Small-Angle X-Ray Scattering

Small-Angle X-Ray Scattering

Proceedings
of the conference held at
Syracuse University, June 1965

Edited by
H. Brumberger
Department of Chemistry
Syracuse University

GORDON AND BREACH, SCIENCE PUBLISHERS
New York   London   Paris

Copyright © 1967 by GORDON AND BREACH, Science Publishers, Inc.
150 Fifth Avenue, New York, N. Y. 10011

Library of Congress Catalog Card Number: 67-26573

Editorial Office for the United Kingdom:

Gordon and Breach, Science Publishers Ltd.
8 Bloomsbury Way
London W.C. 1

Editorial Office for France:

Gordon & Breach
7–9 rue Emile Dubois
Paris 14ᵉ

Distributed in France by:

Dunod Editeur
92 rue Bonaparte
Paris 6ᵉ

Distributed in Canada by:

The Ryerson Press
299 Queen Street West
Toronto 2B, Ontario

Printed in the United States of America

PREFACE

On June 24, 25 and 26, 1965, a conference on Small-Angle X-Ray Scattering was held at Syracuse University under the joint sponsorship of the American Crystallographic Association, the Army Research Office, the National Science Foundation, and the University. Nearly one-hundred and thirty scientists from eleven countries, representing disciplines ranging from polymer chemistry through biophysics and metallurgy, took part in the deliberations. This volume contains the papers presented at the conference. It is the hope of the editor that it will serve, as did the meeting, to describe the state of the art in the interesting and challenging field of small-angle scattering, and also to initiate the newcomer to this area of scientific enquiry. To this end, several papers contain some review material as well as the most recent findings. In some measure, then, these proceedings should complement the monograph of Guinier, Fournet, Walker and Yudowitch.

It has been difficult to place the papers into sharply defined categories; in some instances, a heavily theoretical presentation nevertheless appears in a specialized section (such as metallurgy) because of its relatively narrow scope. In any event, the editor has attempted to preserve the individual character of the papers to the greatest possible extent.

It is a pleasant task to acknowledge the immense help I received from so many sources. The assistance of the Crystallographic Association, and in particular of Dr. H. T. Evans, was much appreciated. The financial support provided by the U. S. Army Research Office, National Science Foundation and Syracuse University is gratefully acknowledged on behalf of myself and the conference participants.

Thanks are due to the Conference Committee—W. W. Beeman, A. Guinier, R. Hosemann, O. Kratky, P. W. Schmidt and W. O. Statton—for invaluable help in preparing and organizing the program. Professors R. H. Marchessault and V. Weiss were most helpful in reviewing some of the manuscripts. I wish to express my thanks also to the graduate students in the Department of Chemistry at Syracuse University who assisted me, to the staff of the University College Conference Office, headed by Miss Eleanor Ludwig, and to the conference secretary, Mrs. Thelma McHale. Finally, I acknowledge a debt of gratitude to my wife for her patience and help throughout the conference.

H. Brumberger

Syracuse, New York

TABLE OF CONTENTS

Preface   v

Part I
THE THEORY OF SMALL-ANGLE X-RAY SCATTERING

1. Determination of General Parameters by Small-Angle X-Ray Scattering   1
   G. Porod
2. The Small-Angle X-Ray Scattering from Filaments   17
   P. W. Schmidt
3. Scattering Functions of Polymer Coils   33
   R. G. Kirste

Part II
EXPERIMENTAL TECHNIQUES

4. Adaptation of the Technique of Diffuse Small-Angle X-Ray Scattering to Extreme Demands   63
   O. Kratky
5. A New Tool for Small-Angle X-Ray Scattering and X-Ray Spectroscopy: the Multiple Reflection Diffractometer   121
   U. Bonse and M. Hart

Part III
APPLICATIONS TO THE STUDY OF MACROMOLECULAR SYSTEMS

6. Optical Analogs of Small-Angle X-Ray Diffraction from Drawn Fibers   131
   P. Predecki and W. O. Statton
7. Small-Angle X-Ray Scattering by Crystalline Polymers   145
   A. Peterlin
8. Small-Angle Equatorial Diffraction in Fibrous Proteins   157
   S. Krimm
9. Combined Analysis of Small-Angle Scattering of Reflections (000) and (hkl) 169
   R. Hosemann

10. Some Experimental Procedures and Recent Results on Liquids, Solutions and Globular Proteins ... 197
    W. W. Beeman
11. A Low-Angle X-Ray Diffraction Study of γ-Globulin ... 213
    W. R. Krigbaum and R. T. Brierre, Jr.
12. Structure of Nucleoproteins Studied by Means of Small-Angle Scattering ... 221
    A. Nicolaïeff
13. X-Ray Scattering from Small RNA Viruses in Solution ... 243
    J. W. Anderegg
14. Myoglobin: Low-Angle X-Ray Scattering Properties as Calculated from the Known Structure ... 267
    H. C. Watson

Part IV
APPLICATIONS TO THE STUDY OF METALS, ALLOYS AND CRYSTAL DEFECTS

15. Application of Small-Angle X-Ray Scattering to Problems in Physical Metallurgy and Metal Physics ... 277
    V. Gerold
16. On the Determination of the Metastable Miscibility Gap from Integrated Small-Angle X-Ray Scattering Data ... 319
    R. W. Hendricks and B. S. Borie
17. Local Atomic Configurations in a Gold-Nickel Alloy ... 335
    S. C. Moss and B. L. Averbach
18. X-Ray Scattering by Point-Defects ... 351
    A. M. Levelut and A. Guinier
19. A New Source of Small-Angle X-Ray Scattering ... 373
    R. G. Perret and D. T. Keating
20. Small-Angle Scattering from Dislocation Rings in Crystals, and Its Relation to the Depleted Zones in Neutron-Irradiated Metals ... 383
    A. Seeger and P. Brand

Part V
CRITICAL PHENOMENA

21. Critical Scattering of X-Rays from Binary Alloys ... 401
    A. Münster
22. Small-Angle X-Ray Scattering from the Perfluoroheptane-Isooctane System in the Critical Region ... 425
    G. W. Brady, D. McIntyre, M. E. Myers, Jr., and A. M. Wims

Part VI
MISCELLANEOUS APPLICATIONS

23. Small-Angle X-Ray Study of Metallized Catalysts — 449
    G. A. Somorjai, R. E. Powell, P. W. Montgomery and G. Jura
24. A Small-Angle X-Ray Scattering Study of Radiation Damage to Silica Gels — 467
    B. C. Larson and H. D. Bale
25. Small-Angle Scattering by Solutions of Complex Ions — 477
    A. Hyman and P. A. Vaughan
26. X-Ray Scattering Study of Bone Mineral — 493
    E. D. Eanes and A. S. Posner

Determination of General Parameters by Small-Angle X-Ray Scattering

G. POROD

Physikalisches Institut, Universität Graz, Austria

INTRODUCTION

The theoretical methods for the evaluation of small-angle scattering (SAS) of dilute systems containing identical particles are well-developed now. For a great variety of simple bodies such as ellipsoids, parallelopipeds and the like the exact scattering functions have been calculated, and good results have been obtained by comparison with the experimental data, especially with proteins.

But in most cases the colloidal systems are of a more complicated nature. The particles may vary in size and shape. Then the direct method will not work. We must be content with finding some parameters giving a general characterization of size, polydispersity and shape. In the present paper we shall only be concerned with systems of this kind.

The most general approach to SAS is made by the use of the well-known correlation function (C.F.) introduced in the fundamental paper by Debye and Bueche (1). The scattering curve is determined unambiguously by that function, if we leave aside the question of *absolute* intensity, which is of quite a different nature and will not be considered here. On the other hand, any information which can possibly be drawn from the X-ray pattern must be contained implicitly in the C.F. In particular, we may choose some parameter by integration over the C.F. in a special way and regard it as a measure of particle size. The comparison of two such "integral" parameters can give a hint as to the shape and polydispersity of the particles. It will be shown later, though, that a strict separation is not possible without additional information.

There is still another type of parameter, which may be termed "differential" parameter for the following reason: obviously the C.F. can be represented by a power series (at least for small distances r):

$$G(r) = 1 - ar + br^2 + cr^3 \ldots \ldots \qquad 1.$$

Here the coefficients $a, b, c, \ldots$ correspond to the first, second, etc., derivatives of the C.F. They determine the course of the C.F. at small distances, that means near the particle surface. So we are led to the suggestion that the differential parameters will give us some information regarding the surface structure. If we wish to distinguish between a sphere and a cube, for instance, the integral parameters are of little use. In fact, it will be seen later that the distinction has to be made by means of b and c. Our problem will be how to determine these parameters from the scattering function.

In principle, all questions can be treated by the use of the C.F. But it will turn out to be useful, partly for mathematical convenience, partly for the sake of intuition, to introduce also another function, the statistical distribution of "intersects." Particularly the asymptotic behavior at larger angles is shown more clearly in that way and the formulae for the differential parameters can be derived more directly. The "intersect" distribution will therefore be widely used in the present paper.

INTEGRAL PARAMETERS

For systems containing separated particles of arbitrary sizes and shapes, but all of the same uniform electron density, embedded in a medium of different but also uniform density, the C.F. has a more special meaning. If we choose at random a point within a particle, the probability of finding another point at a distance r also within the particle is equal to $G(r)$. This function may, therefore, be imagined as a probability distribution in space. Obviously it must vanish for distances larger than the largest diameter of the particles. It is easy, therefore, to define some average size l_c, f_c and v_c by taking the integral in one, two, three dimensions respectively:

$$l_c = 2\int_0^\infty G(r)\,dr; \qquad f_c = 2\pi\int_0^\infty rG(r)\,dr; \qquad v_c = 4\pi\int_0^\infty r^2 G(r)\,dr \qquad 2.$$

Here the index c stood for "characteristic" in the original paper (2), but it seems better to read "correlation" or "coherence" now. For a single particle v_c is identical with the volume. The other parameters cannot be given such a direct meaning.

We might, of course, equally well take the average of r or of some power of it as an integral parameter. In calculating such averages we must only take care that the frequency of a distance r in space is proportional to $r^2G(r)$ rather than to $G(r)$. Of that type is the well-known radius of gyration R introduced by Guinier. Expressed in terms of the C.F. it becomes:

$$2R^2 = \langle r^2 \rangle = \int_0^\infty r^4 G(r)\, dr \bigg/ \int_0^\infty r^2 G(r)\, dr \qquad 3.$$

Its most prominent advantage is the easy way in which it can be obtained from the slope of the well-known Guinier-plot, based on the fact that, to a first approximation, the scattering curve should be of the Gaussian type and determined by the radius of gyration only. The procedure is so widely used now that it needs no further discussion.

The other parameters can also be obtained directly from the scattering function, but in a more complicated way. For that purpose it is convenient to define first a standard form of the intensity $j(h)$. Here h means the variable $h = (4\pi \sin \theta)/\lambda$, where 2θ is the scattering angle. The standard intensity should obey the following relation (2):

$$\int_0^\infty h^2 j(h)\, dh = 2\pi^2 \qquad 4.$$

This invariant integral can at the same time be used to find the standard form from the measured intensity. We need only multiply by a constant factor such that 4 is fulfilled.

In most experimental work a slit-defined primary beam is used rather than a pinhole beam. In that case the X-ray pattern is distorted or "smeared." Though methods have been given to correct for that distortion (3, 4), it is sometimes more convenient to evaluate the smeared intensity $\tilde{j}(h)$ directly. Therefore, the formula will be given for both cases. First the definition of the standard form:

$$\int_0^\infty h\tilde{j}(h)\, dh = 4\pi^2 \qquad 5.$$

The standard intensities are related to the C.F. in a simple way:

$$j(h) = 4\pi \int_0^\infty dr \cdot r^2 G(r) \sin(hr)\bigg/ hr; \quad \tilde{j}(h) = 4\pi^2 \int_0^\infty dr \cdot r G(r) J_0(hr) \qquad 6.$$

where J_0 stands for the Bessel-function of the first kind. In principle the C.F. can, according to 6a, be obtained from the standard intensity by Fourier-transform. Unfortunately that method does not work in practice. This is a question of accuracy only. We need a more direct way of evaluation. In (2) it could be shown that the integral parameters are given by:

$$l_c = \int_0^\infty hj(h)\, dh \Big/ 2\pi; \qquad f_c = \int_0^\infty j(h)\, dh \Big/ \pi; \qquad v_c = j(0) \qquad 7.$$

$$l_c = \int_0^\infty \tilde{j}(h)\, dh \Big/ 2\pi^2; \qquad f_c = \tilde{j}(0)/2\pi \qquad\qquad 8.$$

Unfortunately there is no simple way to derive v_c and R, the most intuitive parameters, from the smeared scattering curve. Nevertheless, in most cases no great error is involved if we use the Guinier-plot with $\tilde{j}$ instead of j (about 3% for a sphere). With an exact Gaussian curve there would be no error at all.

STATISTICS OF INTERSECTS

The definition of the integral parameters in terms of the C.F. is somewhat abstract. We shall try to give them a more intuitive meaning by the introduction of the concept of "intersects." The argument is as follows: Any straight line intersecting a body will cut out a piece of length l, which may be called an "intersect." If we do this everywhere and in all directions, we obtain a whole statistical assembly of intersects from zero up to the largest possible diameter of the body. It may be described by a distribution function $A(l)$ such that the probability of finding a length in the interval from l to $l + dl$ is given by $A(l)\,dl$. It seems plausible that the body should be determined by this function just as well as by the C.F., and that, therefore, a definite relation between the two must exist. This is easily shown. Any intra-particle distance r must necessarily be within an intersect $l > r$. The number of distances r, which can be placed within l, can be measured by the free margin $(l - r)$. Taking into account the correct normalization we have:

$$G(r) = \int_r^\infty (l - r) A(l)\, dl \Big/ \bar{l}; \quad \text{with } \bar{l} = \int_0^\infty l A(l)\, dl \qquad 9.$$

where $\bar{l}$ is the mean intersect (number average). It is equal to the reciprocal of the parameter a.

By differentiating 9 twice with respect to r and then changing r for l we obtain:

$$A(l) = \bar{l}(d^2 G(r)/dr^2)_{r=l} \qquad 10.$$

From this it is seen that the intersect distribution must have a clear relation to the differential parameters, which will be of use later. For the moment we are concerned with the integral parameters. They can be expressed by combination of 2 and 9:

$$l_c = \langle l^2 \rangle / \bar{l}; \quad f_c = \pi \langle l^3 \rangle / 3\bar{l}; \quad v_c = \pi \langle l^4 \rangle / 3\bar{l} \qquad 11.$$

Let us take for example a sphere of radius a. The calculation of $A(l)$ is here particularly simple as all directions are equivalent. Pure geometry shows:

$$\text{sphere:} \quad A(l) = l/2a^2; \quad \text{for } 0 < l < 2a \qquad 12.$$

from which we obtain, by inserting into 11:

$$\text{sphere:} \quad \bar{l} = 4a/3; \quad l_c = 3a/2; \quad f_c = 4\pi a^2/5; \quad v_c = 4\pi a^3/3 \qquad 13.$$

These values are already known and may be regarded as a verification of our argument concerning the intersect distribution.

In a similar way the average of any power of r can be expressed in terms of l:

$$\langle r^n \rangle = 12 \langle l^{n+4} \rangle / \langle l^4 \rangle (n+3)(n+4) \qquad 14.$$

and especially for the radius of gyration:

$$R^2 = \langle l^6 \rangle / 5 \langle l^4 \rangle; \quad \text{sphere } R^2 = 3a^2/5 \qquad 15.$$

which again for a sphere yields the well-known result.

From the above formulae it is seen that all integral parameters can be expressed in terms of mean powers of intersects. As different powers are involved, the ratio of two parameters will depend on the distribution function, that means on the shape of the particles. Evidently the sphere is an extreme case in this respect. Its intersect distribution always increases to the maximum value of l, i.e. the diameter of the sphere. The reason is obviously the high symmetry. With a sphere the maximum intersect can be obtained by shooting through in any direction, while with an elongated particle only one direction is favorable. Any ratio of integral parameters, if formed so as to be dimensionless, will therefore be independent of size and characteristic of shape only. The choice of these structural numbers is rather arbitrary. Volume and radius of gyration were used (5) by forming the ratio of the true R and a fictive one, which would correspond to a sphere of equal volume to that of the particle in question. In another paper (6), $l_c/\bar{l}$ was introduced as a structural number. An evaluation of this kind in terms of particle shape offers no intrinsic difficulty and will not be discussed further here. The only question to be considered in more detail is how to use the structural numbers in order to distinguish between shape and polydispersity.

POLYDISPERSITY

Systems containing identical particles of simple, well-defined shape are an exception. Most colloidal substances consist of more irregular particles with varying sizes and shapes. We shall now examine what can be done about systems of that kind. The problem has often been attacked and is still open.

A first rough comparison of different scattering curves is best made by means of a Guinier-plot will all curves reduced to the same radius of gyration and the same initial value. In that plot all curves coincide in the innermost part and show greater or smaller deviations from one another with increasing angle. This is shown in Fig. 1. The sphere

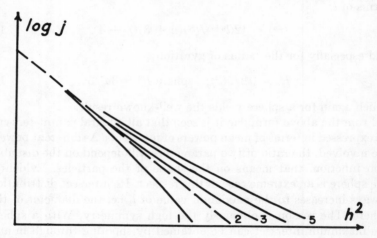

Fig. 1. Guinier-plot of scattering functions of prolate ellipsoids. Axial ratio = 1, 2, 3, 5.
––––– Ideal Guinier-type

is an extreme case again. Any deviation from the spherical shape tends to bend the curve upwards. This behavior is best studied by the use of a power series expansion (7):

$$j(h) = v_c(1 - R^2h^2/3 + \langle r^4\rangle h^4/5! - \langle r^6\rangle h^6/7! \ldots) \qquad 16.$$

An equivalent representation is gained by replacing the "moments" $\langle r^n \rangle$ by the mean powers of l using 14:

$$j(h) = \pi/3\bar{l} \cdot (\langle l^4\rangle - \langle l^6\rangle h^6/15 + 3 \cdot 4!\langle l^8\rangle h^8/8! - \cdots) \qquad 17.$$

The smeared scattering function can be treated in a similar way by developing 6b:

$$\tilde{j}(h) = 2\pi f_c/\langle r^{-1}\rangle \cdot (1 - \langle r\rangle h^2/2^2 + \langle r^3\rangle h^4/(2\cdot 4)^2 - \cdots) \quad 18.$$
$$\tilde{j}(h) = 2\pi^2/3\bar{l} \cdot (\langle\langle l^3\rangle\rangle - 2\cdot 3 \cdot \langle l^5\rangle h^2/4\cdot 5\cdot (2)^2 + \cdots) \quad 19.$$

From the above series it is clear that the deviation from the straight line in the Guinier-plot is due to the higher-power averages of r or l. The more anisometric a particle, the more will the larger distances predominate particularly in the higher moments. Obviously, the same effect can be caused by polydispersity. In fact, if we draw a diagram for systems containing only spheres, but with increasing polydispersity, it would look just like Fig. 1. But the effects are not only similar, but identical in a certain sense. It is a fact, long known and used, for instance, by Guinier (8) and Hosemann (9), that the scattering function of an ellipsoid is strictly identical with that of an assembly of spheres with a special size distribution. The same must obviously be true for any mixture of ellipsoids. Furthermore, we may presume that the scattering of every other body is also equivalent to that of some appropriate mixture. For, according to 16–19, the scattering function is uniquely determined by the moments of r or l. For any desired degree of accuracy a finite number of these coefficients is sufficient. It seems possible that with a given set of moments for a certain body one can always find a special polydispersity function such as to adjust the moments to the values of the scattering function of a different body. Though it will turn out that this argument is not *exactly* true in every case, it is clear that at least in practical evaluation many equivalent systems have to be considered. In short: the determination of shape or polydispersity from the X-ray pattern cannot be unambiguous. The methods developed so far, for instance that of Hosemann (10), are based on the assumption of spherical shape.

In spite of this negative result there is still a possibility of getting *plausible* information at least. The assembly of spheres, equivalent to a certain ellipsoid, shows a very improbable size distribution. It is restricted to the range between the shortest and the longest diameter and exhibits peaks of infinite height. It is hard to imagine a natural process which might produce such an assembly. If for a given system we make the reasonable assumption that the shape of the particles can be sufficiently approximated by some simple body, and that the size distribution is a "normal" one, we may well have a chance of separating the two effects. This was shown in a recent paper by Mittelbach and the author (11).

The argument is as follows. The shape is approximated by an ellipsoid of revolution. Thus it can be defined by *one* number, the axial ratio, only. With the same degree of accuracy we can assume that the polydispersity is sufficiently defined by one single number (the standard deviation is most convenient). In addition, an integral parameter is needed in order to determine the mean size, but this is independent of our question and can be left out of consideration for the present. Under these assumptions the whole variety of our systems represents a two-dimensional manifold. Therefore, two structural numbers should suffice in order to determine a special system. The choice of the structural numbers is rather arbitrary, but it seems best to use numbers which differ as much as possible in their dependence on the system structure. This is the case if, for instance, the one number depends primarily on the large distances, and the other on the small ones.

The method may be illustrated by one example taken from the paper cited above. The particles are assumed to be uniaxial ellipsoids with axial ratio v. The size distribution with the standard deviation σ is represented by a function $g(p)$ (number frequency) of the following type:

$$g(p) \sim p^n \exp(-p); \quad \text{with } \sigma^2 = 1/(n+1) \qquad 20.$$

That type is most convenient for mathematical reasons and seems at the same time to be a natural representation of a normal distribution. As absolute size is not the question here, the variable p means the linear size relative to some arbitrary particle of reference.

The structural numbers K_1 and K_1^* for a single particle are:

$$K_1 = 8l_c/9\bar{l}; \quad K_1^* = 2.08R/\sqrt[3]{v_c} \qquad 21.$$

where the numerical factors have been introduced in order to make the structural numbers equal to unity for the sphere. The corresponding values for the polydisperse systems are found by taking the average of the integral parameters l_c, $\bar{l}$, R^2 and v_c over the size distribution, 20. In the original paper, the calculation of these averages turned out to be a delicate task, where great care must be taken to avoid error. We shall now follow a more systematic way by the use of 11 and 15, so that all parameters are expressed in terms of moments of the intersect. Given such a term $\langle l^m \rangle$ for the particle, it will be

p^m-times as large for some other particle. The latter factor has to be averaged. Here care must be taken that the statistical weight is given by $p^2 g(p)$ rather than $g(p)$, as the total number of intersects is proportional to the cross-section of the particle. Thus we have:

$$\langle\langle l^m\rangle\rangle = \langle l^m\rangle \langle p^{m+2}\rangle / \langle p^2\rangle; \quad \langle p^m\rangle = (n+m)!/n! \qquad 22.$$

The latter formula for $\langle p^m\rangle$ is easily derived from 20. Inserting into 11 and 15, and then into 21, we get the structural numbers for the polydisperse system:

$$K = K_1 \cdot \langle p^4\rangle \langle p^2\rangle / (\langle p^3\rangle)^2; \quad K^* = K_1^* \cdot \sqrt{\langle p^8\rangle} \sqrt[3]{\langle p^3\rangle} / \sqrt{\langle p^6\rangle} \sqrt[3]{\langle p^6\rangle} \qquad 23.$$

The result is plotted in Fig. 2. The diagram represents the variety of

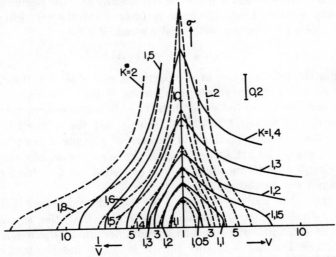

Fig. 2. Diagram of characteristic numbers for a polydisperse system of ellipsoids of revolution (Mittelbach and Porod).

    Abscissa ... axial ratio
    Ordinate ... standard deviation
    ———— $K = 8l_c/9\bar{l}$ = constant
    - - - - $K^* = 2.08R/\sqrt[3]{v_c}$ = constant.

shape and polydispersity, with the axial ratio as abscissa and the standard deviation as ordinate. The full lines mean systems with the same structural number K, and the dashed lines stand for the same K^*. If these numbers are determined from the X-ray pattern, we find the state of the system as the coordinates of the crossing-point of the corresponding full and dashed curves.

Finally, it should be stressed once more that the method explained above rests on special assumptions and cannot be regarded as unambiguous. But it may be reasonable and useful in some cases where other methods are not available.

DIFFERENTIAL PARAMETERS

As was shown independently by Debye (12), van Nordstrand and Hach (13), and the author (2), the scattering curve, when multiplied by h^4, should asymptotically reach a constant value, which is proportional to the internal surface. A similar relation holds for the smeared scattering curve, where h^4 has to be replaced by h^3. The specific surface is closely connected with the mean intersect $\bar{l}$ and the first differential parameter a for purely geometrical reasons. We have:

$$1/a = \bar{l} = 4V/S \qquad 24.$$

V being the total volume of the particles, and S their surface. Generally speaking, the asymptotic behavior of the scattering curve is determined by the initial slope of the C.F. It is convenient, therefore, to use the expansion 1 in order to carry out the integral in 6. It then follows that only the odd powers of r make a contribution, yielding terms like h^{-4}, h^{-6}, etc., while the even powers of r only give rise to oscillating terms which will be cancelled out to a large extent with a not too regular system.

By the use of the power series 1 for the C.F. only small distances are involved. We may conclude, therefore, that the asymptotic course of the scattering function is characteristic of the region near the surface. Further it follows that it is nearly independent of the arrangement of the particles, as far as no direct contact is involved. In the following we assume this to be exactly true. The proof will be given in another paper in connection with the discussion of densely packed irregular systems.

So we have good reason to attribute the differential parameters to the fine structure of the surface. Starting from that idea, Kirste and the author (14) tried to derive a formula for b and c, taking into account the curvature of the surface. The result was, for smooth particles:

$$b = 0; \quad c = (\langle k_1 k_2 \rangle/12 + \langle (k_1 - k_2)^2 \rangle/32) \cdot S/4Vw_1w_2 \qquad 25.$$

where k_1 and k_2 mean the principal curvatures, and w_1, w_2 the volume fractions of particles and voids respectively.

Formula 25 was derived under the assumption that no edges or corners are present. That case seemed too difficult for a general treatment. But from the corresponding two-dimensional problem, for which a special solution could be found, it became rather clear that the parameter b will no longer vanish with angular bodies. It seems to be a criterion for angularity.

Before attacking the general problem we shall first give a special verification of this suggestion. To that purpose let us regard a rectangular parallelopiped. As we are only interested in the beginning of the C.F., a method can be used which was proposed by the author in an early paper (7). We imagine the body (A, B, C) shifted for a distance r in some direction (α, β, γ). The distance probability is equal to the ratio of the volume cut out by the "ghost" to the total volume. The C.F. is the average of the distance probability taken over all directions. Thus we have:

$$G(r) = \langle (A - r\cos\alpha)(B - r\cos\beta)(C - r\cos\gamma) \rangle / ABC \qquad 26.$$

Carrying out the product and using the averages

$$\langle \cos\alpha \rangle = \langle \cos\beta \rangle = \langle \cos\gamma \rangle = \tfrac{1}{2}$$
$$\langle \cos\alpha \cos\beta \rangle = \ldots = 2/3\pi$$
$$\langle \cos\alpha \cos\beta \cos\gamma \rangle = 1/4\pi \qquad 27.$$

we finally have:

$$G(r) = 1 - r(1/A + 1/B + 1/C)/2$$
$$+ r^2(1/AB + 1/BC + 1/CA)2/3\pi - r^3/4\pi ABC \qquad 28.$$

This expression is valid only for distances smaller than the minor of A, B, C. For larger r not all directions of shift are equally possible and the above argument therefore does not hold. For our present problem this means no restriction.

From a comparison of 25 and 28, we may generalize the following conclusion. Smooth and angular particles are distinguished by the value of the parameter b (zero or non-zero), and by the sign of c (plus or minus). The latter criterion has a direct bearing on the scattering curve. The term r^3 of the C.F. gives rise to a term h^{-6} of the asymptotic scattering function. If we multiply by h^4, that means that in the case of smooth particles the constant limiting value is approached from above, and with angular bodies from below. Unfortunately, this criterion is not good for practical evaluation. In the first place, a real colloidal system may be a mixture of the two extreme cases. Then the actual behavior will depend on what influence predominates. Secondly, if the system is not quite irregular, the presence of oscillating terms can overshadow and distort the normal asymptotic slope.

Let us now see how to determine the parameter b. Here it turns out that the problem becomes much more simple and acquires a more intuitive meaning by the use of the intersect distribution. The connection with the C.F. is given by 9 and 10. Using for $C(r)$ the series 1, the function $A(l)$ takes the form:

$$A(l) = \bar{l} \cdot (2b + 6cl + \ldots); \quad A(0) = 2b\bar{l} \qquad 29.$$

So the parameter b is closely related to the initial value $A(0)$. That the latter must vanish for smooth particles and have a finite value for angular ones can now be easily shown without the use of mathematics. It is obvious that very small intersects can only be obtained with lines cutting the particle near the surface nearly tangentially. In the case of a curved surface the length of the chords must necessarily increase with the second power of the distance from the surface inwards. Therefore, an interval dl corresponds to a bundle of intersecting lines small to the second order. That means $A(0) = 0$. In the presence of an edge the relation between intersect length and the distance from the edge is clearly linear. Thus it follows that $A(0)$ must have a finite value.

As the meaning of the parameter b is more distinctly shown in $A(l)$ than in $G(r)$, it seems better suited to our problem to express the scattering function in terms of the intersect distribution. This can be done by inserting 9 into 6 and transforming into a single integral. The calculation shall not be given in detail here. It leads to the result:

$$j(h) = 8\pi/\bar{l}h^4 \cdot \left[1 - \int_0^\infty A(l)\, dl(\cos hl + hl/2 \cdot \sin hl)\right] \qquad 30.$$

$$\tilde{\jmath}(h) = 4\pi^2/\bar{l}h^3 \cdot \left\{1 - \int_0^\infty A(l)\, dl[R(hl) + hlJ_0(hl)]\right\} \qquad 31.$$

where J_0 means the Bessel-function of the first kind and first order, and the function $R(x)$ is defined by

$$R(x) = \int_x^\infty J_0(t)\, dt \qquad 31a.$$

The representation 30 and 31 seems particularly suited to our present problem, as it shows in a direct manner the asymptotic behavior of the scattering function. If we multiply by h^4 or h^3 respectively, the first term is constant and characteristic of $\bar{l}$ or the specific surface. The second term, i.e., the integral, accounts for the deviation from that final value. We shall now prove the following statement: The deviation integral is proportional to $A(0)$ or b; that means it must vanish with smooth particles and have a positive value with angular ones.

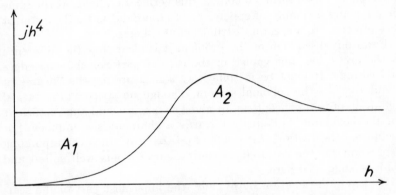

Fig. 3. Asymptotic behavior of the scattering function (schematic).

$\mathbf{A}_1$, $\mathbf{A}_2$ ... Areas characteristic of the angularity of the particles (cf. Eq. 32).

The situation is shown in Fig. 3. The deviation integral in question is the difference of the two areas $\mathbf{A}_1 - \mathbf{A}_2$:

$$\mathbf{A}_1 - \mathbf{A}_2 = 8\pi \Big/ \bar{l} \cdot \iint_0^\infty A(l)\, dl(\cos hl + hl/2 \cdot \sin hl)\, dh \qquad 32.$$

This double integral can first be carried out in a closed form with respect to h, the upper limit being taken as a very large value H tending to infinity. So we have:

$$\mathbf{A}_1 - \mathbf{A}_2 = 4\pi \Big/ \bar{l} \cdot \int_0^\infty A(l)\, dl(3 \sin Hl/l - H \cos Hl) \qquad 33.$$

The second term represents a rapidly oscillating function whose contribution tends to zero in the limit of H toward infinity. The first term is well known as a sort of Dirac delta-function, so that only $A(0)$ appears in the result. We find:

$$\mathbf{A}_1 - \mathbf{A}_2 = 6\pi^2 A(0)/\bar{l} = 12\pi^2 b \qquad 34.$$

The corresponding case of the smeared scattering function can be treated in a quite analogous manner (h^3 instead of h^4), so that we need not carry out the argument in detail. The result is:

$$\tilde{\mathbf{A}}_1 - \tilde{\mathbf{A}}_2 = 8\pi^2 A(0)/\bar{l} = 16\pi^2 b \qquad 35.$$

In addition, it may be mentioned that 34 and 35 were verified also for the special case of a sphere, for which the scattering function is available in closed form. Of course, this verification holds at the same time for any assembly of spheres or of ellipsoids, as the latter must be equivalent to a certain distribution of spheres.

From our discussion of the problem it is clear that the difference of the areas does not vanish in the case of particles showing edges and corners. It must be in some way a measure for the "degree of angularity" of the colloidal system. Though no strict mathematical definition of that concept can be given so far, it would seem sufficient for the beginning to define arbitrarily a characteristic number. For instance, the ratio $(\mathbf{A}_1 - \mathbf{A}_2)/(\mathbf{A}_1 + \mathbf{A}_2)$ or the corresponding expression for the smeared scattering curve seems well suited and shall be suggested here.

By the multiplication by h^4 or h^3, the innermost part of the scattering curve is largely suppressed, so that chiefly the values at relatively large angles will contribute to our characteristic number. This must be so as it is the question of a differential parameter. On the other hand, the experimental determination becomes a delicate task. At present, no direct evidence is available. Therefore, it cannot be said now whether or not there will be a practical use for the method presented here.

Finally, it should be stressed once more that the above calculations were based on the assumption of separate particles. The case of densely packed systems needs a special investigation. In a subsequent paper it will be shown that the conclusions concerning the parameter b remain essentially unchanged (but for a slight modification). For the parameter c, on the contrary, we must expect a serious influence of dense packing.

SUMMARY

The distinction between integral and differential parameters is introduced, the former being related to the size of the colloidal regions and the latter to the structure of the internal surface. The parameters are expressed in terms of the correlation function and of the intersect distribution function. Their determination from the scattering curve is discussed.

The influence of particle shape and of polydispersity is expressed by means of characteristic numbers. A special method for the separation of the two effects is considered in detail.

It is shown that there is a clear distinction between smooth and angular particles in the asymptotic behavior of the scattering curve. With the aid of the intersect distribution function, criteria for the two cases are derived.

BIBLIOGRAPHY

1. Debye, P. and Bueche, A. M., J. Appl. Phys., **20,** 518 (1949).
2. Porod, G., Kolloid-Z., **124,** 83 (1951).
3. Guinier, A. and Fournet, G., Nature, **160,** 501 (1947). Dumond, J. W. M., Phys. Rev., **72,** 83 (1947).
4. Kratky, O., Porod, G. and Kahovec, L., Z. Elektrochem., **55,** 53 (1951). Kratky, O., Porod, G. and Skala, Z., Acta Phys. Austriaca, **13,** 76 (1960).
5. Kratky, O., Progr. Biophys., **13,** 105 (1963).
6. Kahovec, L., Porod, G. and Ruck, H., Kolloid-Z., **133,** 16 (1953).
7. Porod, G., Acta Phys. Austriaca, **2,** 255 (1948).
8. Guinier, A., Ann. Phys., **12,** 161 (1939).
9. Hosemann, R., Z. Physik, **113,** 751 (1939).
10. Hosemann, R., Ergeb. exakt. Naturw., **24,** 142 (1951).
11. Mittelbach, P. and Porod, G., Kolloid-Z., **202,** 40 (1965).
12. Debye, P., Anderson, H. R., and Brumberger, H., J. Appl. Phys., **28,** 679 (1957).
13. Van Nordstrand, R. A. and Hach, K. M., Am. Chem. Soc. meeting, Chicago, Sept. 1953.
14. Kirste, R. and Porod, G., Kolloid-Z., **184,** 1 (1962).

The Small-Angle X-Ray Scattering from Filaments*

P. W. SCHMIDT

Physics Department, University of Missouri, Columbia, Mo.

INTRODUCTION

In small-angle X-ray scattering studies of the structure of colloids, it is desirable to know as much as possible about the relationship between the sample structure and the angular distribution of the radiation scattered by the sample, in order to develop techniques for interpretation of scattering data and to widen the range of applicability of these methods. Although this relation is too complicated to be discussed in detail without making some restrictions and assumptions about the sample, a number of studies of small-angle X-ray scattering theory have dealt with properties of the scattering process which, though not completely general, make relatively few restrictions about the sample.

One approach to this problem has been to assume as general as possible a class of particle shapes and then consider the properties of the scattering pattern expected for this class of particles. By dealing with a class of particles instead of with particles of a single shape, more general properties of the scattered intensity can be discovered than could be obtained by studying the intensity from particles of only one shape. For example, the author has recently published a study of some of the properties of the scattered intensity from right cylinders of arbitrary cross section (1). In this investiga-

---

* Work supported by the National Science Foundation.

tion, however, it was not possible to express the final results in terms of the shape of the cross-section. Instead, the answers were given in terms of a function, called the characteristic function, which can, in principle, be determined for any given cross-section shape but which at present, at least, cannot be computed for a cross-section of arbitrary shape and which has, in fact, been explicitly calculated only for a few very simple cross-sections, such as circles and rectangles.

As the problem of a three-dimensional solid appeared to be too complicated to discuss explicitly in terms of the particle shape, thin rods with the form of a smooth plane curve were chosen for this investigation. Many properties of the scattering curves for these particles, which will be called filaments, were found to be directly expressible in terms of the geometrical properties of the curve generating the filaments. This class of particles was studied primarily because it permitted a relatively general treatment of the scattered intensity.

Since few experimental samples can be approximated by assemblies of identical, non-interacting filaments, the results probably will not often be directly applicable to the analysis of experimental scattering data. Instead, this study will contribute in two ways. First, some of the properties of the scattering curves for filaments may also hold, either exactly or with some modification, for particles which comprise the samples ordinarily encountered. Also, the techniques developed for filaments may be applied to calculations of the scattering from other particles.

THE CHARACTERISTIC FUNCTION FOR FILAMENTS

The particles in the sample will be assumed to be identical rods which have a negligible cross-section and uniform electron density, and which are bent in the form of a plane curve. At all points on this curve, an arbitrary number of the derivatives of the radius of curvature will be assumed to be continuous, and the radius of curvature will be assumed to be single-valued. The particle thus will have no sharp bends or corners, and the curve will not cross or touch itself at any point, except that for a closed curve, or loop, the two ends of the curve will be joined together.

The location of points on the curve will be specified by the arc length t from the point, which will be called "point t," to one end of the curve, which will be taken as a reference point. The values of t will range from 0 to L, where L is the total arc length of the curve.

The particles will be considered to be randomly oriented, with the observed scattering being the average over all particle orientations. The particles will be assumed not to interact with each other, and therefore only the scattering from a single particle need be considered.

The expression for the average scattered intensity $I(h)$ for a filament can be obtained by writing the one-dimensional analogue of the intensity of the scattering from a randomly oriented three-dimensional particle (2). The resulting expression is

$$I(h) = \frac{1}{L^2} \int_0^L dt_1 \int_0^L dt_2 \frac{\sin hs}{hs} \qquad 1.$$

where the t_1 and t_2 integrations both extend over the length of the particle, $s(t_1,t_2)$ is the distance between points t_1 and t_2, $h = 4\pi(\lambda)^{-1} \sin \theta$, λ is the X-ray wavelength, and 2θ is the scattering angle. (In the small-angle region, h is essentially proportional to the scattering angle.)

If s is taken as one of the variables of integration, Eq. 1 can be expressed by

$$I(h) = \frac{2}{L} \int_0^{s_m} ds \gamma(s) \frac{\sin hs}{hs} \qquad 2.$$

where s_m is the maximum distance between two points of the particle. The function $\gamma(s)$, which is called the characteristic function, contains all information about the particle obtainable from scattering data. As $\gamma(s)$ almost never has a simple relation to the particle shape, one of the main purposes of this discussion is to point out some features of this relationship and to develop approximations for $\gamma(s)$ for filaments which are valid in the neighborhood of certain values of s.

Emphasis will be given to the form of $I(h)$ for large h—specifically, for $hD \gg 1$, where D is the smallest dimension characterizing a feature of the form of the filament. For large h, integrals like the integral in Eq. 2 can be evaluated by the method of asymptotic expansion of Fourier integrals which has been discussed by Erdélyi (3), who showed that the asymptotic expansion is determined by the behavior of $\gamma(s)$ and its derivatives at the end points of the interval of integration and at any interior points at which $\gamma(s)$ or its derivatives have discontinuities. When the behavior of $\gamma(s)$ is known in the neighborhood of these points, the asymptotic expansion can be obtained from Erdélyi's equations. The calculation of the asymptotic expansion of $I(h)$ thus reduces to the problem of finding approximate expressions for $\gamma(s)$ in the vicinity of $s = 0$, $s = s_m$, and any points at which $\gamma(s)$ or its derivatives are discontinuous.

The asymptotic expansion of $I(h)$ can be conveniently obtained by first considering the function

$$J(h) = \frac{d}{dh}[hI(h)] = \frac{1}{L^2} \int_0^L dt_1 \int_0^L dt_2 \cos hs \qquad 3.$$

This function can also be expressed in the form

$$J(h) = \frac{2}{L} \int_0^{s_m} ds\gamma(s) \cos hs \qquad 4.$$

The scattered intensity $I(h)$ can be found from $J(h)$ by the relation

$$I(h) = \frac{1}{h} \int_0^h dk J(k) = \frac{1}{h} \int_0^\infty dk J(k) - \frac{1}{h} \int_h^\infty dk J(k) \qquad 4a.$$

Because of the properties of Fourier transforms,

$$\int_0^\infty dk J(k) = \frac{2}{L} \int_0^\infty dk \cos ku \int_0^{s_m} ds\gamma(s) \cos ks \bigg|_{u=0} = \frac{2}{L} \frac{\pi}{2} \gamma(0) \qquad 4b.$$

For filaments the characteristic function has the property that $\gamma(0) = 1$, just as for the three-dimensional case (4). Thus

$$I(h) = \frac{\pi}{hL} - \frac{1}{h} \int_h^\infty dk J(k) \qquad 5.$$

As the integral in Eq. 5 involves only large values of h, the integrand can be approximated by an asymptotic expansion. Thus, once the asymptotic expression for $J(h)$ is calculated, the asymptotic expansion of $I(h)$ can be obtained from Eq. 5.

APPROXIMATIONS FOR $\gamma(s)$ IN THE NEIGHBORHOOD OF CERTAIN VALUES OF s

As $s(t_1,t_2) = s(t_2, t_1)$, Eq. 3 can be written

$$J(h) = \frac{2}{L^2} \int_0^L dt_1 \int_0^{t_1} dt_2 \cos hs \qquad 6.$$

The region of integration in Eq. 6 will be called Region R. It is bounded by the lines $t_1 = t_2$, $t_1 = L$, and $t_2 = 0$. In Region R, $t_2 \leq t_1$.

If the variables of integration in Eq. 6 are changed from t_1 and t_2 to s and r (usually r will equal t_1 or t_2), $\gamma(s)$ can be expressed as

$$\gamma(s) = \frac{1}{L} \sum_{i=1}^{j(s)} \int_{a_i(s)}^{b_i(s)} dr \left| \frac{\partial t_1}{\partial s} \frac{\partial t_2}{\partial r} - \frac{\partial t_1}{\partial r} \frac{\partial t_2}{\partial s} \right| \qquad 7.$$

The limits $a_i(s)$ and $b_i(s)$ and the number $j(s)$ of terms in the sum depend on the form of $s(t_1, t_2)$ and on the value of s.

Jones and Kline (5) have treated the general problem of asymptotic expansion of double Fourier integrals, including integrals like the integral in Eq. 6. They showed that the asymptotic expansion, instead of being determined by the integral over the entire region R, depends only on the integrals over certain domains of this region. These domains are the neighborhoods of points and arcs called critical points and critical arcs. Jones and Kline list the critical points and arcs which contribute to the asymptotic expansion, and they show how these points and arcs can be identified from the properties of $s(t_1,t_2)$. They develop series expansions for $\gamma(s)$ in the neighborhood of these critical arcs and points and then use Erdélyi's theorem (3) to find the contribution to the asymptotic expansion from each critical point and arc.

The procedures developed by Jones and Kline can be used to identify and locate the critical points and arcs in Region R in Eq. 6. (The discussion will be limited to the critical points and arcs likely to be encountered with filaments with the properties which have been assumed above.) After the critical points and arcs have been located, however, the general formulas of Jones and Kline for $\gamma(s)$ and the asymptotic expansion of $J(h)$ will not be used, since for this discussion, $\gamma(s)$ can be found more easily by approximate evaluation of Eq. 7 for s values corresponding to critical points and arcs of Region R. Except for s values near $s = 0$, only the first term of each approximation will be calculated. Erdélyi's theorem will then be employed to find the resulting contribution to the asymptotic expansion of $J(h)$, with the asymptotic expression for $I(h)$ then being obtained from Eq. 5.

Taylor expansions of $s(t_1,t_2)$ for t_1 and t_2 near points T_1 and T_2, respectively, will be needed for evaluating Eq. 7. Let $\mathbf{r}_j$ be the vector from point T_i to point t_i. Then according to the Frenet-Serret formulas of differential geometry (6),

$$\mathbf{r}_i = \alpha_i r_{\alpha i} + \beta_i r_{\beta i} \qquad 8.$$

where

$$\left.\begin{aligned}
r_{\alpha i} &= \sum_{j=1}^{\infty} p_{\alpha j}(t_i - T_i)^j \\
r_{\beta j} &= \sum_{j=1}^{\infty} p_{\beta j}(t_i - T_i)^j \\
p_{\beta 1} &= 0 \\
p_{\alpha 1} &= 1 \\
p_{\alpha 2} &= 0 \\
p_{\alpha 3} &= -(\tfrac{1}{6})R(T_i)^{-2} \\
p_{\alpha 4} &= -\frac{1}{16} \frac{d}{dT_i}\left[\frac{1}{R(T_i)}\right]^2 \\
p_{\beta 2} &= (\tfrac{1}{2})[R(T_i)]^{-1} \\
p_{\beta 3} &= \frac{1}{6} \frac{d}{dT_i}\left[\frac{1}{R(T_i)}\right] \\
p_{\beta 4} &= -\frac{1}{24}\frac{1}{[R(T_i)]^3} + \frac{1}{24}\frac{d^2}{dT_i^2}\left[\frac{1}{R(T_i)}\right]
\end{aligned}\right\} \quad 8a.$$

and where $R(t)$ is the radius of curvature at point t, $\boldsymbol{\alpha}_i$ is a unit tangent vector in the direction of increasing t_i, and $\boldsymbol{\beta}_i$ is a unit normal vector which points in the direction in which the curve is concave. The radius of curvature is always taken to be positive.

When s is near 0, t_1 is near t_2, and an approximate expression for $s(t_1,t_2)$ can be obtained by letting $\mathbf{r}_i = \mathbf{r}_2$ and $T_i = T_2 = t_1$. Then $s^2 = \mathbf{r}_2 \cdot \mathbf{r}_2$, and from Eq. 8 one obtains

$$s^2 = (t_2 - t_1)^2 - \frac{(t_2 - t_1)^4}{12[R(t_1)]^2} - \frac{(t_2 - t_1)^5}{24}\frac{d}{dt_1}\left[\frac{1}{R(t_1)}\right]^2 + \cdots \quad 9.$$

This expression can be used to find the approximate dependence of t_2 on s and t_1. The result can be written in the form

$$t_2 = u_2(s,t_1)$$

Since in Region R, $t_2 \leq t_1$, the values of t_2 given by $u_2(s,t_1)$ must satisfy the condition that $t_2 \leq t_1$. An expression analogous to Eq. 9 can be written to express s^2 in powers of $(t_1 - t_2)$, with the coefficients being functions of t_2. When this relation is solved for t_1, the result can be expressed

$$t_1 = u_1(s,t_2)$$

These t_1 values must satisfy the condition $t_1 \geq t_2$.

To find $\gamma(s)$ for small s for a particle which is not a loop, let $r = t_2$ in Eq. 7. Then $\gamma(s)$ can be written

$$\gamma(s) = \frac{1}{L} \int_{u_1(s,0)}^{L} dt_1 \left| \frac{\partial t_2}{\partial s} \right| \qquad 10.$$

By use of Eq. 9, $\gamma(s)$ can be expressed as follows:

$$\gamma(s) = 1 - \frac{s}{L} + \frac{s^2}{8L} \int_0^L \frac{du}{[R(u)]^2} - \frac{s^3}{12L} \left\{ \frac{1}{[R(L)]^2} + \frac{1}{[R(0)]^2} \right\} + \cdots \qquad 10a.$$

For a loop, the domain in Region R in the neighborhood of $t_1 = L$, $t_2 = 0$ corresponds to small s values. When account is taken of the contribution of this domain, for a loop the expression for $\gamma(s)$ is found to be

$$\gamma(s) = \frac{1}{L} \int_0^L dt_1 \left| \frac{\partial t_2}{\partial s} \right| \qquad 10b.$$

When Eq. 9 is used to evaluate $\partial t_2/\partial s$, for a loop, $\gamma(s)$ can be shown to be given by

$$\gamma(s) = 1 + \frac{s^2}{8L} \int_0^L \frac{du}{[R(u)]^2} + \cdots \qquad 11.$$

In Eq. 11 the next non-vanishing term is proportional to s^4.

When s is not near 0, a different expansion must be used instead of Eq. 9. Let $\mathbf{d}$ be the vector from point T_1 to point T_2. Then

$$s^2 = d^2 + |\mathbf{r}_1 - \mathbf{r}_2|^2 + 2\,\mathbf{d} \cdot (\mathbf{r}_2 - \mathbf{r}_1) \qquad 12.$$

where $d = |\mathbf{d}|$. When $|\mathbf{r}_2 - \mathbf{r}_1| \ll d$, s can be approximated by

$$s = d + \mathbf{d}_0 \cdot (\mathbf{r}_2 - \mathbf{r}_1) + \frac{|(\mathbf{r}_2 - \mathbf{r}_1) \times \mathbf{d}_0|^2}{2d} + \cdots \qquad 13.$$

where $\mathbf{d}_0 = \mathbf{d}/d$. When $\mathbf{r}_1$ and $\mathbf{r}_2$ are evaluated from Eq. 8, s can be approximated by the Taylor series

$$s(t_1,t_2) = \sum_{i=0}^{2} \sum_{j=0}^{i} \frac{s_{i-j,j}}{j!(i-j)!} (t_i - T_i)^{i-j} (t_2 - T_2)^j \qquad 14.$$

where $s_{00} = d$, and

$$\left.\begin{array}{l} s_{10} = -(\mathbf{d}_0 \cdot \boldsymbol{\alpha}_1) \qquad\qquad s_{01} = (\mathbf{d}_0 \cdot \boldsymbol{\alpha}_2) \\ s_{20} = \frac{(\mathbf{d}_0 \cdot \boldsymbol{\beta}_1)^2}{d} - \frac{(\mathbf{d}_0 \cdot \boldsymbol{\beta}_1)}{R(T_1)} \qquad s_{02} = \frac{(\mathbf{d}_0 \cdot \boldsymbol{\beta}_2)^2}{d} + \frac{(\mathbf{d}_0 \cdot \boldsymbol{\beta}_2)}{R(T_2)} \\ s_{11} = -\frac{1}{d}[(\boldsymbol{\alpha}_1 \cdot \boldsymbol{\alpha}_2) - (\mathbf{d}_0 \cdot \boldsymbol{\alpha}_1)(\mathbf{d}_0 \cdot \boldsymbol{\alpha}_2)] \end{array}\right\} \quad 15.$$

Jones and Kline show that the critical points will either be stationary points, at which both s_{10} and s_{01} are zero, or they will be non-stationary points or arcs on the boundary of Region R. Stationary points can occur either on the boundary or within the region. There will be critical non-stationary boundary points or arcs at points or arcs on the boundary at which the curves of constant s are tangent to the boundary. Also, critical nonstationary boundary points will occur at points where the boundary has a sharp corner.

At all points on the boundary line $t_1 = t_2$ of Region R, $s = 0$, and thus the entire boundary line is tangent to the curve $s(t_1, t_2) = 0$ and therefore is a critical non-stationary arc. However, as the form of $\gamma(s)$ for small s has already been calculated, this critical arc need not be considered again. For filaments satisfying the conditions assumed in this discussion, no other critical arcs will ordinarily be expected.

At the point $t_1 = L$, $t_2 = 0$, the boundary of Region R has a corner. When the particle is not a loop, the corner point will be a critical point, and the contribution of this domain of Region R must be calculated from Eq. 7. For a loop, Eq. 7 includes the contribution of this domain, and thus it need not be considered separately.

The curves of constant s will be tangent to the line $t_2 = 0$ at points on this line at which s_{10} vanishes. These points thus will be critical non-stationary boundary points. Similarly, there will be critical non-stationary boundary points on the line $t_1 = L$ at points at which $s_{01} = 0$. Critical non-stationary points for which s_{10} or s_{01} is zero occur when the distance from one end of the curve to a point on the curve has a maximum or minimum value as the point is moved along the curve.

At stationary critical points, both s_{10} and s_{01} are zero, and $s(t_1, t_2)$ ordinarily has either a maximum, a minimum, or a saddle point.

When the stationary point is a maximum, $s_{20} < 0$, $s_{02} < 0$, and $(s_{11}^2 - s_{20}s_{02}) < 0$. Assume that the stationary point occurs at a point in the interior of Region R at which $t_1 = T_1$, $t_2 = T_2$. Then only points in this neighborhood will contribute to the integral in Eq. 7. By letting $t_1 = r$, Eq. 7 can be written

$$\gamma(s) = \frac{1}{L} \sum_{i=1}^{j(s)} \int_{a_i(s)}^{b_i(s)} dt_1 \left| \frac{\partial t_2}{\partial s} \right| \qquad 16.$$

For $s \leq s_{00}$, from Eq. 14 the relation

$$|\partial t_2/\partial s| = [2s_{02}(s - s_{00}) - (s_{02}s_{20} - s_{11}^2)(t_1 - T_1)^2]^{-\frac{1}{2}} \qquad 17.$$

can be obtained. The integration in Eq. 7 ranges over all allowed values of t_1 at a given value of s in the neighborhood of s_{00}. Then $j(s) = 2$, and

$$b_1(s) = b_2(s) = 2T_1 - a_1(s) = 2T_1 - a_2(s) = T_1 + b(s) \qquad 18.$$

where

$$b(s) = [2s_{02}(s - s_{00})(s_{02}s_{20} - s_{11}^2)^{-1}]^{\frac{1}{2}} \qquad 19.$$

Thus, when s is near s_{00}, the limiting form of $\gamma(s)$ is given by

$$\gamma(s) = (2\pi/L)[s_{02}s_{20} - s_{11}^2]^{-\frac{1}{2}} + \cdots \qquad 20.$$

for $s \leq s_{00}$. For $s > s_{00}$, $\gamma(s) = 0$.

Similarly, if $s(t_1,t_2)$ has a minimum at $t_1 = T_1$, $t_2 = T_2$ in the interior of Region R, in the neighborhood of s_{00}, $\gamma(s)$ can be approximated by

$$\gamma(s) = (2\pi/L)(s_{02}s_{20} - s_{11}^2)^{-\frac{1}{2}} \qquad 21.$$

for $s \geq s_{00}$, and $\gamma(s) = 0$ for $s < s_{00}$.

At an interior stationary point of Region R at which

$$(s_{11}^2 - s_{20}s_{02}) > 0,$$

there will be a saddle point. Equation 16 again can be used to find $\gamma(s)$, which can be written in the form

$$\gamma(s) = \gamma_c(s) + \gamma_d(s) \qquad 22.$$

where $\gamma_c(s)$ is a function with all derivatives continuous at $s = s_{00}$. According to Erdélyi's theorem, $\gamma_c(s)$ will not contribute to the asymptotic expansion and thus need not be considered. For $s_{02}(s - s_{00}) > 0$, by use of Eq. 16 $\gamma_d(s)$ can be written

$$\gamma_d(s) = \frac{1}{L} \sum_{i=1}^{j(s)} \int_{T_1-T}^{T_1+T} dt_1 \left| \frac{\partial t_2}{\partial s} \right| \qquad 23.$$

where $j(s) = 2$ and T is some convenient value of t_1 such that $[2s_{02}(s - s_{00}) + (s_{11}{}^2 - s_{02}s_{20})T^2] > 0$. Thus

$$\gamma_d(s) = \frac{2}{L[s_{11}{}^2 - s_{02}s_{20}]^{\frac{1}{2}}}$$
$$\times \left[2 \log \frac{(s_{11}{}^2 - s_{02}s_{20})^{\frac{1}{2}} T + [s_{02}(s - s_{00}) + (s_{11}{}^2 - s_{02}s_{20})T^2]^{\frac{1}{2}}}{[2s_{02}]^{\frac{1}{2}}} \right.$$
$$\left. - \log |s - s_{00}| \right] \quad 24.$$

For $s_{02}(s - s_{00}) \leq 0$, the same expressions are obtained for $\gamma_c(s)$ and $\gamma_d(s)$. As all derivatives of

$$\log \frac{(s_{11}{}^2 - s_{02}s_{20})^{\frac{1}{2}} T + [s_{02}(s - s_{00}) + (s_{11}{}^2 - s_{02}s_{20})T^2]^{\frac{1}{2}}}{[2s_{02}]^{\frac{1}{2}}} \quad 25.$$

are continuous at s_{00}, the contribution of the saddle point to the asymptotic expansion comes from the quantity

$$\gamma_{dd}(s) = -(2/L)(s_{11}{}^2 - s_{02}s_{20})^{-\frac{1}{2}} \log |s - s_{00}| \quad 26.$$

This expression holds for all s in the neighborhood of s_{00}. In an appendix to their paper, Jones and Kline (7) discuss the asymptotic expansion of Fourier integrals with logarithmic discontinuities.

The case $(s_{11}{}^2 - s_{02}s_{20}) = 0$ must be treated differently and will not be discussed.

When there is a boundary critical point which is a maximum, a minimum, or a saddle point, the contribution to the asymptotic expansion can be shown to be exactly half that given by Eqs. 20, 21, and 26, respectively.

There will be a critical non-stationary boundary point on the line $t_2 = 0$ at a point $t_1 = T_1$ on this line at which $s_{10} = 0$. In the neighborhood of this point, Eq. 14 can be solved for t_2, giving the approximation

$$t_2 = (s_{01})^{-1}[(s - s_{00}) - (s_{20}/2)(t_1 - T_1)^2] \quad 27.$$

This equation can be used to obtain $a_i(s)$ and $b_i(s)$ for the integral in Eq. 7. For a given value of s, the allowed t_1 values are those which make t_2 positive in Eq. 27. If $s_{01} > 0$ and $s_{20} > 0$,

$$\gamma(s) = 2(Ls_{01}s_{20})^{-1}[2s_{20}(s - s_{00})]^{\frac{1}{2}} \quad 28.$$

for $s_{20}(s - s_{00}) \geq 0$, and $\gamma(s) = 0$ for $s_{20}(s - s_{00}) \leq 0$. For $s_{01} > 0$, $s_{20} > 0$, $\gamma(s)$ can be written in the form $\gamma(s) = \gamma_c(s) + \gamma_d(s)$, with

$\gamma_d(s)$ being given by Eq. 28 and with $\gamma_c(s)$ being a function which has all derivatives continuous at $s = s_{00}$ and which thus will not contribute to the asymptotic expansion. Thus for $s_{10} > 0$ the contribution to the asymptotic expansion can be calculated from Eq. 28 regardless of the sign of s_{20}. This equation can also be shown to apply for $s_{01} < 0$.

A similar procedure can be used to show that for a nonstationary critical point at which $s_{01} = 0$ for $t_2 = T_2$ on the boundary line $t_1 = L$, the contribution to the asymptotic expansion is given by

$$\gamma(s) = -2(Ls_{10}s_{02})^{-1}[2s_{02}(s - s_{00})]^{\frac{1}{2}} \qquad 29.$$

for $s_{02}(s - s_{00}) \geq 0$, with $\gamma(s) = 0$ for $s_{02}(s - s_{00}) \leq 0$.

At the corner point of Region R, where $t_1 = L$ and $t_2 = 0$, assume that neither s_{10} nor s_{01} is zero. (Special treatment is required when one or both of these quantities is zero at the corner point.) In the vicinity of this point, the lines of constant s have the approximate form

$$s - s_{00} = s_{10}(t_1 - L) + s_{01}t_2 \qquad 30.$$

If $s_{10} < 0$ and $s_{01} > 0$, from Eq. 7 the limiting form of $\gamma(s)$ can be shown to be given by

$$\gamma(s) = -(s_{01}s_{10}L)^{-1}(s - s_{00}) \qquad 31.$$

for $s \geq s_{00}$, with $\gamma(s) = 0$ for $s \leq s_{00}$. A similar calculation shows that regardless of the signs of s_{10} and s_{01}, the contribution to the asymptotic expansion is given by Eq. 31.

THE ASYMPTOTIC EXPANSION OF THE INTENSITY

After the critical points have been located, the s_{ij} can be calculated from Eq. 14, and by use of the methods of the preceding section, the approximate form of $\gamma(s)$ can be found in the neighborhood of each value s_i which contributes to the asymptotic expansion. The convention will be used that $i = 0$ corresponds to $s = 0$, and thus $s_0 = 0$. For $i \geq 1$, the s_i represent s values at which $\gamma(s)$ or its derivatives have discontinuities. In the neighborhood of one of the s_i, $\gamma(s)$ can be expressed by

$$\gamma(s) = \gamma_{ci}(s)$$
$$+ \frac{B_{i0}}{\pi} \log |s - s_i| + \sum_{j=0}^{N} A_{ij} 2^{-\alpha_{ij}}[|s - s_i| + (-1)^i(s - s_i)]^{\alpha_{ij}} \qquad 32.$$

where $\gamma_{ci}(s)$ is a function with all derivatives continuous at $s = s_i$. In this equation, for odd i the factors multiplying the A_{ij} are zero for $(s - s_i) \geq 0$, while for even i these factors vanish for $(s - s_i) \leq 0$. For $i \geq 1$, only the A_{i0} have been calculated in the preceding section. The first A_{0j} can be found from Eq. 10a or Eq. 11. Note that one value of s_i can correspond to more than one critical point in Region R.

The contributions of Eq. 32 to the asymptotic expansion of $J(h)$ can be obtained by Erdélyi's theorem (3) and the theorem of Jones and Kline (7). From Eq. 5, the dominant terms in the asymptotic expansion are found to be

$$I(h) \sim \frac{\pi}{hL} + \frac{2}{L} \sum_{j=0}^{\frac{N-1}{2}} \frac{(-1)^j A_{0,2j+1}}{(2j+1)h^{2j+2}}$$

$$+ \frac{2}{L} \sum_{i=1}^{I} \left[\frac{A_{i0}\Gamma(\alpha_{i0}+1)\sin(hs_i + \beta_i)}{s_i h^{\alpha_{i0}+2}} - \frac{B_{i0}\sin hs_i}{s_i h^2} \right] \quad 33.$$

where $\beta_i = (-1)^i (1 + \alpha_{i0})(\pi/2)$.

ELLIPTICAL FILAMENTS

The procedure outlined above can be used to compute the asymptotic expansion of the intensity scattered by an elliptical filament with semi-minor axis a and semi-major axis va. If the angles θ_1 and θ_2 are used to describe the positions of two points on the filament, with the x and y coordinates of the points being given by

$$x_i = a \cos \theta_i \qquad y_i = va \sin \theta_i$$

then the distance s between the two points has the value

$$s = 2a[\sin(\tfrac{1}{2})(\theta_1 - \theta_2)][1 + (v^2 - 1)\cos^2(\theta_1 + \theta_2)/2]^{\frac{1}{2}} \quad 34.$$

The intensity $I(h)$ can then be expressed

$$I(h) = \frac{2a^2}{L^2} \int_0^{2\pi} d\theta_1 \int_0^{\theta_1} d\theta_2 \frac{\sin hs}{hs}$$
$$\times [1 + (v^2 - 1)\cos^2 \theta_1]^{\frac{1}{2}} [1 + (v^2 - 1)\cos^2 \theta_2]^{\frac{1}{2}} \quad 35.$$

There is a stationary interior critical point for $\theta_1 = 3\pi/2$, $\theta_2 = \pi/2$, and there are boundary stationary critical points for $\theta_1 = \pi$, $\theta_2 = 0$

and for $\theta_1 = 2\pi$, $\theta_2 = \pi$. The interior stationary point is a maximum, and the boundary stationary points are saddle points. Also, for $v^2 > 2$, on the line $\theta_2 = 0$ there are critical non-stationary points for the two θ_1 values for which $\cos\theta_1 = -(v^2-1)^{-1}$, and on the line $\theta_1 = 2\pi$ there are critical non-stationary points for the two θ_2 values for which $\cos\theta_2 = -(v^2-1)^{-1}$. The s_{ij} for the stationary and non-stationary critical points can be found from Eq. 14. Then, from Eq. 33 the asymptotic expansion of Eq. 35 is found to be

$$I(h) \sim \frac{\pi}{hL} + \frac{8\pi a^2 v^2}{L^2(v^2-1)^{\frac{1}{2}}} \frac{\sin 2ha}{(2ha)^2} - \frac{8\pi a^2 v}{L^2(v^2-1)^{\frac{1}{2}}} \frac{\cos 2hav}{(2hav)^2} \qquad 36.$$

The contributions from the non-stationary critical points cancel each other and thus do not appear in Eq. 36.

By a change of variables and of the order of integration, Eq. 35 can be expressed

$$I(h) = \frac{2}{L}\int_0^{2va} ds \frac{\sin hs}{hs}\gamma(s) \qquad 37.$$

where

$$\left.\begin{aligned}
\gamma(s) &= \frac{4a}{L}\int_0^{g(s)} dw\, G(s,w) \\
g(s) &= \frac{\pi}{2} \qquad\qquad\qquad\qquad\qquad 0 \le s \le 2a \\
g(s) &= \cos^{-1}\left[\frac{s^2 - 4a^2}{4a^2(v^2-1)}\right]^{\frac{1}{2}} \qquad 2a \le s \le 2va \\
G(s,w) &= \left[1 - \frac{s^2}{4a^2} + (v^2-1)\cos^2 w\right]^{-\frac{1}{2}} \\
&\qquad\qquad \times [1 + (v^2-1)\cos^2 w][H(s,w)]^{\frac{1}{2}} \\
H(s,w) &= \left\{1 - \frac{(v^2-1)\dfrac{s^2}{4a^2}}{[1+(v^2-1)\cos^2 w]^2}\right\}^2 \\
&\qquad + \frac{4\left(\dfrac{s}{2a}\right)^2 (v^2-1)\sin^2 w}{[1+(v^2-1)\cos^2 w]^3}
\end{aligned}\right\} \qquad 38.$$

This equation gives an explicit expression for $\gamma(s)$ for an elliptical filament. Direct calculation of the asymptotic expansion of Eq. 37 reproduces Eq. 36.

DISCUSSION

In Eqs. 10a and 11 the coefficients of s^2 are quantities averaged over the entire filament. In the former equation, the coefficient of s^3 depends on the values of the radius of curvature at the ends of the filament, and the term in s occurs because of the properties of the ends of the filament. There are no terms proportional to s and s^3 in Eq. 11.

These results suggest that in the expression for $\gamma(s)$ in the neighborhood of $s = 0$, the coefficients of odd powers of s are determined by properties of the ends of the filament, while the coefficients of even powers of s involve averages over the entire filament. Then, as a loop can be considered as a filament without ends, the coefficients of all odd powers of s would vanish. This suggestion is supported by the results obtained for a few higher-order terms which have been calculated, but which are not shown in Eqs. 10a and 11.

If for a loop the A_{0j} in Eq. 32 vanish for odd j, the only non-oscillatory term in the asymptotic expansion for $I(h)$ for a loop will be the term proportional to h^{-1}.

Luzzati and Benoit (8) studied some properties of $\gamma(s)$ for filaments for s in the neighborhood of $s = 0$. Without restricting their discussion to particles which did not have sharp bends and did not cross themselves, they calculated the coefficient of the term proportional to s in the expansion of $\gamma(s)$ for s near $s = 0$. Their calculations suggest that for a filament without bends, ends, or points where the particle crosses itself, the asymptotic expression for the intensity is $\pi/(hL)$, just as was found in this discussion of the scattering from filaments. When a minor error is corrected in the right-hand bracket of Luzzati's and Benoit's Equation 12, their Equation 13 becomes equivalent to our Eq. 11.

An analogous association of certain powers of s with boundary discontinuities may occur in the series expansion of $\gamma(s)$ near $s = 0$ for plane laminae and three-dimensional solids, for which Kirste and Porod (9) computed approximate expressions for $\gamma''(s)$ in the neighborhood of $s = 0$. For both types of bodies, $\gamma''(0)$ was found to vanish when the boundary surface or curve was smooth, with no corners. For a generalized cylinder (1), however, $\gamma''(0) \neq 0$. As generalized cylinders always have corners where the ends of the cylinder join the sides, the non-vanishing value of $\gamma''(0)$ for generalized cylinders suggests that for two- and three-dimensional particles, a non-zero value of $\gamma''(0)$ possibly is associated with the presence of corners or other discontinuities in the curvature of the boundary.

The asymptotic expression, Eq. 36, for the intensity from an elliptical filament is not applicable to circular filaments, for which $v = 1$. In this case, however, the asymptotic expansion can be found after letting $v = 1$ in Eq. 37. The dependence on h is different from that given by Eq. 36. This behavior suggests that a circle is too symmetric a particle to provide a representative illustration of the properties of the scattering from filaments. A similar result is found for two- and three-dimensional particles, since the asymptotic behavior of the scattering from circles and spheres differs from the asymptotic scattering from generalized cylinders (1), which are less symmetrical particles.

In Eq. 32 the exponents α_i are either integers or half-integers. The results of Jones and Kline show that the relatively simple values of these exponents are a fundamental property of the asymptotic expansion of this type of Fourier integral and that more complicated exponents would rarely be expected. Simple exponents might also be expected to occur in the non-logarithmic discontinuities in the corresponding approximations for $\gamma(s)$ for two- and three-dimensional particles, since in all cases where $\gamma(s)$ has been calculated for these particles, the exponents are either integers or half-integers.

The methods developed in the calculations of the scattering from filaments should be useful for computing the asymptotic expansion of the intensity for two- and three-dimensional particles. For example, analogues of Eqs. 7 and 14 might be developed for two- and three-dimensional particles, and the results of Jones and Kline indicate that the concept of critical points should be applicable to triple and higher-order multiple integrals.

REFERENCES

1. Schmidt, P. W., J. Math. Phys., **6,** 424 (1965).
2. Guinier, A., et al., *Small-Angle Scattering of X-Rays*, J. Wiley and Sons, New York, 1955; Eq. (15), p. 10.
3. Erdélyi, A., *Asymptotic Expansions*, Dover Publications, New York, 1956, p. 49.
4. Reference 2, p. 15.
5. Jones, D. S., and Kline, M., J. Math. and Phys., **37,** 1 (1958).
6. Widder, D. V., *Advanced Calculus*, Prentice-Hall, New York, 1947, p. 84.
7. Reference 5, p. 27.
8. Luzzati, V., and Benoit, H., Acta Cryst., **14,** 297 (1961).
9. Kirste, R., and Porod, G., Kolloid-Z., **184,** 1 (1962).

Scattering Functions of Polymer Coils

R. G. KIRSTE

Institut für physikalische Chemie, Universität Mainz, Germany

INTRODUCTION

There are already a number of papers (1–8), which deal with the scattering of light and X-rays from coiled, chainlike molecules. The various scattering functions calculated in these papers differ in part from one another, because different approximations were used in the calculations. But obviously it was the intention of the authors to calculate a universal scattering function for polymer coils.

In reality, a universal scattering function for coils does not exist, because the scattering behavior depends on individual features of the threads. For instance, distinct qualitative differences were found between the scattering functions of syndiotactic and isotactic polymethyl methacrylate (9–10), though both form coils with nearly the same ratio of radius of gyration to chain length in solution (11). Among the previously calculated theoretical scattering functions there are none which would permit such differences. In this paper it will be attempted to fill this gap.

DEBYE'S SCATTERING FUNCTION FOR THE IDEALIZED RANDOM COIL AND POSSIBILITIES FOR A REFINEMENT OF THE CALCULATION

Let an unbranched chain molecule of N atoms be given. Let the bond length be l and the valence angle be α. The position of a bond in relation to the next but one may be characterized by the angle of rotation φ (see Fig. 1). By the statement of all angles of rotation an instantaneous shape of the molecule can be described. Now, the instantaneous shape is neither recognizable nor essential. Only the mean shape is of any interest.

In order to see how one can illustrate the mean shape of coiled molecules, let us regard the random coils. A coil may be termed as random, if the position of two bonds relative to one another is random with regard to direction unless the length L of that part of the thread which connects the two bonds, is shorter than a certain finite value characteristic for the whole molecule. For $L \to 0$, the transition from random to regular orientation shall take place according to a fixed law, which again is valid for the whole thread.

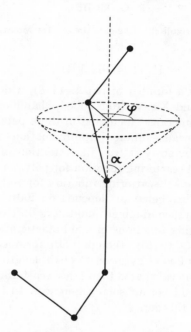

Fig. 1. The characterization of a conformation of a chain with valence angles. α = valence angle, φ = angle of rotation.

The distance r_{ij} between the atoms i and j of a polymer coil varies with time. In random coils, predictions can be made (12) about the mean square $\langle r_{ij}^2 \rangle$ and about the distribution $W(r_{ij})$. Let L_{ij} be the length of the thread between the atoms i and j. Then for $L_{ij} \to \infty$ the relation

$$\langle r_{ij}^2 \rangle = b \cdot L_{ij} \qquad 1.$$

is valid. The factor b is a constant of the thread and is termed the statistical chain element. Equation 1 is strictly valid for $L_{ij} \to \infty$, within the limits of error, however, even if $r_{ij} \ll L_{ij}$, i.e., if the part

L_{ij} of the thread already resembles a coil rather than a rod. Liable to the same restriction as Eq. 1 is the expression for $W(r_{ij})$

$$W(r_{ij}) = (3/2\pi\langle r_{ij}^2\rangle)^{3/2} \cdot 4\pi r_{ij}^2 \cdot \exp\left(-3r_{ij}^2/\langle 2r_{ij}^2\rangle\right) \qquad 2.$$

Equation 1 can be verified experimentally on θ-solutions, i.e. solutions with ideal osmotic pressure. Therefore it is generally accepted that in θ-solutions the coils are random. For good solvents one obtains instead of Eq. 1

$$\langle r_{ij}^2\rangle = \text{const.}\ L_{ij}^{1+\epsilon} \quad \text{with } 0 < \epsilon < 0.5 \qquad 1^*.$$

The coil is expanded in comparison with the random coil.

P. Debye was the first to perform a calculation about electromagnetic scattering of coils (1). The model used is an idealized random coil. Equations 1 and 2 are assumed to be valid for all pairs of atoms in the coil. The well-known result of the calculation is

$$i(r_g h) = 2/(r_g h)^4 \{\exp[-(r_g h)^2] + (r_g h)^2 - 1\} \qquad 3.$$

with

$$h = (4\pi/\lambda)\sin\vartheta$$

r_g is the root mean square of the radius of gyration of the coil, λ is the wavelength and ϑ half of the scattering angle. The scattering function $i(r_g h)$ is normalized in such a manner that $i(0)$ is put equal to unity.

For light-scattering measurements Eq. 3 is valid in all cases of practical interest, but in small-angle X-ray scattering, greater values of the argument occur and therefore deviations are observable.

The investigations which were carried out in order to refine the function of Debye can be divided into two groups. First group (2–6): An approximation to real conditions is performed in the range of small values of L_{ij}. The random character of the coils is however maintained in these papers. Second group (7): Those deviations from the random character of the coiling are taken into account which are caused by the effect of the excluded volume in good solvents. However the details of the distance distribution in the range of small L_{ij} are ignored.

The investigations of the first group led to a modification of the scattering functions in the range of relatively large values of the argument $r_g h$. For small $r_g h$ (light-scattering range) the new calculated functions become identical with the function of Debye. In the functions which were calculated in the second group, it is just reverse: in the light-scattering range they emerge modified in comparison with Debye's equation. For larger values of the argument, however, the functions become unrealistic, like Eq. 3.

From a rough contemplation of this matter it follows that the effect of a non-random behavior of the coil on the scattering function must vanish for sufficiently high values of the argument. Unfortunately only one investigation has been made of this problem until now (8). It did not lead to a general clarification because the molecular weights taken into consideration were too low.

The present paper belongs to the first of the above-mentioned groups. The influence of the excluded volume is being neglected as is usual in this group, though experimental results in a very good solvent are involved. In justification of this it may be stated that: 1) The comparison between experimental results and calculations will be performed for relatively large scattering angles 2) The general course of the experimental scattering functions is probably not affected by the second virial coefficient. The observed influence of the solvent is more probably based on the electron density.

A further deviation from the random character of coil molecules will be neglected in the present paper: the dependence of the probability for a segment to stay at a point on the position of that point in relation to other segments. This effect, though closely related to the excluded volume effect, doesn't vanish in θ-solutions. If one assumes that besides the geometrical volume of the segments there are certain areas in the neighborhood of the segments which are forbidden or less easy of access to other segments, then in θ-solutions a greater part of the segments must sit in the potential trough of other segments (13). Such aggregates of segments lead to a decrease of intensity for greater scattering angles.

Finally, it may be mentioned that only high molecular weights will be involved in the calculations.

THREADS WITH PERSISTENCE OF DIRECTION AND PERSISTENCE OF CURVATURE

The intention of the calculations of the present paper was the interpretation of measurements on syndiotactic and isotactic polymethyl methacrylate (PMMA) in solution. Let us first look at the calotte-models of both the polymers in Fig. 2. The models derive from X-ray investigations on the solid polymers (14). In the model of syndiotactic PMMA we recognize a helix-like structure, whereas in isotactic PMMA, matter is more equally distributed around the axis. The chain C-atoms in both the models form a 5.2-helix. The discernible helix of the syndiotactic PMMA in Fig. 2 is a 10.1-helix.

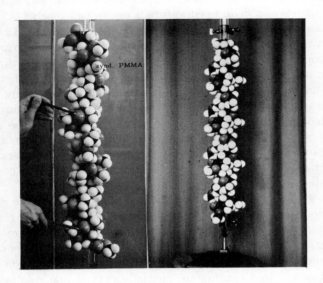

Fig. 2. Calotte models of isotactic and syndiotactic PMMA. On the left syndiotactic, on the right isotactic.*

It is realized by the arrangement of the side groups. The chain C-atoms in the model are virtually hidden. For the scattering function it is obviously not the position of the chain-C-atoms which is decisive, but the X-ray-optical center line. By this we shall understand the connection line of the X-ray-optical centers of gravity of the cross-sections of the thread. In the calotte model of the syndiotactic PMMA the center line is approximately a helix. In the model of the isotactic PMMA it is a complicated line travelling about in the thread.

In solution the polymer thread is coiled. The helix structure has been collapsed. Certainly smaller or greater fragments of the helix will be maintained, however. From the radius of gyration (11), the mean length of the intact helical sequences may be estimated. It

* We thank Dr. G. Schröder (Röhm u. Haas GmbH, Darmstadt) for permitting reproduction of this photograph.

appears that these sequences in solutions of PMMA must be smaller than an identity period of the syndiotactic 10.1-helix. The periodicity of the helix therefore cannot play any part in the coil. The curvature of the helix on the other hand can be dominating in the coil even if the helix fragments are small. For the coiled isotactic PMMA the predominance of a definite curvature is not to be expected.

Consequently for isotactic PMMA a model thread should be used, the curvature of which fluctuates randomly from one point to another. This property is possessed by the model of the persistent thread of Kratky and Porod (2). Syndiotactic PMMA on the other hand should diffract X-rays like a thread in which the curvature of the 10.1-helix of the solid polymer is favored. As an extension of the terminology of Kratky and Porod such a model may be named "thread with persistence of curvature." The persistence defined by Kratky and Porod one would have to name accordingly "persistence of direction." A thread with a persistence of curvature has of course a persistence of direction too.

Now the question is how to construct such threads. One can use the same procedure as for the construction of a valence angle chain according to Fig. 1. N vectors $\mathbf{l}_i$ of constant length l are strung together, the position of the vectors relative to each other being determined by the valence angle α and the angles of rotation φ_i. An approximation to a continuous, spatially curved line results, if the length of the bond vectors and, together with this, the valence angle, are made small. A preference for a definite curvature results, if in the angles of rotation the *cis* position is favored. For given values of α and l of course only one value of preferred curvature can be produced, whereas the degree of preference of the curvature will be determined by the degree of the hindrance to rotation.

The formalism of the construction is the same as in the case of a real polymer chain. One has to keep in mind, however, that the lines constructed in this way are "center lines" in the sense defined above. The vectors $\mathbf{l}_i$ and the angle α have nothing do to with a real bond length and a real valence angle. The mean end-to-end distance of the center line is identical with that of the molecule. The length of the center line may be a little greater than the length of the corresponding molecule chain just as a helical line is longer than the corresponding axis. Accordingly the mass per unit length of the center line may be somewhat smaller than the mass calculated from the chemical formula of the polymer and the length of the connection line of the actual chain atoms.

THE CONSTRUCTION OF RANDOM THREADS
Relations between the Parameters of the Coil

The connection between the end-to-end distance r_{ee} and the parameters l, α and φ for sufficiently long threads is given by

$$\langle r_{ee}^2 \rangle = nl^2 \cdot \frac{1 + \cos \alpha}{1 - \cos \alpha} \cdot \frac{1 + \eta}{1 - \eta} \quad \text{with } \eta = \langle \cos \varphi \rangle \qquad 4.$$

n is the number of vectors $\mathbf{l}_i$. The equation is valid if $\langle \sin \varphi \rangle$ vanishes and no correlations between different rotation angles are to be considered (12). Further we have

$$L = n \cdot l \qquad 5.$$
$$\langle r_{ee}^2 \rangle = b \cdot L \qquad 6.$$

L is the total length of the thread and b is the statistical chain element. From Eq. 4 to 6 it follows that

$$b = \langle r_{ee}^2 \rangle \Big/ L = l \cdot \frac{1 + \cos \alpha}{1 - \cos \alpha} \cdot \frac{1 + \eta}{1 - \eta} \qquad 7.$$

Threads with different degrees of preference for the cis-position were constructed with the aid of a random generator. In order to be able to compare the obtained threads with each other, care was taken for the statistical chain element to be of the same length in all constructed threads. We put $b = 20l$. Equation 7 changes with this into

$$\frac{1 + \cos \alpha}{1 - \cos \alpha} \cdot \frac{1 + \eta}{1 - \eta} = 20 \qquad 7a.$$

φ was allowed to be within a range of $\pm \varphi^*$ symmetrical to the cis-position.

The connection between $\eta = \langle \cos \varphi \rangle$ and φ^* arises from the following simple calculation ($\varphi = \varphi_0$ is the midpoint of the allowed range):

$$\langle \cos \varphi \rangle = (1/2\varphi^*) \cdot \int_{\varphi_0 - \varphi^*}^{\varphi_0 + \varphi^*} \cos \varphi \, d\varphi = (\sin \varphi^* / \varphi^*) \cdot \cos \varphi_0 \qquad 8.$$

In the present case $\varphi_0 = \pi$ and consequently

$$\cos \varphi = - \sin \varphi^* / \varphi^* \qquad 8a.$$

Thread Construction

From Eqs. 7a and 8a some pairs of parameters α and φ^* were calculated. They are listed in Table I. Each pair of values defines a type of random coil with the chain element $b = 20$. (The uneven values for α and φ^* arise from the fact that they originally were derived from another parameter which is of no interest in this connection.)

TABLE I
The parameters α and φ^* of the models used.

Model number	α	$2\varphi^*$
1 (free rotation)	0.43998	2π
2	0.18118	2.7324
3	0.08060	1.2413
4	0.04030	0.6235

Each of the vectors l_i can be calculated from the preceding one according to

$$l_i = l_{i-1} \cdot \cos \alpha + (\mathbf{u} \cos \varphi + \mathbf{v} \sin \varphi) \cdot l \sin \alpha \qquad 9.$$

$\mathbf{u}$ is an unit vector orthogonal to l_{i-1} and pointing in the cis-direction relative to l_{i-2}. $\mathbf{v}$ is an unit vector orthogonal to $\mathbf{u}$ and orthogonal to l_{i-1}. φ must be determined by a random number z (see appendix). If z is placed with equal probability anywhere between 0 and 1 then φ is to be calculated from

$$\varphi = (z - 0.5) \cdot 2\varphi^* \qquad 10.$$

As the start of the chain it is most convenient to choose the origin of the coordinate system. By successive calculation of the l_i and addition to the last of the calculated thread points a special conformation of the respective type of coil is formed. In Figs. 3–6, projections are shown of one conformation of each of the types of coils which are characterized by the values in Table 1. Each thread consists of 1000 vectors l_i, i.e., of 50 statistical chain elements. In Fig. 5 two projections of the same thread are shown. They are orthogonal to each other. The thread in Fig. 3 has no persistence of curvature.

On viewing Figs. 3–6 one has to consider the perspective shortening by which corners and peaks can be simulated. The plotted lines illustrate instantaneous conformations, but all essential characteristics in which they differ from each other are independent on the instantaneous shape. As a result of these experiments one can say: Threads with equal statistical chain elements can be constructed, of which

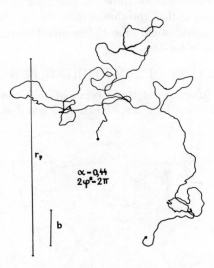

Fig. 3. Projection of a random conformation of thread model no. 1. $b =$ statistical chain element, $r_p =$ mean length of a projection of the end-to-end distance ($= 0.7854 \cdot \langle r_{ee}^2 \rangle^{1/2}$).

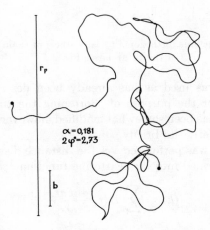

Fig. 4. Projection of a random conformation of thread model no. 2. For the notation see Fig. 3.

nevertheless the mean shapes differ in principle from one another. Especially it is possible that the thread exhibits any chosen curvature on the length of a statistical chain element. The threads in Fig. 2 are only one example for this.

CALCULATION OF SCATTERING FUNCTIONS

The scattering functions of those types of coils were calculated of which special conformations are shown in Figs. 3–6. Virtually the

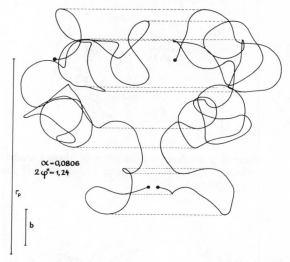

Fig. 5. Two projections (orthogonal to each other) of chain model no. 3. For the notation see Fig. 3.

same procedure was used as has already been described by other authors (4, 6). For the purpose of shortening the time needed for the calculation it has been somewhat modified, however. These modifications may be referred to briefly.

The calculation was performed via the distance distribution $H(r_i)$ which was transformed into the scattering function according to

$$i(h) = \sum_i H(r_i) \frac{\sin hr_i}{hr_i} \qquad 11.$$

By proceeding in such a manner the calculation of a high number of sine values has been avoided.

The distribution $H(r_i)$ was calculated via $H(r_i{}^2)$. By this means the calculation of a large number of square roots was avoided.

$H(r_i{}^2)$ was obtained as a step function in 550 intervals. The lengths of the intervals were arranged in the following way:

[0, 0.1]; [0.1, 0.2]; [0.2, 0.3]............................ [9.9, 10];
[10, 11]; [11, 12]; [12, 13] .... [99, 100];
[100, 110]; [110, 120]; ....
....
[100 000, 110 000] .... [990 000, 1 000 000].

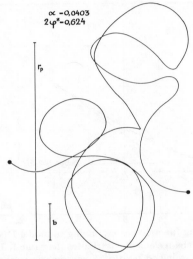

Fig. 6. Projection of a random conformation of thread model no. 4.
For the notation see Fig. 3.

In this kind of graduation, it can be determined easily from the first two digits whether a distance square belongs to a certain interval. The last 90 intervals remained empty in general. They served only for the event that extremely extended conformations would have been formed.

For the transformation of $H(r_i{}^2)$, the square roots of the boundaries of the intervals had to be extracted. For application of Eq. 11, the 550 intervals were intersected further, the range from the 191st up to the last non-empty interval being divided uniformly into 2500 intervals, by which altogether 2690 intervals were formed.

For the determination of $H(r_i{}^2)$ for each type of coil, 10 conformations were constructed. Each conformation consisted of 2016 scattering centers which correspond with 100.8 statistical chain elements. Let P_i and P_j be two scattering centers of a conformation. For $|i - j| \leq 40$ all distance squares were determined. For $41 \leq |i - j| \leq 95$ the centers were collected into groups of three, for $|i - j| \leq 96$ into groups of nine. Thus for each conformation 116,721 distance squares had to be calculated. Without this gathering into groups one would have had 2,031,120 distances per conformation.

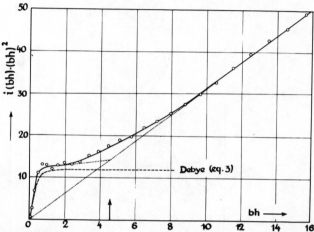

Fig. 7. The scattering function corresponding to model 1 (see Fig. 3). Normalization: For $i(0)$ every chain element contributes the value 1. The arrow designates the junction point according to Kratky. For comparison, the function of Debye (Eq. 3) is also shown (more detailed elucidation will be found in the paper and the appendix).

The calculation of a scattering function according to the method described requires 14 hours on a computer with a word time of 90 μsec. A further improvement of the calculation program seems to be possible.

In Figs. 7–10 the results of the calculations are shown. The usual plot for coil functions was chosen, that is $i(bh) \cdot (bh)^2$ vs. bh with b = statistical chain element. The functions are normalized in such a manner that $i(0)$ is equal to the number of statistical chain elements of the coil. For comparison the function of Debye (Eq. 3) also has been drawn in the graphs.

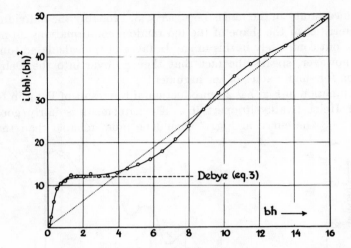

Fig. 8. The scattering function corresponding to model 2 (see Fig. 4). For the notation see Fig. 7.

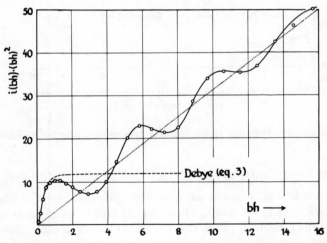

Fig. 9. The scattering function corresponding to model 3 (see Fig. 5). For the notation see Fig. 7.

In principle the values of a scattering function calculated with the aid of the Monte Carlo method are approximate. The accuracy depends on the number of conformations taken into account. From the fluctuation of the values in Fig. 7–10 one can perceive a relatively great uncertainty in the range $0.5 < bh < 3$. Probably the individual characteristics of the shape of the ten random conformations do not average out completely in this range. In favor of the calculated functions, however, speaks the fact that they go over into the Debye function for small values of the argument.

The function in Fig. 7 is a reproduction of the result of Peterlin (4) and of Heine-Kratky-Roppert (6). The agreement is fairly good. For $bh > 3$ the curve in Fig. 7 should be more reliable than that

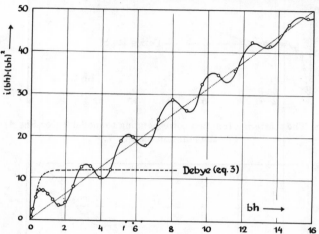

Fig. 10. The scattering function corresponding to model 4 (see Fig. 6). For the notation see Fig. 7.

published earlier, because it contains a greater number of calculated values.

DISCUSSION OF THE CALCULATED FUNCTIONS

The arrow in Fig. 7 marks the so-called "junction point" (Kratky and Porod). This is the point, where the asymptote of the scattering curve and the tangent through the inflection cross each other. According to Kratky and Porod (2, 5, 6, 15) one can derive the persistence length a from the abscissa of the junction point:

$$ah^* = \text{const.} \qquad\qquad 12.$$

For the constant, Kratky gave the value 2.3 in the last paper of the series. $ah^* = 2.3$ means for the model underlying Fig. 7, for which $a = b/2$, that $bh^* = 4.6$. The tangent of inflection has been drawn accordingly in Fig. 7, though other positions would also have been compatible with the calculated values.

Kratky and Porod defined the persistence length as the mean length of the projection of the (sufficiently long) thread on its starting direction. The author of the present paper considers it more convenient to define the persistence length as the length of a part of the thread, which on the average exhibits a change in direction of arc cos $(1/e) = 68.4°$, which for many coil models is identical with the original definition. Instead of "change in direction" one might speak of "curvature." That it would be more suitable to consider junction point and persistence length as measures of the mean curvature may be seen from Figs. 7–10.

If one goes forward from Fig. 7 to Fig. 10, the functions increasingly acquire a new character, though the persistence length in the original sense of the word is the same in all cases. The latter will be proved in the appendix. What does change from Fig. 7 to Fig. 10 is the curvature behavior of the threads.

For statistically fluctuating curvature, a scattering function of the kind of Fig. 7 emerges. The mean curvature in this case is given by the persistence length.

If a certain curvature in the coil is favored, then waves in the scattering curve emerge. From the position and the amplitude of the waves the radius of the preferred curvature and the amount of preference of the curvature can be estimated.

TABLE II

The radius (r_c) of preferred curvature of the model threads and the position of the first "submaximum" of the corresponding scattering functions. (The unit of length is $b/20$.) The values for the scattering functions of helices recently calculated may serve as a comparison.

Model number	r_c	bh_{max}	$r_c h_{max}$
1	—	—	—
2	5.52	11.0	3.0
3	12.40	5.5	3.4
4	24.80	2.9	3.6
Helix $r_a/t = 0.2$	—	—	3.50
Helix $r_a/t = 0.3$	—	—	3.54

The scattering functions of coils with preferred curvature resemble the scattering functions of helices. The latter were calculated in a preceding paper (16), and are shown again in Fig. 11. In Table II are listed the abscissae of the first "submaximum" of the scattering functions for the coils and their connection with the radius of preferred curvature. Those points which in Figs. 7–11 exceed the asymptote by the largest amount are considered the maxima.

INTERPRETATION OF SMALL-ANGLE X-RAY MEASUREMENTS ON TACTIC POLYMETHYL METHACRYLATES*

In Figs. 12–14, experimental scattering curves of tactic PMMA fractions in various solvents are shown. The dashed lines are the

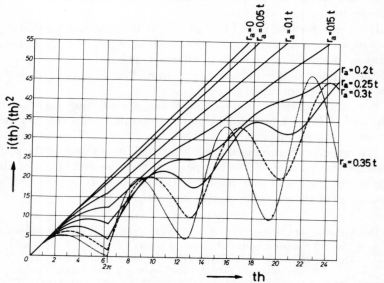

Fig. 11. Scattering functions of helices. H = identity period.
$x = r_a \cdot \cos \psi, \; y = r_a \cdot \sin \psi, \; z = t\psi/2\pi.$

functions which would result in the respective solvent, if the thread were completely stretched and had an infinitesimally small cross-section. These functions can be determined in correspondence with a given measurement, if the scattered intensity has been determined on an absolute scale. The determination of the absolute intensity was carried out according to O. Kratky (15), with the help of a

* A detailed description of the experimental technique will be found in a preceding paper (10).

rotating disk. If one intends to compare the scattering intensities in the several solvents with one another, one has to take the dashed lines in Figs. 12–14 as reference. A direct comparison is impossible, because the electron densities and the absorption coefficients of the various solvents differ from one another.

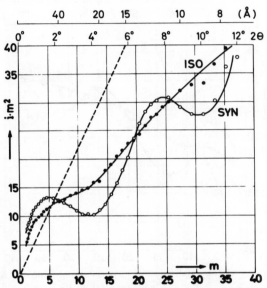

Fig. 12. Intensity of the small-angle X-ray scattering from PMMA in benzene. ● isotactic, ○ syndiotactic. m is the distance between the counting tube position and the primary beam in mm. $m = 179 \cdot \sin 2\theta \approx 179(\lambda/2\pi) \cdot h$; 179 mm = distance between sample and plane of registration. Dimension of the ordinate: counts $\cdot \text{min}^{-1}$ cm^2. On the uppermost scale are the distances calculated from Braggs equation.

In discussing the effects of the solvent on the scattering functions, one has to distinguish between the effect of the electron density and the effect of the solvent power. The solution of PMMA in benzene is nearly athermic, which means that benzene is a very good solvent (17) for PMMA. Acetone is of moderate solvent power and endothermic. Methylisobutyrate is a θ-solvent. In terms of the electron density, acetone has an extreme position among the three solvents (see Table III).

In a recent paper it was shown that the cross-section factor of a polymer coil depends rather strongly on the electron density of the solvent (18). With the exception of the curve SYN in Fig. 14 the different slopes of the curves in Figs. 12–14 for $m > 10$ may be attributed to this effect. (Strong cross-section factor depression in acetone, weaker in benzene and in methylisobutyrate.)

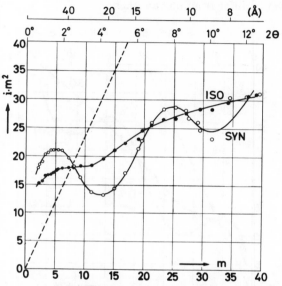

Fig. 13. Intensity of the small-angle X-ray scattering from PMMA in acetone. For the notation see Fig. 12.

A systematic influence of the solvent power on the curves is observed only on the extreme left part of the curves. The scattering intensity there is the greater the poorer the solvent. The effect can be interpreted by the greater coil density in poor solvents.

TABLE III

The electron densities of PMMA, benzene, acetone and methylisobutyrate at 20°C. ($n = z - v.d$, n = number of surplus electrons of PMMA per monomer unit; z = total number of electrons per monomer unit in PMMA = 54; v = partial molecular volume of the monomer unit = 80 cm³/6.023 · 10²³; d = electron density of the solvent).

Solvent	vd	n
Acetone	34.7	19.3
Benzene	37.8	16.2
Methylisobutyrate	39.0	15

A special interpretation is necessary for syndiotactic PMMA in methylisobutyrate (Fig. 14). In this solvent no stable solutions could be obtained. The solutions had to be heated briefly during the measurements at intervals of about two hours in order to prevent coagulation. On the scattering curve a lower intensity at large scattering angles is observed in comparison with the values for isotactic PMMA. This indicates that partial aggregation has taken place, as does the steep course of the scattering curve at low angles. According

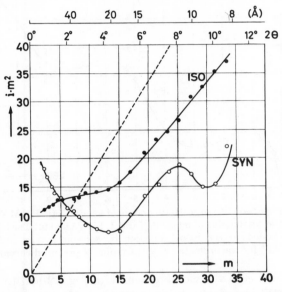

Fig. 14. Intensity of the small-angle X-ray scattering from PMMA in methylisobutyrate. For the notation see Fig. 12.

to experiments which were made with the same fraction earlier (11), the light-scattering should not be affected to an appreciable extent.

What do the measured curves predict for the shape of the segments in the coil? A rough comparison with the functions of Figs. 7–10 shows that isotactic PMMA scatters as a thread with random, syndiotactic PMMA as a thread with non-random curvature. For the comparison between theoretical and experimental scattering curves only the measurements in benzene were used, because the cross-section factor probably can be neglected here (18). First let us discuss the scattering behavior of syndiotactic PMMA.

1. *Discussion of the Scattering Curve of Syndiotactic PMMA*

The function in Fig. 9 is very similar to the scattering curve of syndiotactic PMMA in benzene. By linear transformation of the coordinates the experimental curve can be fitted to the theoretical one (Fig. 15). From this one can determine the amount of preferred curvature:

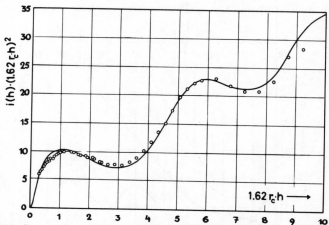

Fig. 15. Comparison of the scattering curve of syndiotactic PMMA in benzene with the scattering function of model 3 (Table I, Figs. 5 and 9). The experimental data were adapted to the model function by suitable factors (0.247 for the abscissa and 0.65 for the ordinate). r_c is the radius of preferred curvature.

The abscissa m of the measured curve in Fig. 12 had to be multiplied by 0.247 in order to bring it into accordance with the model function of Fig. 9. The abscissa of the model function is $bh = 20h = 1.62 r_c \cdot h$ where r_c = radius of preferred curvature. (Compare Table II: $r_c \cdot h = 3.4$ for $bh = 5.5$). From this it follows that

$$m \cdot 0.247 = r_c \cdot h/0.62$$
$$r_c = 0.62 \times 0.247 \, m/h \qquad\qquad 13.$$

The ratio m/h is given by the experimental arrangement. In the present case we have $m = (1.54/2\pi) \cdot 179h$. Putting this into Eq. 13 one obtains $r_c = 6.7$ Å.

This value can be interpreted as the radius of curvature of the helix which is formed by the line of X-ray optical centers of gravity in the solid syndiotactic PMMA. The identity period of the helix is $t = 21.1$ Å. Let r_a be the axial radius of the helix; then t, r_a and r_c are connected by the relation

$$r_a^2 + t^2/4\pi^2 - r_a r_c = 0 \qquad 14.$$

or

$$r_a = r_c/2 \pm (r_c^2/4 - t^2/4\pi^2)^{1/2}$$

Taking the values found for r_c and t one obtains $r_a = 3.35$ Å which is a reasonable value.

The significance of the agreement between the model function and the measured scattering function, however, should not be overestimated; it indicates only that those features of the model conform to the measured behavior which are important in small-angle X-ray scattering. A difference between the model and the molecule is given for instance in the radius of gyration and in the statistical chain element. In the model the relation between b and r_c is $b = 1.6r_c$, in syndiotactic PMMA on the other hand it is $b \approx 2.4r_c$, where b had been determined from light-scattering measurements.* It should be possible, however, to expand the model coil in such a way that the experimental value of the radius of gyration is reached, whereas the X-ray scattering in the considered range still remains the same.

2. *The Interpretation of the Scattering Curve of Isotactic PMMA*

The attempt to fit the scattering curve of the isotactic PMMA with one of the calculated model functions did not succeed. For this, only model No. 1 came into question and, as Fig. 16 shows, the scattering function of model 1 resembles the measured one only qualitatively.

To interpret the scattering curve of the isotactic PMMA two additional model functions may be considered.

Model 5: Chain with valence angle $\alpha = 1.24$ (= supplement to the tetrahedral angle) and $\varphi^* = 1.50$. The *trans* position may be favored. Then a chain results in which $b = 10l$. A short part of a conformation of this type of coil is shown in Fig. 17.

Model 6: Chain with the same valence angle and the same chain element as in the case of model 5. The rotation states are *l-gauche* favored ($\langle\varphi\rangle = 1.05$, $\varphi^* = 1.54$). Since $\langle\sin\varphi\rangle \neq 0$ the calculation of the chain element has to be done according to the following equation:

* This is described in reference (11). In Table 4 of this reference persistence lengths are listed. According to the present paper, these persistence lengths are to be considered as half statistical chain elements.

$$b = l \cdot \frac{1 + \cos \alpha}{1 - \cos \alpha} \cdot \frac{(1 + \eta)^2 + \epsilon^2}{1 - \eta^2 - \epsilon^2} \quad \text{with } \eta = \langle \cos \varphi \rangle = \cos \langle \varphi \rangle \frac{\sin \varphi^*}{\varphi^*}$$

$$\epsilon = \langle \sin \varphi \rangle = \sin \langle \varphi \rangle \frac{\sin \varphi^*}{\varphi^*} \quad 15.$$

A hint at the calculation of this formula will be found in the appendix. A part of a conformation of model 6 is shown in Fig. 17.

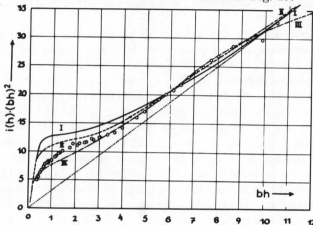

Fig. 16. Comparison of the scattering curve of isotactic PMMA in benzene with model functions. Curve 1: model 1 (Table I, Figs. 3 and 7). Curve 2: trans favored valence angle chain (see page 53 and Fig. 17). Curve 3: gauche-favored valence angle chain (see page 53 and Fig. 17).

The measured scattering function in Fig. 16 is—roughly speaking—placed in the middle between the model functions 5 and 6 (curves II and III). The question is, what has one to conclude from this with respect to the molecule? If we look at Fig. 17 (right), the conception is supported that the real "scattering line," that is the center line in the case of isotactic PMMA, is a line randomly walking about the axis of the thread. Presumably one need only have luck to find a model which gives total agreement between the calculated and the measured functions.

Common to all functions in Fig. 16 is the existence of a junction point in the sense of Kratky. True, it is scarcely perceptible in curve III. The persistence length a can be calculated from the position of the experimentally determined junction point according to Eq. 12 or from the factor with which the abscissa of the measured values were multiplied, in order to be fitted to Fig. 16. The second way is to be favored over the first in general, because the junction point is determinable neither in model functions nor in experimental curves with satisfactory accuracy. Both ways of calculation may be applied to the reported values:

(a) Persistence length from the experimentally determined junction point: We take from Figs. 12 and 13 for isotactic PMMA in benzene and acetone as the abscissa of the junction point $m^* = 12$

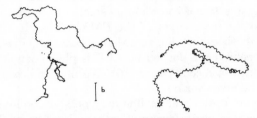

Fig. 17. Projections of random conformations of two coil models. Chain length $L = 156$. On the left: trans favored angles of rotation. On the right: gauche favored angles of rotation (see page 53).

and from Fig. 14 (θ-solution) $m^* = 14$. With Eq. 12 and $m = (\lambda/2\pi) \cdot 179h = 43.9h$ it follows that

$$h^* = 0.274 \text{ resp. } 0.320; \quad a = 2.3/h^*;$$
$$\underline{a = 8.4 \text{ Å in acetone and benzene}}$$
$$\underline{a = 7.2 \text{ Å in the } \theta\text{-solvent.}}$$

(b) Persistence length from the factor used in fitting the model curve: The factor is equal to 0.315: $bh = 0.315m$. For the models, $a = b/2$ holds.

$$bh = 0.315 \, m = 2 \, ah$$
$$a = 0.315 \, m/2h = 0.315 \times 0.5 \times (\lambda/2\pi) \times 179$$
$$\underline{a = 6.9 \text{ Å}}$$

As average value one obtains for isotactic PMMA in benzene and acetone $a = 7.7$ Å. For the θ-solution, an analogous fitting to Fig. 16 cannot be performed because of the beginning association. One can assume, however, that a calculation of a in the second way would have led to a smaller value in this case too. With this assumption we get as the most probable value for isotactic PMMA in the isobutyric acid ester $a = 6.6$ Å.

In all experimental values an error of ± 1 Å must be taken into account. Therefore the diminution of the persistence length from

7.7 Å to 6.6 Å in the transition from benzene to the θ-solvent is still within the limits of error.

(c) The problem of the absolute intensity of the measured scattering curves: To the right of the junction point respectively to the right of the first minimum in the representation $i(bh) \cdot (bh)^2$ vs. bh (respectively im^2 vs. m) the absolute scattering intensity of chain models is proportional to the mass cover per unit length of the thread. But in all the measurements presented here the scattering intensity in the range in question is considerably smaller than that which would correspond to the mass cover of the PMMA-thread: The measured scattering intensities in the right-hand parts of Figs. 12–14 lag considerably behind the dashed lines. This phenomenon has already been discussed in a recent paper (10). We may indicate again what reasons may be given for it:

(1) The X-ray optical center line may possibly be longer than the junction line of the chain-forming C-atoms. Because of the responsibility of the center line for the scattering behavior, the mass cover corresponding to this line has to be taken into consideration.

(2) An additional cross-section factor may be involved.

(3) By segment aggregation a part of the scattering intensity may appear at smaller scattering angles.

FINAL REMARKS

The results of this paper demonstrate that it is possible to construct an innumerable number of different types of random coils. Each of these models has its own scattering function. By comparison of the experimental scattering functions with the model functions one can find a model for the real molecule. Unfortunately not much is known about the uniqueness of a model detected in such a manner. A one-to-one correspondence exists between the scattering function and the distance distribution of the particles:

$$i(h) = \frac{1}{h} \int_0^\infty \frac{\psi(r)}{r} \sin rh \, dr \qquad \text{16a.}$$

$$\psi(r) = \frac{2r}{\pi} \int_0^\infty h \cdot i(h) \sin rh \, dh \qquad \text{16b.}$$

$\psi(r)$ is the distance distribution of the particle. It is normalized in such a manner that

$$\int_0^\infty \psi(r) \, dr = 1 \qquad \text{17.}$$

One can calculate the distance distribution uniquely from the shape of the particle. For the reverse, until now neither the existence nor the uniqueness nor the non-uniqueness of a solution has been proved. There is also no way to find any solution systematically. The author found a class of very specially shaped bodies, for each of which one can find a second, differently shaped body with the same distance distribution. I suppose there exist for each distance distribution only a finite number of particle shapes, provided that the particles are monodisperse. A comprehensive investigation of these interrelations would be desirable, because our imperfect knowledge of them is now an appreciable hindrance for advances in the area of small-angle X-ray scattering.

ACKNOWLEDGEMENT

I wish to express my sincerest thanks to my teacher, Prof. Dr. G. V. Schulz, for his participation and interest in my work.

I also thank the *Deutsche Forschungsgemeinschaft* for providing the apparatus as well as a scholarship.

MATHEMATICAL APPENDIX

The Random Procedure Used for the Monte Carlo-Calculations

For the generation of random numbers a modified mean square method was used. The algorithm may be presented in ALGOL-formulation:

```
'PROCEDURE' RANDOM (X1,X2) RESULT: (Y1,Y2);
    'VALUE'X1, X2;
    'INTEGER'X1, Y1;
    'REAL'X2, Y2;
'BEGIN'
    'INTEGER'X11, EX 12, REST;
    'REAL'X22, UX12, X12;
    C: = C + ₁₀5 × C + 0.1234567891
    C: = C − ENTIER(C);
    X2: = X2 + C; X2: = X2 − ENTIER(X2);
    X11: = X1 × X1; X22: = X2 × X2; X12: = X1 × X2;
    EX12: = ENTIER(X12); UX12: = X12 − EX12;
    X22: = X22 + 2 × UX12; REST: = ENTIER(X22);
    Y2: = X22-REST;
```

```
X11: = X11 + 2 × EX12 + REST;
Y1: = X11 − ₁₀5 × ENTIER (X11 × ₁₀ − 5);
'IF'Y1 'LESS'1000 'THEN'Y1: = Y1 + 31415;
'END'
```

Let $X1$ be an integer between 0 and 99999, $X2$ and C real numbers between 0 and 0.9999999999, then according to this procedure the random numbers $Y1$ and $Y2$ are calculated. They are of the same type and lie in the same interval as do $X1$ and $X2$. As a random number will be taken properly only $Y2$. C is a global parameter of the procedure. It is not a random number but has the function of making the cycles of the random numbers large ($\geq 10^{10}$).

Proof of the Identity of the Persistence Length According to the Original Definition in the Coil Models 1–4 (Table I, Figs. 3–6)

One has to calculate the mean length a of the projection of the end-to-end distance on the starting direction of the thread for $L \to \infty$

$$a = \frac{1}{2} |\mathbf{l}_1| + \left\langle \sum_{i=2}^{\infty} |\mathbf{l}_i| \cos(1,i) \right\rangle \qquad \text{A1.}$$

$\cos(1,i)$ is the cosine of the angle between $\mathbf{l}_1$ and $\mathbf{l}_i$. The $\mathbf{l}_i$ are the vectors forming the thread ("bonds"). The first vector is only considered half of its length.

$$a = l \left[\frac{1}{2} + \sum_{i=2}^{\infty} \langle \cos(1,i) \rangle \right] \qquad \text{A2.}$$

Let each vector $\mathbf{l}_i$ be connected with a coordinate system, the z-axis of which has the direction of $\mathbf{l}_i$. Let A_j be the matrix of the cosines of the angles between the axes of the coordinate systems belonging to $\mathbf{l}_{j-1}$ and $\mathbf{l}_j$. Then the following relation is valid:

$$A_j = \begin{pmatrix} -\cos \varphi_j \cos \alpha & -\sin \varphi_j & \cos \varphi_j \sin \alpha \\ \sin \varphi_j \cos \alpha & -\cos \varphi_j & -\sin \varphi_j \sin \alpha \\ \sin \alpha & 0 & \cos \alpha \end{pmatrix} \qquad \text{A3.}$$

α is the angle between $\mathbf{l}_{j-1}$ and $\mathbf{l}_j$ (= "valence angle"), φ_j is the angle of rotation (Fig. 1).

The matrix

$$A_{1,i} = \prod_{j=2}^{i} A_j$$

represents the relative position of the coordinate systems belonging to l_1 and l_i with respect to one another and $\{A_{1,i}\}_{33}$ is the $\cos(1,i)$ wanted for Eq. A2. We get further

$$\langle \cos(1,i) \rangle = \{\langle A_{1,i}\rangle\}_{33} = \prod_{j=2}^{i} \langle A_j \rangle_{33} = \{\langle A_j \rangle^{i-1}\}_{33} \quad \text{A4.}$$

Substituting in Eq. A2 with $\langle A_j \rangle = \bar{A}_j$ gives

$$a = l\left[\frac{1}{2} + \sum_{i=2}^{\infty} \{\bar{A}_j{}^{i-1}\}_{33}\right] = l\left[\frac{1}{2} + \left\{\sum_{i=2}^{\infty} \bar{A}_j{}^{i-1}\right\}_{33}\right] \quad \text{A5.}$$

The sum is a geometrical series:

$$\sum_{i=2}^{\infty} \bar{A}_j{}^{i-1} = \bar{A}_j(J - \bar{A}_j)^{-1} \quad \text{with } J = \text{unit matrix} \quad \text{A6.}$$

$$a = l[\tfrac{1}{2} + \{\bar{A}_j(J - \bar{A}_j)^{-1}\}_{33}] = l[-\tfrac{1}{2} + \{(J - A_j)^{-1}\}_{33}] \quad \text{A7.}$$

For symmetrical hindrance of rotation ($\langle \sin \varphi \rangle = 0$), one obtains from Eq. A3

$$\bar{A}_j = \begin{pmatrix} -\eta \cos \alpha & 0 & \eta \sin \alpha \\ 0 & -\eta & 0 \\ \sin \alpha & 0 & \cos \alpha \end{pmatrix} \quad \text{A8.}$$

Calculation of $(J - \bar{A}_j)^{-1}$ and substituting in Eq. A7 leads to

$$a = \frac{1}{2} l \cdot \frac{1 + \cos \alpha}{1 - \cos \alpha} \frac{1 + \eta}{1 - \eta} \quad \text{A9.}$$

The comparison with 7a yields

$$a = 10l = b/2 \quad \text{A10.}$$

for the models 1 to 4.

Derivation of Eq. 15

For statistical valence angle chains without correlation between the rotation angles the following general expression can be employed (12):

$$b = l[1 + 2\{\bar{A}_j(J - \bar{A}_j)^{-1}\}_{33}] \qquad \text{A11.}$$

l is the bond length, J the unit matrix and

$$\bar{A}_j = \begin{pmatrix} -\eta \cos \alpha & -\epsilon & \eta \sin \alpha \\ \epsilon \cos \alpha & -\eta & -\epsilon \sin \alpha \\ \sin \alpha & 0 & \cos \alpha \end{pmatrix} \qquad \text{A12.}$$

If one compares A11 with A7, one obtains $a = b/2$. Substitution of Eq. A12 into A11 leads to Eq. 15.

The Plotting of the Scattering Function of Debye in Figs. 7–10

To the scattering functions in Figs. 7–10 belong threads with 100.8 statistical chain elements. Consequently because of the normalization method used, $i(0) = 100.8$. With this factor the right part of Eq. 3 is to be multiplied. For the radius of gyration r_g we have

$$r_g^2 = \frac{1}{6} \times (100.8)^2 \times \frac{2016}{100.8} = 6720 \qquad \text{A14.}$$

The abscissa in Figs. 7–10 is $bh = 20\ h$; $(bh)^2 = 400\ h^2$

$$r_g^2 h^2 = 16.8\ b^2 h^2 \qquad \text{A15.}$$

With this Eq. 3 changes into

$$100.8\ i(r_g h) = 100.8\ \frac{2}{(16.8\ b^2 h^2)^2} [\exp(-16.8\ b^2 h^2) - 1 + 16.8\ b^2 h^2]$$

$$100.8\ i(r_g h) \cdot b^2 h^2 = \frac{2 \times 100.8}{16.8^2\ b^2 h^2} [\exp(-16.8\ b^2 h^2) - 1 + 16.8\ b^2 h^2]$$

$$= 12 - \frac{0.7143}{b^2 h^2} [1 - \exp(-16.8\ b^2 h^2)] \qquad \text{A16.}$$

This function was plotted in Figs. 7–10 as the Debye function.

BIBLIOGRAPHY

1. Debye, P., J. Phys. Coll. Chem., **51**, 18 (1947).
2. Kratky, O. and Porod, G., Rec. Trav. Chim. Pays-Bas, **68**, 1106 (1949).
3. Hermans, J. and Ullman, R., Physica, **18**, 951 (1952). Porod, G., J. Polymer Sci., **10**, 157 (1953). Hermans, J., Jr. and Hermans, J.J., J. Phys. Chem., **62**, 1543 (1958)
4. Peterlin, A., J. Polymer Sci., **47**, 403 (1960).
5. Heine, S., Kratky, O., Porod, G., and Schmitz, P. J., Makromol. Chem., **44-46**, 682 (1961).
6. Heine, S., Kratky, O. and Roppert, J., ibid., **56**, 150 (1962).

7. Ptitsyn, O. B., J. physic. Chem. (USSR), **31,** 1091 (1957). Benoit, H., Compt. Rend., **245,** 2244 (1957). Loucheux, C., Weill, G. and Benoit, H., J. Chim. Phys., **55,** 540 (1958). Hyde, A. J., Ryan, J. H., Wall, F. T., and Schatzki, T. F., J. Polymer Sci., **33,** 129 (1958). Ptitsyn, O. B., Soviet Physics Usp., **2,** 797 (1960).
8. Heine, S., Makromol. Chem., **71,** 86 (1964).
9. Kirste, R. G., and Wunderlich, W., ibid., **73,** 240 (1964).
10. Wunderlich, W. and Kirste, R. G., Ber. Bunsenges. physik. Chem., **68,** 646 (1964).
11. Schulz, G. V., Wunderlich, W. and Kirste, R. G., Makromol. Chem., **75,** 22 (1964).
12. Volkenstein, M. V., *Configurational Statistics of Polymeric Chains*, J. Wiley and Sons, New York, 1963.
13. Kirste, R. G., Z. physik. Chem. (Frankfurt), **39,** 20 (1963).
14. Stroupe, J. D. and Hughes, R. E., J. Am. Chem. Soc., **80,** 2341 (1958).
15. Kratky, O., Kolloid-Z., **182,** 7 (1962).
16. Kirste, R. G., Z. physik. Chem. (Frankfurt), **42,** 358 (1964).
17. Schulz, G. V. and Doll, H., Ber. Bunsenges. physik. Chem., **56,** 248 (1952); **57,** 841 (1953).
18. Kirste, R. G., Z. physik. Chem. (Frankfurt), **42,** 351 (1964).

Adaptation of the Technique of Diffuse Small-Angle X-Ray Scattering to Extreme Demands

O. KRATKY

Institut für physikalische Chemie der Universität Graz, Graz, Austria

REQUIREMENTS FOR ADAPTATION TO EXTREME DEMANDS

The adaptation of the small-angle technique to extreme demands requires that the following prerequisites be met:

(1) Sufficient freedom from parasitic scattering in the whole angular range, but particularly at extremely small angles.

(2) High accuracy in the measurement of the scattered intensity, particularly if the effect proper is obtained by taking the difference between the total scattering and a background scattering (solvent scattering etc.), and the difference amounts to only a fraction of both intensities measured; the experimental accuracy of the individual measurements must then necessarily be very much higher than that required for the result. For a solution, the above-mentioned difference will decrease with increasing dilution, and with decreasing size of the dissolved particles (the last statement will be substantiated in detail later). The prerequisite for obtaining the necessary high accuracy by taking correspondingly high counts in the individual measurements, is a sufficiently high constancy of the primary energy. A possibility for eliminating fluctuations of the primary energy by reference measurements of the latter is pointed out.

(3) Since the majority of measurements of the diffuse small-angle X-ray scattering are made with a primary beam of line-shaped cross-section for the sake of a sufficiently high energy, the elimination of the collimation error caused thereby is necessary. This can be done exactly only when the intersection of the primary beam with the plane of registration has a constant intensity over a sufficiently large range in its length direction.

(4) Suitable mechanization of measurements must be provided. In this way the total time needed by the experimenter is kept within

reasonable limits, despite the long counting times required for the requisite accuracy of the measurements.

(5) Provision must be made for the performance of very exact absolute measurements, that is simultaneous exact determinations of the primary energy.

THE EXPERIMENTAL METHOD

X-Ray Unit, Air Conditioning, Cooling System, X-Ray Tube

It is evident that an X-ray unit with highest stabilization is required. The commercially available modern units will meet the demands perfectly. In most cases, a copper target tube with line focus is used. It is true that an X-ray unit with rotating anode tube, as developed in the laboratory of Beeman (1), will bring considerable advantages because of the energy gain due to the high load capacity. However, construction and operation of such a unit present a difficult electronic problem, particularly if high dose constancy is required. Therefore this device is, for the time being, not generally accessible.

Even with the very best stabilization of the unit, simultaneous airconditioning of the laboratory is necessary, so that temperature fluctuations will not exceed $\pm 1°C$. For work with the line focus in particular (that is the broad side of the focus), a room temperature fluctuation of several degrees may cause a variation in the intensity of several percent, mainly due to the thermal motion of the collimator relative to the focal spot. Very much smaller is, according to our experience, the influence of temperature for work at the square focus, that is at the narrow side of the focus. If there is no possibility of installing a sufficiently powerful air-conditioning unit, then the camera should be mounted at the narrow side of the focus. However, this will bring about a considerable loss in intensity (cf. next section).

Regulation of humidity is also required, though here a considerably larger tolerance may be accepted. In consideration of the experimenters, and because of the danger of drying out the cables, too low a humidity is not desirable. Too high a humidity may cause annoying corrosion and electrical phenomena. The acceptable mean would be a relative humidity of about 60%. Regulation to $\pm 5\%$ will suffice.

One more source of error, mostly neglected, must be pointed out. The operation of an X-ray tube expected to give constant output intensity also requires constant temperature of the cooling water. This fact may in certain regions necessitate the installation of a cooling system with regulated temperature. As H. Leopold in our laboratory has found, a temperature variation of 3°C in the cooling water will shift the line focus by about 1 μ. This can change the scattered intensity appreciably, namely up to 10%.

Primary Beam Shape

Now we deal with the practically most important problem (mentioned in the first chapter among the prerequisites)—production of a primary beam which has a uniform distribution of intensity in its length direction within a sufficiently long zone in the plane of registration. The maximum length of this zone may amount to several centimeters for a 20 cm distance between sample and plane of registration.

We assume that the focus has an area of 1×10 mm. This corresponds to many commercial tubes. Further, the radiation shall be taken off at an angle of 6° to the plane of the focus. The projection of the focus onto plane F (Fig. 1) intersecting the center of the tube and normal to the direction of radiation, will then have an area of 1×1 mm for take-off at the narrow side (namely 1 mm in the vertical direction, 1 mm normal to the plane of the paper); for take-off at the broad side an area of 10×0.1 mm (namely 0.1 mm in the vertical direction, 10 mm normal to the plane of the paper). In the first case, we speak of a "square focus," in the second of a "line focus." The diameter of the tube windows (lying in the plane W) is, as a rule, about equal to the length of the focus; thus we shall assume 10 mm.

Before we approach the discussion, we need to become familiar with the setup of the camera used in our laboratory, and to be described in detail later on. The primary beam coming from the focal plane F is led past three blocks D_1, D_2, and D_3, placed normal to the plane of the paper. The resolution is thereby given by the breadth of the entrance slit Sl. The dimensions in the vertical direction in Fig. 1 are enlarged by a factor 100 as compared with those in the horizontal direction. The scattering produced in the irradiated sample Sa is measured in the plane of registration R. The problem is now to determine the intensity distribution of the beam in the plane of registration in the direction normal to the plane of the paper. For its discussion we must make use of a section coinciding with the plane of the flat beam, that is a section in the direction of the beam, but normal to the plane of the paper in Fig. 1.

Figure 2, using such a section, shows the conditions for work with the square focus. The distance of the tube window D from the focal spot has been assumed to be 40 mm, the distance tube window-plane of registration to be 380 mm, both in accordance with Fig. 1. With a tube window diameter of 10 mm, a uniform length of the

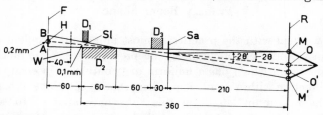

Fig. 1. Distance ratios in the vertical section of the camera, parallel to the direction of the beam, and normal to the plane of the beam. The dimensions in the horizontal direction are in mm. The dimensions in the vertical direction are enlarged relatively by a factor of 100. F: focal plane, W: plane of the tube window, D_1, D_2, D_3: blocks, arranged normal to the plane of the paper and determining the vertical beam geometry. Sl: entrance slit between D_1 and D_2; Sa: sample; R: plane of registration. For explanation of further symbols, see text.

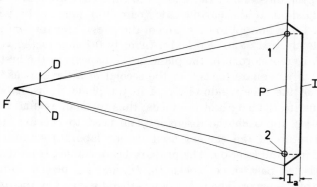

Fig. 2. Geometry of the beam in the beam plane for work with the square focus. F: focus, D: pair of stops in the plane of the tube window, whose opening is equal to the diameter of the tube window. P: primary beam, I: intensity distribution in the primary beam, 1-2: uniform portion of the primary beam.

primary beam in R of 9.5 cm can be obtained. A shortening is readily possible by confinement at the window, or at any site between window and sample (cf. Fig. 1). The entrance slit Sl of 0.1 mm, as assumed in Fig. 1, will correspond to a smallest scattering angle 2θ, for which a measurement will be possible, of 1.19×10^{-3} radians, and a corresponding maximum equivalent Bragg spacing of about 1300 Å. This is a resolution sufficient for most cases. The projection of the effective portion of the focus in F (Fig. 1) has in this case an area of $0.2 \times 1 = 0.2$ mm². Thus, only $\frac{1}{5}$ of the total effective area of 1 mm² is utilized. This is why, as a rule, the line focus, which allows the attainment of a considerably greater "efficiency," is used. We shall now turn our attention to this case.

If the beam is allowed to emerge without confinement in the length direction (Fig. 3a), the uniform portion 1–2 of the primary beam in the plane of registration will possess a length equal to that of the focus or the diameter of the window, that is with the above assumptions, 1 cm. An enlargement of the exit window is not possible with commercial tubes; the only possibility of variation is a narrowing. If such is done, e.g. according to Fig. 3b, the uniform length 1–2 is shortened; for the arrangement in Fig. 3c it is diminished to zero: the whole primary beam will then consist only of two penumbral regions, rising in intensity towards the middle. If, however, the exit window is even more confined, then, as shown in Fig. 3d, a uniformly illuminated region 1–2 is again produced in the plane of registration, since from each point along the line 1–2 of the primary beam in R one "sees" a portion of the focus of equal length, namely a portion of the length I–II. This method of the "crossed beam," which is required due to the construction of the tube, has, however, two very essential disadvantages:

(1) To the same degree to which the length of the uniform portion of the primary beam is enhanced, the irregularities of the focal spot will be portrayed more and more in the plane of registration.

(2) Intensity is lost, since at any one point of the region 1–2 of the primary beam (in the plane of registration) the energy of a piece of the focus will be found which is the shorter the longer one makes the uniform portion of the primary beam. The comparison of Fig. 3a, 3d and 3e indicates directly the shortening of the distance I–II in the focus enforced by the elongation of the primary beam. In the illustrated profiles of the primary beam, the relative intensities have been indicated. As a very simple calculation will show, the intensity will decrease linearly if the uniform portion of the primary beam exceeds 1 cm and will reach the value zero (for the dimensions given in Fig. 1) at a length of this portion of 9.5 cm.

If we assume the same resolution, that is the same breadth of the entrance slit as shown in Fig. 1, and as was used above for the work with the square focus, then the collimation system will now no longer be fully illuminated in breadth, since the projection of the focus onto plane F is only 0.1 mm broad. For full illumination, however, a breadth of 0.2 mm would be required.

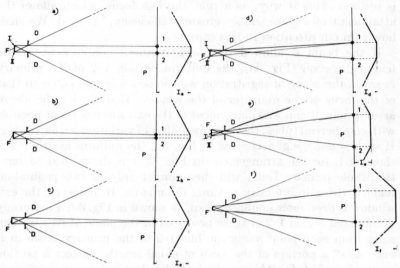

Fig. 3. Geometry of the beam in the beam plane for work with the line focus. The symbols correspond to those in Fig. 2. Only the pair of stops D has an opening varying from case to case, which determines the beam geometry. (a) The diameter of the tube window, resp. the opening of the stops D, is equal to the length of the focus. The uniform portion of the primary beam has only the length of the focus. (b) The opening of D is somewhat smaller, therefore the uniform portion of the primary beam is shorter. (c) The opening of D has been made sufficiently small to reduce the uniform portion of the primary beam to zero length. (d) In order to make the uniform portion of the primary beam just equal to the 2.5-fold length of the focus, the opening of D must be made still smaller as compared with case c. This will lower the intensity I_d of the uniform portion of the primary beam, as compared with I_a. (e) For making the uniform portion of the primary beam equal to the five-fold length of the focus, a still stronger confinement at D is required. This will lower the intensity I_e along the uniform portion of the primary beam still more. (f) The focus length is shorter than the window. Now a uniform portion of the primary beam of multiple length of the focus is obtained without confinement and loss in intensity.

It is of interest to reflect on the loss in intensity which is thereby brought about, in comparison with full illumination. Imagine the whole vertical distance AB in Fig. 1 as the source of radiation; then each point will send a bundle into the collimation system whose breadth will increase along AB approximately linearly. Then the triangular area ABC in Fig. 4 represents a measure for the energy entering the collimation system. If now, for instance, only the distance DE would emit radiation, then the area DEFG will be a comparable measure for the radiated energy. If the focus projection is only half as broad as AB, as is the case in our example with line focus, then the camera is best adjusted in such a way, that the focal spot projection occupies the upper half of distance AB, that is distance HB.

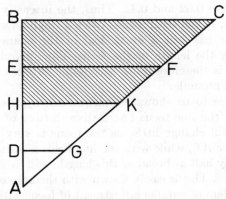

Fig. 4. Scheme for illustrating the intensity as a function of the breadth of the entrance slit.

Such a position is obtained automatically, if, by means of raising and lowering the entrance slit, adjustment for maximum intensity of the primary beam is made. Now let us compare the square focus, fully illuminating the entrance slit Sl (because its projection is five times as broad as AB), with the line focus for optimum adjustment. The radiated energies (per unit length of the focus, of course), will be related to each other as area ABC to HBCK, that is as $1:\frac{3}{4}$.

On the other hand, with square focus, only a horizontal length of 1 mm is effective (see Fig. 2), while for instance in case 3d a length of 7.5 mm, in case 3e one of 5 mm will be effective. Therefore we obtain the relative intensities as $7.5 \times (\frac{3}{4}) \times 0.2 = 1.1$, resp. $5 \times (\frac{3}{4}) \times 0.2 = 0.75$, if we put the intensity of the square focus at 0.2. We must, however, consider that with such "oblique" radiation from focus BH the center of the beam in the plane of registration will no longer be found at O (Fig. 1), but rather at O'.

Although the boundaries of the primary beam in plane R then are given by O and M', the measurement of the scattering will now also only be possible above point M, due to the parasitic scattering extending up to this point (cf. following section). Therefore the smallest accessible scattering angle is $2\theta'$; the resolution will be considerably diminished and corresponds to a Bragg value of 1300 × $\frac{2}{3} \approx 870$ Å. In order to get back to the old value, the entrance slit must be reduced to 0.067. This will lower the intensity by a factor 0.56, as a simple consideration will show. Instead of the values 1.1 and 0.75, we have 0.62 and 0.42. Thus, the intensity is in the first case three times, in the second case twice as high as with the square focus. Then one has lost a great deal of the chance of intensity enhancement by the length of the focus. For a required very large focus length, it is therefore quite acceptable to work with a square focus on several grounds:

(1) The square focus shows, as mentioned, a lower temperature sensitivity than the line focus (A relative shifting of both entrance slit and focus will change little, as the focus is very much broader than the distance AB, while with the line focus a shifting within AB of the focus, only half as broad as this length, will cause considerable intensity changes. This is easily shown with the help of Fig. 4.)

(2) The problem of parallel adjustment of focus with the direction of the entrance slit is eliminated.

(3) On lowering the resolution (enlargement of the entrance slit Sl in Fig. 1), the intensity will rise more than with the line focus.

The above-mentioned difficulties are met at once, if the line focus is considerably shorter then the exit window (Fig. 3f), since in this case a very long, uniform primary beam, an "opening beam," will be available in the plane R without confinement at the tube window. Both disadvantages are gone: the intensity of radiation in the uniform portion of the primary beam is large since at each point the energy of the total length of the focus is assembled. For the same reason no irregularities of the focus will be portrayed.

One might object that with given diameter of the tube window, the focus must be shorter than in tubes of ordinary design. This is true, but one must consider that by shortening the focal spot without changing the total load the breadth can be raised correspondingly. So for instance with a shortening from 10 to 5 mm, a broadening from 1 mm to 2 mm will be possible. In the projection, that is if the radiation emerges at 6°, the beam will no longer appear with a breadth of 0.1 mm, but 0.2 mm. Apart from the case of extremely high resolution, the collimation system will not yet be completely illuminated with a 0.1 mm broad primary beam, so that a broadening to 0.2 mm can be used fully. If we again refer to the intensity of the square focus as 0.2, we now have $0.2 \times 5 = 1$, that is roughly twice the value as in the case of the crossed beam.

These considerations were already stressed in 1960 (2), but only recently have we succeeded in convincing a major company to produce a tube with such a short focal spot.

Summarizing, then, we want to emphasize again: With a primary beam length of 3–5 cm, using an "opening beam," as can be done with a shortened line focus, will give rise to an energy gain by a factor of 2 compared to an ordinary tube with a 10 mm focus. Thus a 1.5 kW unit with an appropriately constructed tube will yield (with the assumed resolution of 1300 Å) the same primary energy as a 3 kW unit would give with a tube of usual design. Furthermore, the troublesome source of error due to the portraying of irregularities of the focal spot with crossed path of the beam is eliminated, and adjustment is greatly facilitated. Therefore we may state that the X-ray tube with shortened line focus will meet extreme demands for quality (i.e. uniformity of the primary beam) and intensity to a much higher degree than ordinary tubes.

Practically the same effect could be achieved with a focal spot of normal length in combination with a tube window of correspondingly greater diameter.

Monochromatization of the Radiation and Measurement of the Scattered Intensity

The measurement of the scattered intensity can be done either by photographic film, or pointwise with a counter. Since photographic measurement does not meet the highest demands for accuracy, we shall not deal with it in this contribution, and base further discussion only on counting techniques. Since on good grounds the $CuK\alpha$ wavelength has so far nearly exclusively been used for small-angle X-ray measurements, we shall confine ourselves in the following to the discussion of methods for the monochromatization of copper radiation.

For obtaining the monochromatic scattering curve, there is then, first, the use of a primary beam gotten by Bragg reflection or total reflection, second the filter difference method of Ross (3), and third the use of a pulse height discriminator in conjunction with a proportional counter or a scintillation counter.

For studies of diffuse small-angle scattering, primary beams of line-shaped cross-section are usually employed for intensity reasons. We shall therefore discuss only methods which have the purpose of producing such a primary beam.

In the well-tested method of Guinier (4), a bent and ground Johansson crystal (5) is used, which exerts a focussing action at its surface by Bragg reflection of a primary beam with line-shaped cross-section. For the same purpose, Jagodzinski and Wohlleben (6) use an arrangement in which one reflecting Johansson monochromator is combined with a penetration monochromator described by Cauchois (7). Total reflection at a plane glass plate has been recommended recently by Damaschun (8) for obtaining a line-shaped primary beam. This author discusses in a very general manner the monochromatizing effect of a combined application of total reflection and a nickel filter. Older investigations of total reflection at an elastically bent optical flat go back to Ehrenberg and Franks (9) and Franks (10). There is no doubt that the possibilities of using total reflection for the purpose of small-angle scattering are by no means exhausted. A completely new, obviously very important way for obtaining a high-intensity monochromatic primary beam is the multiple reflection interferometer described by Bonse and Hart during this meeting (11).

The intensity losses encountered with any type of reflection are avoided with the filter difference method (3), but part of this gain is lost by the double measurement required for each scattering angle. Besides, every difference measurement brings about an enhancement of the statistical error.

For the time being, therefore, the most suitable method of monochromatization for the $CuK\alpha$ line seems to be the proportional counter or scintillation counter with pulse height discrimination.

However, the device with proportional counter and pulse height discriminator is considerably more complicated than an arrangement with a Geiger counter, which can be used for the measurement of the beam monochromatized by reflection. Therefore, the question may be posed whether the higher reliability of operation of the Geiger counter will outweigh the disadvantage of the smaller intensity of a reflected beam. Now, developments will without doubt be directed towards greater reliability of complicated electronic equipment. This is augmented by the transition taking place now from the electron tube to the transistor, so the chance of a very high intensity gain as encountered with the proportional counter and using the direct beam, should not be abandoned in favor of the simpler Geiger counter instrumentation. Even if for many problems the longer time of measurement connected with the use of the monochromatized beam may be bearable, one must not forget that the higher intensity of the direct radiation will make it possible to measure a larger number of counts during the same time, and therefore a greater accuracy is obtained. An additional advantage is, that the dead time of the proportional counter lies below that of the Geiger counter by two or more orders of magnitude. This will greatly facilitate the measurement of high intensities, e.g., of the inner portions of the scattering curves of fibers or powders.

During the measurements, the constancy of the unit should be checked often, by measuring the scattered intensity for any sample at a given scattering angle several times in succession. If the differences of the measured quantities exceed the statistical error, and if temperature fluctuations of the room or the cooling water can be excluded as a source of disturbance, the first suspicion will be directed towards the counting device. Therefore it is important to have a quick way for checking its performance. In our laboratory, the use of a radioactive calibration sample has proved very good for this purpose, namely Zn 65, which can easily be obtained from any reactor station. This isotope undergoes transition to Cu by electron capture from the K-shell, and then emits the ordinary monochromatic Cu-spectrum, since an empty hole has been left in the K-shell. The half-life of Zn 65 is 285 days, so no notable decrease of emission will take place within the duration of the check. The electron capture is connected with the emission of a quantum of about 1 MeV. It is appropriate to procure about 1 millicurie of the isotope (but not more!), which can be handled without danger. The commercial Zn-salt is best fixed to an aluminum platelet by means of some lacquer, covered with a thin foil, and put into a small lead casing. The Cu-radiation leaves the casing through a hole, and enters the counter of the device to be tested. Such a sample needs to be replaced by a new one after 1 or 2 years.

If the measuring time for a certain number of counts, say 10^5, proves constant within the statistical error for several successive measurements (average error $\pm 0.3\%$), the performance of the counting device can be regarded as satisfactory. In this way, in case of an inconstancy occurring during a measurement, a "differential diagnosis" will be possible at once: whether the X-ray unit, poor adjustment of the camera, temperature fluctuations or the counting device is the source of the trouble.

In this connection, we want to return to the statement made at the beginning, that one can get rid of the difficulties caused by the fluctuations of the primary intensity of the X-radiation by means of

Fig. 5. Double beam camera.

a comparison method: namely by comparing a reflection of the primary beam at the primary beam stop with the scattered radiation. Cu-sheeting has proved good for this purpose. The basic advantage of this method, it is true, is purchased only with certain disadvantages. First, the double measurement will enlarge the statistical error; second, the counting device is still more complicated; and finally, the stationary primary beam stop—required to give an absolutely constant reflected intensity—fits the total construction of our camera less well than the previous arrangement, in which the primary beam stop was moved together with the counter (see later). We have such a double beam camera in operation (Fig. 5), and are presently collecting comparative experience.*

* This investigation will be reported elsewhere in more detail with H. Leopold.

The Small-Angle Camera

We shall confine ourselves to the description of an apparatus developed by our group (12), which we know to guarantee sufficient freedom from parasitic scattering even for smallest angles due to the special construction of its collimation system, and which allows combination with automated measuring equipment in a very simple way.

The entire device is shown schematically in Fig. 6. The collimation system $C.S.$ allows the passage of a plane bundle of rays lying in the plane normal to that of the figure (comp. Fig. 1). It passes through the sample mounted at Sa, and is finally absorbed in the primary

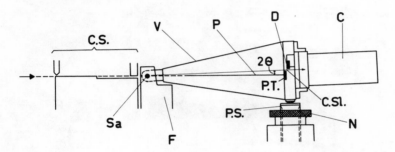

Fig. 6. Complete schematic of the small-angle camera. $C.S.$: collimation system, Sa: sample, F: fork, V: evacuated tube, P: primary beam, 2θ: scattering angle, D: slit jaw, $P.T.$: primary beam stop. D and $P.T.$ together form the counter slit $C.Sl.$, C: counter, $P.S.$: precision spindle. N: screw nut.

beam stop $P.T.$ The evacuated tube V is mounted in a fork F bracketing the sample, and can therefore be rotated around the sample axis Sa. The evacuation of V as well as of the portion of the collimation system facing the sample is necessary to avoid parasitic air scattering. Up to an air pressure of 2 Torr, readily obtainable with any oil pump, no measurable effect will be observed according to our experience. But in order to avoid a slow increase in pressure beyond this limit, caused by small leaks and the permeability of the foils sealing the evacuated space, we have installed an automatic control device. We prefer such a partial evacuation to the use of a complete vacuum camera, as it has the advantage of free accessibility of the sample. Repeated changing of samples, temperature control by using an appropriate cuvette, and eventual use of a flow device during the exposure can all be done more easily with a sample mounted outside the vacuum chamber. Besides, the danger of an explosion if using thin-walled Mark capillaries as sample containers is avoided. Further since the length of the air column between the exit from the collimation system and the entrance into the vacuum tube need not be

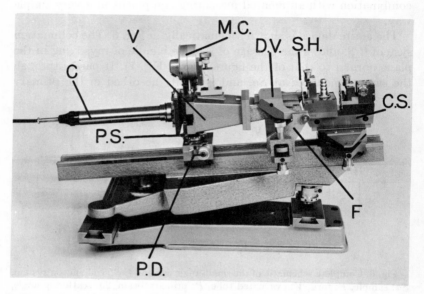

Fig. 7. Small-angle camera with distance sample—plane of registration $a \approx 20$ cm. The symbols have the same meaning as in Fig. 6. Further: $S.H.$: sample holder, $P.D.$: precision drive for the precision spindle $P.S.$, $D.V.$: dovetail, which makes it possible to adjust the rotation axis of the vacuum tube exactly to the center of the sample for samples (resp. cuvettes) of varying thickness. $M.C.$: dial gauge.

more than half a centimeter, the remaining air scattering will be insignificant, and hardly measurable.

The primary beam stop is followed by a movable slit jaw D situated close to the outside of the vacuum tube. $P.T.$ and D form a variable slit, the counter slit $C.Sl.$, which confines a narrow portion of the radiation coming from sample Sa and entering the counter C. A screw nut N, mounted on the optical bench carrying the collimation system and other components of the camera, can be rotated and thereby moves the precision-threaded spindle *P.S.* vertically, to vary the scattering angle 2θ by rotation of *V* around the axis *Sa*.

Figures 7 and 8 are photographs of the complete camera, which represents a further development of the model described elsewhere. Figure 7 shows a type with shorter vacuum tube (the distance from sample to the plane of registration is about 20 cm as shown in Fig. 1). Figure 8 has a vacuum tube of twice that length. The dial gauge *M.C.* arranged at the end of the vacuum tube serves for exact adjustment of the opening of the counter slit.

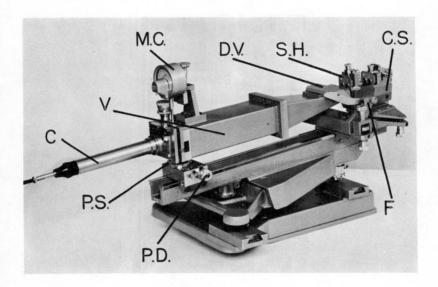

Fig. 8. Small-angle camera with distance sample—plane of registration $a \approx 40$ cm. The symbols have the same meaning as in Fig. 7.

A separate discussion is required concerning the construction of the collimation system. As already shown more than a decade ago (12), it is possible to eliminate the parasitic scattering originating at the collimation system extensively in a simple way without loss of resolution. It is correct that for certain samples no particular measures for suppressing the parasitic scattering need be taken, since the scattering effect to be measured is very intense. This case prevails e.g. with activated charcoal, often used as a test sample. Now, the performance of an apparatus with such samples will tell nothing about its applicability to samples which are really poor scatterers, e.g. a 0.1% protein solution of molecular weight 10^5, and a similarly dilute solution of a dye of molecular weight 10^3. One must consider that the intensity of small-angle scattering, with two-phase systems, will be proportional to the square of the electron density difference of the two phases and, for not too dense systems, also to the volume

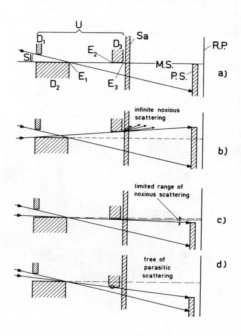

Fig. 9. Collimation system composed of three blocks D_1, D_2, and D_3. $M.S.$: principal section, $P.T.$: primary beam stop, E_1, E_2, E_3: edges, Sl: entrance slit, Sa: sample, $R.P.$: plane of registration. (a) normal arrangement, (b) block D_3 lifted on purpose, (c) block D_3 lowered on purpose, (d) block D_3 lowered and ground to a circular radius at the edge E_2 facing the focus (for an explanation of this measure see text).

concentration. Let us compare a coal powder of, say, 30% volume concentration with a 0.1% protein solution, supposing the coal particles to be of a size similar to that of the protein molecules. Since the electron density difference between coal and air is at least 10 times larger than that between protein and solvent, we have first a factor of 100. In addition there is a factor of $30/0.1 = 300$ due to the concentration. Thus the scattering of the coal will be higher in intensity by a factor 3×10^4 compared to the protein solution. Similarly unfavorable are the conditions for diluted solutions of chain molecules, while solid fibers take an intermediate place (the electron density difference between the two phases, crystalline and amorphous, is similarly small, but the volume fractions are of the order of 50%). Further, the intensity is lowered under otherwise equal conditions with particle size. That is, a 0.1% solution of dye molecules of molec-

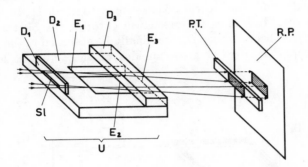

Fig. 10. Realization of the collimation principle shown in Fig. 9a by means of a U-shaped carrier and a bridge D_3 resting on its arms. The symbols have the same meaning as in Fig. 9.

ular weight 1000 will scatter more poorly than coal by a factor of about 3×10^6 (comp. the later discussion of scattering by various dyes).

The discussion of the newest experience in the field of the construction of collimation systems practically free of parasitic scattering shall be prefaced by a brief explanation of older results with the help of Fig. 9a. In the arrangement of D_1 and the two blocks D_2 and D_3, it is essential that the upper surface of D_2 and the lower surface of D_3 lie in the same plane. Then it is easily understood that no parasitic scattering will occur above the "principal section" $M.S.$ since the scattering originating at the edges E_1 and E_2 can only pass along the perfectly plane lower limiting face of D_3, but cannot suffer any scattering which could direct it upwards from the principal section. The principle of this arrangement is perhaps understood best by discussing the effect of deviations. Imagine the block D_3 somewhat lifted, as indicated in Fig. 9b. The direct radiation will strike the edge E_3 opposite to the focus, and give rise there to an unlimited parasitic scattering. On the other hand, if D_3 is lowered below the extended tangential plane to D_2, a parasitic radiation will arise at the edge E_2 facing the focus. It will be limited, but nevertheless fill a space above the primary beam itself. Therefore, an exact coincidence of the tangential planes to D_2 and D_3 is required. The realization of this idea is accomplished, in principle, in the way sketched in Fig. 10. The central part of the U-shaped body corresponds to block D_2. The "bridge," whose lower face coincides with the upward face of the U-body, and which corresponds exactly to block D_3, rests on the arms of the "U." Thus the beam will enter at slit Sl, travel over edge E_1, pass below bridge D_3, and finally be absorbed in the primary beam stop $P.T$. Fig. 11 gives an idea of the actual scale ratios.

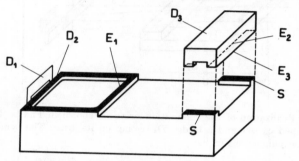

Fig. 11. Collimation system in natural scale ratios. Bridge D_3 is lifted and must be thought of as resting on supports S. The other symbols have the same meaning as in Fig. 9.

The cross-hatched areas are ground and lapped to highest smoothness, and are coplanar. If now bridge D_3, whose lower plane is similarly ground and lapped to highest smoothness, is placed on the supports S, we have the arrangement corresponding to Figs. 9a and 10.

As already mentioned, this principle was invented more than ten years ago (12). It has since proven successful in many investigations. In the first communication on this subject it was already mentioned, however, that a certain parasitic scattering, whose origin is at first not understandable, will still be found above the principal section. It is much less than that caused by a collimation system consisting of ground pairs of slit jaws, but for measurements of very weak scattering effects, such as those discussed above from very dilute solutions in the range of smallest angles, it will nevertheless be disturbing.

ADAPTATION TO EXTREME DEMANDS

In Fig. 12 we show measurements of the primary beam cross-section with a 10 μ slit with a very carefully constructed collimation system. The primary beam intensity, measured with attenuation filters, should reach zero in the principal section $M.S.$ after linear extrapolation. From this abscissa value on, measurements were made without attenuators. Due to the breadth of the counter slit, 10 μ, the end of

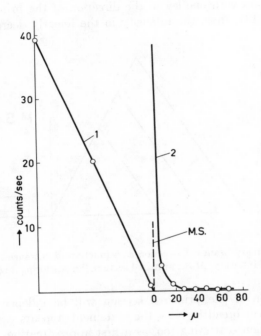

Fig. 12. Measurement of the parasitic scattering with a collimation system according to Fig. 9a. 1 represents the decrease of the primary energy towards the principal section $M.S.$ after appropriate attenuation. 2 is the parasitic scattering measured without attenuation. Entrance slit into the collimation system: 100 μ, counter slit dimensions: 10 × 1500 μ.

the primary beam may be smeared out to 5 μ. Actually, up to 20 μ a measurable scattering is present which exceeds cosmic ray background. It amounts to 310 counts/sec at the theoretical limit, but decreases to a few counts/sec after a further 5 μ. Since a resolution equivalent to a Bragg value of 1000 Å corresponds to a primary beam breadth of about 600 μ the total range of the superimposed scattering of 20 μ will not represent any significant disturbance. The question of the origin of this scattering, weak as it is, was still of interest, however, as we wanted to eliminate this disturbance completely. Investigations together with H. Leopold (13) have led to an extensive clarification.

We start from the statement, that on variation of the entrance slit Sl there will always result triangle-shaped cross-sections of the primary beam with one leg in the direction of the principal section $M.S.$ (Fig. 13); thus, the intensity in the linearly decreasing leg in

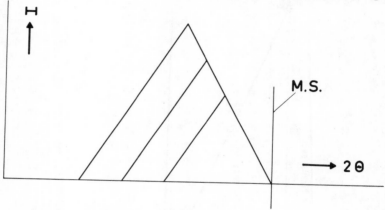

Fig. 13. Primary beam cross-section as a function of entrance slit breadth. I: Intensity, $M.S.$: principal section, 2θ: scattering angle.

the direction of the principal section will be independent of the primary beam breadth. Now the statement appears essential, that the collimation scattering too, as a first approximation, will not depend on the primary beam breadth, but rather on this intensity along the decrease towards the principal section. From this we arrive at the idea, that the slight superimposed radiation is caused by the radiation immediately below area D_3 in Fig. 9a. The fact that the radiation is actually coming from the collimation system is proven by the application of an adjustable slit a larger distance (20 cm) from the collimation system: in this way the parasitic scattering could be sharply cut off. Therefore we may assume that the irregularities of D_3 lying close to edge E_3 represent the scattering centers. They can obviously be excited only by radiation passing along the bridge plane. This may be:

(1) Direct X-ray radiation.
(2) Radiation originating from E_1.
(3) Radiation originating from E_2.

Therefore it should be possible, by lowering of D_3 (as indicated in the scheme of Fig. 9c), to eliminate the disturbance caused by 1. and 2. Measurements with a bridge lowered by a few μ show a considerable reduction of the disturbing scattering, as expected (13). If, besides, the front edge of the bridge is ground cylindrically, as shown in Fig. 9d, then the third cause of the disturbance should be eliminated as well. Even if each site of the cylindrical surface, struck by the primary beam resp. the scattering from edge E_1, would radiate in all directions, this radiation can still never pass along the lower plane face of the bridge. The beam approaching this direction most closely travels along the tangential plane at the lowest point of the cylindrical surface struck by the primary beam.

It must be added, however, that no method could so far be found for the cylindrical grinding which would guarantee good success of this operation in all cases. The requirements in accuracy amount to fractions of a μ. But we did succeed in preparing in this way individual bridges, for which no background scattering at all could be measured (13).

For most purposes it will suffice to lower the bridge by several μ, and to machine very carefully the plane of D_3 facing the focus. The effect of the latter procedure is not presently understood, but appears certain from experiments performed by I. Pilz in our group.

Of course one could object that by the lowering of the bridge an arrangement is in principle produced, in which a limited range of angles must be filled with scattered radiation. Now, a very slight lowering by a few μ must be regarded as a compromise, in view of the impossibility of grinding the bridges to a state which could be regarded as completely plane in relation to the X-ray wave length. It is better to abandon such a small angular range for the measurement, than to have the unlimited scattering due to the irregularities close to the bridge edge E_3 apart from the focus, as caused by direct radiation and radiation at edges E_1 and E_2. One can easily see that for a distance of about 5 cm between E_1 and E_2, a scattering zone of about 10^{-4} radians angular breadth will arise by lowering the bridge by 5 μ. This will correspond to a limiting Bragg value of 15,000 Å. With high intensity of the scattering effect, a larger or smaller part of this zone of parasitic scattering too will be usable for the measurement, so as a rule this compromise will be readily bearable even for extremely high demands of resolution. Practically the lowering is carried out in such a way, that the supporting areas S of the bridge (Fig. 11) are ground appropriately.

We use a chromium steel with about 12% chromium for the collimation system. It is readily workable and has sufficient resistance to corrosion. Although with Cu radiation a good deal of Fe fluorescence radiation is released, the collimation principle works so well that nothing will be observed of this disturbance, even if the channel of the discriminator is opened so wide that the wave length of the Fe radiation could be registered. It may be, that if the very hard molybdenum radiation were to be used (e.g. for the investigation of objects containing heavy metals) it would be necessary to change to materials with high absorption, such as tungsten. For Cu radiation, which we have used exclusively so far, tungsten brings no advantage.

Automation of the Measurement. A Step-Scanning Device

As explained above, during the measurement of the small-angle X-ray scattering the counter with its entrance slit is moved stepwise through the angular range of interest. For each step, the intensity is recorded. The steps must in general be small at the beginning of the measurement, that is at the smallest scattering angles (order of magnitude 1/100 mm), while for larger angles mostly considerably larger steps (up to 1 mm) are appropriate.

In order to relieve the experimenter and to carry out measurements outside the normal working hours also, an automatic device is necessary which moves the counter after the desired count has been attained at each angular position. Since the type of scattering curve will be known from preliminary experiments, one will generally be able to decide on the appropriate angular step sizes prior to the measurement proper. With the simple step-scanning device described below, programming of the step intervals can be done prior to the measurements (14, 15). Figure 14 shows a schematic section of its mechanical part. By turning the nut 1, which rotates in and is supported by the base 2, a threaded spindle 3 (thread height 1 mm) is shifted in its axial direction. This device takes over the function of the precision spindle (Fig. 6) in our ordinary cameras. The nut 1 is rigidly fixed to a circular disk 4, in which small holes of 1 mm diameter are drilled in a spiral sequence, beginning from the periphery. The locus of all holes is a spiral covering 30 × 360 degrees of arc. The radii of two successive holes form an angle of 3.6 degrees or a multiple thereof. Thus, the rotation of the turntable from one hole to the following one (relative to a fixed point) corresponds to a vertical motion of the spindle and counter of 0.01 mm or a multiple of this. The traverse of the entire spiral corresponds to a vertical travel of the spindle of 30 mm. The threaded spindle carries a screw nut 5, which serves for zero adjustment, since it can be moved up and down independently

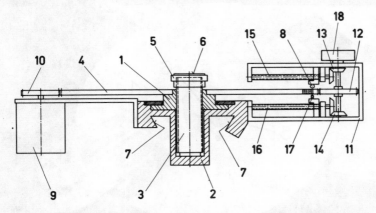

Fig. 14. Section through the step-scanning device. For explanation of the numbers see text.

of the spindle. By turning it, it is possible to set its horizontal limiting plane 6 to any desired position in height; it will consequently serve to set the counter or counter slit to zero angle.

Above the disk 4 there is a light source 8, which is adjusted above the outermost hole (or any other desired starting hole) at the beginning of a measurement. A motor 9 turns the disk by means of a gear 10 which meshes with the toothed rim of the disk. The latter in turn engages gear 12 inside the casing 11. Gear 12 actuates, via gears 13 and 14 and worm drives 15 and 16, the light source 8 and simultaneously the photodiode 17. These move radially, in such a way that the light beam from 8 to 17 will travel along the spiral locus of the holes, traversing all holes in succession as it moves from periphery to center of the disk.

Pre-programming is done by shutting a portion of the holes (whose diameter is 1 mm at the top but offset in a step to a smaller diameter at the bottom) with steel balls 1 mm in diameter. Figure 15 is a photograph of the device. The optical bench, not shown in Fig. 14, is here designated by "19." In Fig. 16, the vacuum tube V, supported in fork F, is resting on the screw nut for zero adjustment. Counter slit and counter have been removed. The operational sequence of the scanner is as follows: The electronic control unit of the step-scanner stops the motor if light strikes the photodiode. It simul-

Fig. 15. Step-scanning device. For explanation of the numbers see text.

taneously delivers a "start" pulse to the counter and timer which record the scattered intensity. After a preset time (or alternatively a preset count) the readings of the counter and timer are printed out. The "print finished" pulse of the printer control unit is used to reset counter and timer and to start the motor again via the control unit, even though the photodiode is illuminated. At the next open hole the motor will be stopped again, and so on. This way the scattering curve will be obtained as a sequential printout (corresponding to the previously programmed scanner positions) of time and count.

Such a device not only renders the operation of the laboratory more effective, but also more rational: the "warming up" of the X-ray

unit to constancy of dose will require several hours. Without the
use of an automatic step-scanner, one can either switch on the unit
in the morning, in which case the actual time for measurement will,
under ordinary conditions, be reduced to but a few hours, or the unit
is kept in operation day and night, but used for measurements only
during the normal working hours, which only amounts to about one
third of the total time of operation. By using the step-scanner, the
X-ray unit is utilized 100%, a notably better constancy is obtained

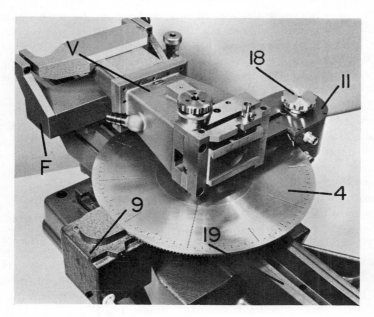

Fig. 16. Step-scanning device with vacuum tube V, turnable in fork F.
For explanation of the numbers see text.

than with daily switching on and off, the measurements themselves
will proceed three times as fast, and all that without manual operation which occupies one person fully. Of course there will be other
ways to solve the problem of automatic recording. J. Jecny (16)
described a system using a teletape reader as control unit. A teletape
carrying the program operates via the reader, controlling scanner,
timer, scaler and teleprinter. In our laboratory H. Leopold developed
an electronically programmed step-scanner. The program is stored
in an electronic memory which allows convenient programming by

setting a number of switches. As this installation is able to record real position and intensity at each point of the scattering curve, it is in addition intended to operate on line with the computer which is used to eliminate collimation errors.

The Absolute Measurement of Intensity

(a) *General*

By "absolute measurement" we will understand the determination of the scattered intensity relative to the primary energy. With dilute solutions, the size of this quotient will lead to a "weighing" of the dissolved particles. For dense and irregular systems too it has become important for the determination of the relative amount of two phases as well as their internal surface (17).

The first measurements of absolute intensity were performed 12 years ago (18). In that paper, the final theoretical basis for their evaluation could already be established. The experimental methods were improved in the course of time, and for the past 5 years we have disposed of precision apparatus (developed in cooperation with Wawra) which allows a routine measurement of this important quantity (19, 20). In recent years, accurate absolute measurements have also been carried out by Luzzati (21) and lately in Beeman's group (22). We refer to summarizing papers (23, 24).

The main difficulty of the absolute measurement lies in the high intensity of the primary beam. The counter devices for measuring the low-intensity scattering are generally not suitable for measuring the primary radiation as well. Fundamentally, there are two ways to overcome this difficulty. Either the primary beam is weakened in a defined way, or a gaseous scattering sample is employed, whose scattering at small angles can be calculated from its composition on the absolute scale. The second way has recently been utilized by Beeman (we refer to his paper in this volume). In the following we shall discuss the first method.

(b) *Defined Attenuation by Filters*

It is obvious that nickel filters may be used for a defined attenuation, as we have done in former years when using photographic techniques (25). Luzzati (21) is using this method in appropriately improved and refined form with counting techniques also, where considerable attenuation factors, namely 10^5–10^6, are required. Lately, Damaschun (26) also has made use of this method. Great care has to be taken in the exact calibration of the sets of filters. A thickness irregularity of but 1 μ will already give rise to an intensity variation of 3%! We cannot judge whether it will thus be possible to attain the desired accuracy of 1%. For us, this method is out of the question in connection with the proportional counter, since it requires monochromatic radiation, such as Luzzati is getting by Bragg reflection according to Guinier (4), Damaschun (8), as mentioned, by total reflection at a plane glass plate in combination with filters. If one would try to weaken polychromatic radiation by means of filters in order to reduce the number of counts, then for the reduction of the short-wave portion by several orders of magnitude one would need filters which will weaken the characteristic portion, considerably longer in wavelength, by several orders of magnitude more, so that practically no more characteristic radiation would be present. Neither will the use of the proportional counter reach our goal, since with appropriately reduced (i.e. by several orders of magnitude) intensity of the Kα-line the short-wavelength portions will be but little attenuated and greatly overload the counting tube. Consequently the discriminator also will no longer be able to make the selection without errors. Therefore, the only way is actually a full monochromatization, which we do not want to employ because of the loss in intensity connected with it.

(c) *The Rotator Method*

Since 1960 (19, 20) we have been using an arrangement which can be regarded as an improved rotating sector for the attenuation of the primary beam without variation of its spectral composition. Immediately in front of the counter is placed a disk, whose plane is normal to the direction of propagation of the X-ray beam, and which shields the counter from the beam (Fig. 17). The disk has a number of small holes—in our measurements four of them—which cross the ribbon-shaped primary beam on rotation. It is necessary, and technically possible, to select the dimensions of the disk and of the holes, and the rate of rotation, in such a way that the number of quanta traversing a hole per passage will on the average be considerably smaller than unity. (If during every passage at least one quantum would travel through the hole, then the pulse counting would only be equivalent to the determination of the rate of rotation.) Of course the time between successive hole passages must exceed the deadtime of the counter.

First, an order-of-magnitude estimate shall show that the stated conditions can actually be met.

Let us regard a primary beam of integral breadth 0.04 cm in the plane of registration. This will correspond to average exposure conditions. Then, a portion of the primary beam of 1 cm length will correspond to an integral area of 0.04 cm². About 2×10^7 quanta/sec

Fig. 17. Rotator device for defined attenuation of the primary beam.
R: rotator, V: vacuum tube, C: counter.

shall intersect this area. This figure represents an approximate average value obtained with commercial X-ray units.

One hole of the rotator has an area of about 10^{-5} cm²; it is, therefore, smaller than the cross-section of the primary beam by the factor 4×10^3. If the hole would rest within the primary beam, $2 \times 10^7/4 \times 10^3 = 0.5 \times 10^4$ quanta/sec would intersect it.

Now, the holes of the rotator are traveling around a periphery of about 20 cm at about 50 rotations/sec. Therefore, the passage of one hole will require $(1/50) \times (0.04/20) = 4 \times 10^{-5}$ sec. During this time interval, under the given conditions, on the average only $0.5 \times 10^4 \times 4 \times 10^{-5} \approx \frac{1}{5}$ quanta will traverse the hole. Thus the condition is fulfilled that the quanta should be removed individually. Since one rotation will last 1/50 sec, and the rotator has four holes, a time of 1/200 sec will elapse between two passages. This time is longer than the deadtime, therefore the apparatus is reset and ready to count at the next hole passage.

The argument leading to the exact calculation of the primary energy is somewhat simplified if we imagine the four holes combined into one hole of size f_4. Let the counting rate (counts/sec) be p. Now we can see that we will obtain the number of counts (counts/sec) corresponding to the primary beam of length 1 cm by multiplying p with two factors:

(α) The reciprocal dwell time: the fraction of the time during which the hole, in its rotation, is within the primary beam, is given by b/U, where b is the integral breadth of the primary beam, and U the circumference of the circle along which the holes travel. It appears clear that multiplication with U/b is necessary.

(β) The quotient of the integral area of the primary beam $b \cdot 1$ cm^2 and the hole area f_4; therefore we have:

$$P_0 = p \cdot \frac{U}{b} \cdot \frac{b}{f_4} = p \cdot \frac{U}{f_4} \qquad\qquad 1.$$

For very exact measurements, various corrections must be applied, and several sources of error have to be considered. We want to refer to the original papers in this regard (20), but the most important corrections shall be discussed. In spite of the few passages of quanta through the holes, there will still occur coincidences for a certain fraction of the passages, that is, a hole will be traversed by two or more quanta during one passage. Since one passage lasts several 10^{-5} sec, as mentioned above, it is doubtful whether for such coincidences a double or multiple counting will take place, or whether the quanta are counted as a single one. Therefore it is essential to apply a pulse-lengthening device, to guarantee that the deadtime of the counter after the input of a quantum will be longer than the duration of the passage. Then we have clear conditions, which allow exact application of a coincidence correction.

First we obtain the number of counts per passage, n, from the number p of counts/sec, by dividing p by the number of holes (four) and the number of rotations per second, u:

$$n = p/4u \qquad\qquad 2.$$

According to well-known principles of statistics (Poisson's formula) we then obtain the number of counts actually occurring per passage, n_{corr}:

$$n_{\text{corr}} = n + \frac{n^2}{2} + \frac{n^3}{3} + \cdots = -\ln(1-n) \qquad 3.$$

As long as $n \ll 1$, n_{corr} is only afflicted with the statistical error occurring during the measurement of n (resp. p). However, if n approaches unity, then n_{corr} will become more and more sensitive to small errors in n, as the above formula immediately demonstrates. If this case takes place, the area of the holes f_4 is obviously too large for the evaluation of the primary beam, and a rotator with smaller holes must be used.

From n_{corr}, one determines first

$$p_{\text{corr}} = n_{\text{corr}} \cdot 4u \qquad 4.$$

and finally obtains the primary energy P_0 by inserting 3 into 4, and 4 into 1.

Of course an exact determination of hole size is necessary. Because of the smallness (edge length about 0.03 mm), the accuracy of microscopic measurement is not quite satisfactory. Therefore we found it necessary to make a second rotator with a considerably larger hole, namely about 0.012 mm², which can be measured with sufficient accuracy. Then the ratio of the areas can be determined by comparing measurements with any primary beam.

So the absolute intensity measurement amounts finally to a length measurement. The repetition of the measurements for many times and different illuminations in the microscope (reflected light of transmitted light), and by different workers has resulted in a reproducibility of $\pm 0.5\%$. In order to obtain the same accuracy with a nickel filter, the thickness had to be defined to $\pm 0.12\ \mu$.

(d) *Use of a Standard Sample*

It is an obvious idea to determine the scattering of a standard sample at an appropriately selected angle on the absolute scale once and for all by means of the rotator method. A comparison of the scattering of a sample under investigation with the test sample will then make it possible to calculate the former's scattering on the absolute scale.

We shall mention briefly, that polyethylene* in the form of about 3 mm-thick platelets has proved very suitable for this purpose (48). It has the advantages of mechanical stability, completely sufficient stability to X-radiation,† homogeneity, and sufficiently intense scattering at the angle corresponding to a Bragg value of 150 Å. In this range, the intensity decreases linearly, and already approaches zero at relatively small angles. Therefore the scattering is practically independent of primary beam length and breadth in a wide range. If the relation between the two samples is established, however, the difference in absorption must be taken into account. While the attenuation by the standard sample can also be determined once and for all, the individual test sample requires an absorption measurement. For this purpose provision is made for inserting the sample, mounted on a carrier, into the center of the collimation system, as discernible especially in Fig. 8. Thus it will weaken the primary beam, but not give rise to any radiation entering the counter. One now inserts any auxiliary sample into the normal position between collimation system and vacuum tube, and measures its scattering at any suitable angle. This measurement is done first with the unattenuated beam, and then a second time after having inserted the test sample into the center of the collimation system as described above. From the quotient of the two scattering measurements, the desired absorption is obtained.‡

Of course other types of standard sample are also conceivable. Thus Kirste and Wunderlich (27) have proposed an exposed and developed photographic film, whose scattering from the colloidal silver is calibrated by means of the rotator method. Beeman (22) has measured the scattering of pure water in the small-angle range, and in this way provided a standard available to everyone. However,

* The sample is the substance "Lupolen" 1811 M of the Badische Anilin- and Soda-Fabrik. We wish to express our thanks for these materials, suitably prepared for our experiments.

† The objection might be raised, that high X-ray doses will alter any plastic material, and therefore sufficient constancy of scattering may be questioned. Without digging deeper into the problem it may be stated that we could not detect any variation in the course of one year even with daily irradiation. Therefore, for practical purposes, where an absolute measurement will hardly be performed every day, one can safely depend on constancy for at least several years.

‡ Since absolute measurements with the rotator are performed currently in the laboratory of the author, he will be glad to supply interested research workers on request with a calibrated sample.

the scattering of this substance is extremely weak and depends on the length-profile of the primary beam, which therefore must be determined carefully.

Remarks Concerning Collimation Corrections

An important and at the same time difficult problem of the evaluation is the elimination of collimation errors. A theoretically correct method by Guinier and Fournet (28) is applicable if the primary beam

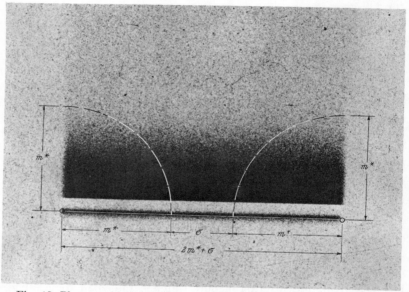

Fig. 18. Photographically recorded small-angle scattering on a greatly enlarged scale. The figure shows the relation between effect length m^*, counter slit length σ and the minimum length required for the uniform portion of the primary beam $(2m^* + \sigma)$.

is of "infinite" length, that is if the primary radiation at the ends of the primary beam cross-section will send no more radiation into the counter slit. As Fig. 18 demonstrates, in this case the effect length, that is the breadth of the scattering range in the plane of registration, must be smaller than the distance m^*, therefore smaller than half the difference between primary beam length and counter slit length σ. Since a long counter slit (about 10 mm) is desirable on intensity grounds, and the effect length is often enough also about 10 mm, a length of the uniform portion of the primary beam of 30 mm would be required in these cases. Since every irregularity of the primary beam will lead to errors in the curve obtained by correction for the collimation error (*"Entschmierung"*), the importance of using X-ray tubes with "opening" beam—as explained previously— is obvious. With extremely large effect lengths, as found for instance with solutions of chain molecules (30 mm and more), it will practically no longer be possible to work with an "infinitely" long primary beam. One must not forget, that the background radiation caused e.g. by the solvent and the sample container (Mark-capillary or cuvette windows) and which has only a small angular dependence, will, at every place of the small-angle scattering field, increase about proportionally to the length of the primary beam. So for large scattering angles and "infinitely" long primary beam it will in this case no longer be possible to subtract the extremely weak effective scattering from the background. We are therefore forced to use a primary beam of "finite" length. In order to be able to apply the sufficiently accurate approximation method of Kratky, Porod and Kahovec (18) for the elimination of the collimation error with finite primary beam, the middle portion of the primary beam must be even more completely uniform, and both ends must decrease linearly in a given fashion.

In our group an electronic program for carrying out the correction for the collimation error has been worked out by Heine (29). Since it will not be possible within the scope of this contribution to discuss in detail the problems connected with collimation in X-ray small-angle measurements, reference shall be made to a summarizing paper (2).

EXAMPLES OF MEASUREMENTS UNDER EXTREME CONDITIONS

Only new measurements are discussed, which in part have not yet been published anywhere else, and for which the above-mentioned experimental techniques have been utilized.

Emulsion of TiO_2 in Linseed Oil, as an Example for the Study of Very Large Particles by Measurements at Extremely Small Angles

[With H. Leopold (30)]

With sufficiently small entrance and counter slits, it was possible to measure the scattering curve up to Bragg values of about 6800 Å, neglecting the background. In the region of smaller angles up to a Bragg value of about 11,000 Å, however, a small correction was applied by subtracting the values of the background readings corrected for absorption by the sample.

Figure 19 shows the log–log plot of two experimental scattering curves for a TiO_2 sample designated by I. There is nearly no difference in the shape of the curves for both given concentrations. From this it may be supposed that in this range of concentrations practically no effects of interparticle interference occur, and curve 1 may be interpreted as undisturbed particle scattering according to Guinier.

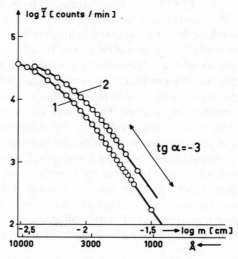

Fig. 19. Log–log plot of the experimental scattering curves of a TiO_2 sample designated by I for 2.12 (curve 1) and 29.50 wt.% (curve 2).

Such behavior is not surprising, as the concentrations by volume are about four times lower than the stated concentrations by weight. In addition, interparticle interference effects are attenuated by polydispersity and particle anisotropy. The figure shows how well the observed points fit on a smooth curve. This is an immediate indication of the accuracy attained in this measurement. The tail end of the curve follows the negative third power of the scattering angle, as required by the theory (17).

After elimination of the collimation error we obtain the curve of Fig. 20 (from curve 1 in Fig. 19). As expected from the theory, the tail end of the log–log curve now drops as the negative fourth power

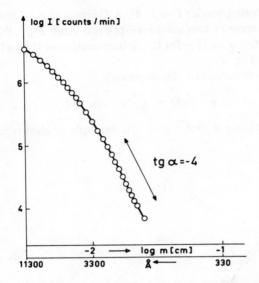

Fig. 20. Log–log plot of the scattering curve for sample I corrected for collimation error; 2.12 wt.%.

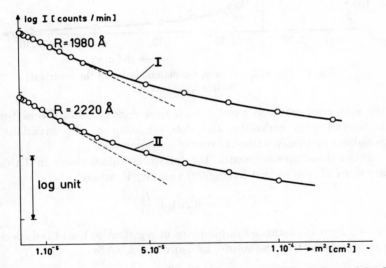

Fig. 21. Guinier plot of scattering curves for two samples, designated by I and II, corrected for collimation error. 2.12 wt.% (curve I), 1.38 wt.% (curve II).

of the scattering angle: $I = k/\theta^4$. A Guinier plot of the collimation-corrected curves of two samples—the one from Fig. 20 and a second one designated with II—for the determination of the radii of gyration, is shown in Fig. 21.

The determination of the invariant

$$Q = \int_0^\infty I \cdot m^2 \cdot dm \qquad 5.$$

requires plotting $I \cdot m^2$ vs. m. An example is shown in Fig. 22. As

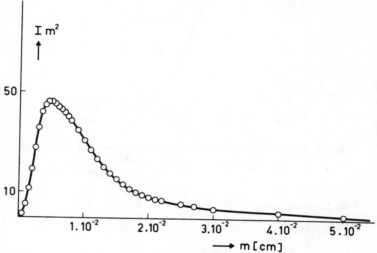

Fig. 22. Plot of $I \cdot m^2$ vs. m for determination of the invariant; Sample I, 2.12 wt.%.

the scattering curve is available up to a region where the m^{-4}-rule is obeyed with certainty, the determination of the invariant is possible practically without error of extrapolation.

From these measurements, the following values listed in Table I are derived: the weight average of volume, V, where

$$V = 0.291 \, a^3 \frac{I_0}{Q} \qquad 6.$$

($a = 22$ cm = distance sample-plane of registration), and the specific inner surface O_s of the solute, measured in $\text{Å}^2/\text{Å}^3$

$$O_s = \frac{12.82 \, w_2}{a} \frac{k}{Q} \qquad 7.$$

(w_2 = volume fraction of the solvent).

TABLE I

	TiO$_2$(I)	TiO$_2$(II)
c[g/ml]	1.89×10^{-2}	1.23×10^{-2}
R[Å]	1980	2220
d_R[Å]	5120	5740
Q	0.667	0.522
k	3.40×10^{-3}	2.60×10^{-3}
V[Å^3]	2.68×10^{10}	3.08×10^{10}
$\bar{M}_v$	6.2×10^{10}	7.12×10^{10}
O_s[Å^2/Å^3]	2.95×10^{-3}	2.88×10^{-3}
V_{max}[Å^3]	7.04×10^{10}	9.87×10^{10}
O_{min}[Å^2/Å^3]	1.17×10^{-3}	1.045×10^{-3}

Assuming the particles to be spheres of uniform size, one can derive a maximum value for the volume

$$V_{max} = \frac{4\pi}{3}\left(\sqrt{\frac{5}{3}}R\right)^3 \qquad 8.$$

as well as a minimum inner surface

$$O_{min} = 3 \bigg/ \sqrt{\frac{5}{3}}R \qquad 9.$$

from the radius of gyration R. The difference between V_{max} and O_{min} on the one hand, and the experimental V and O_s on the other hand, are a manifestation of the anisotropy and polydispersity of the particles. From the experimental volume the weight average of the particle mass $\bar{M}_v$ (in atomic weight units) can also be derived. This value is listed in Table I.

Since considerable multiple scattering takes place, we shall not attempt a calculation of M from the absolute intensity at zero angle.

Measurements With Dye Solutions for the Determination of Degree of Association and Particle Shape, as an Example for the Investigation of Extremely Small Particles

(With H. Ledwinka and I. Pilz)

(a) *General Remarks*

We have shown in the preceding chapter that it is possible to extend the small-angle X-ray method beyond its usual range also to very large particles. In this section we want to discuss measurements, which, on the contrary, represent an extension to astonishingly small particles. In order to obtain reliable results about size, shape, and mass, we have studied diluted solutions. First we ask what experimental difficulties could be expected if the particles were to fall below the order of magnitude of volume and mass usual for instance for proteins. For any particle size and concentration, the scattering curve 2 in Fig. 23 may result; curve 1 is the corresponding blank

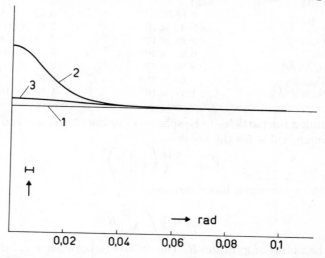

Fig. 23. Scheme of the scattering curves for 0.5% solution of a protein with $M = 40{,}000$ (curve 2), $M = 8{,}000$ (curve 3) and the solvent (curve 1).

exposure, that is the scattering of the solvent, sample container, etc. If we now imagine the weight concentration kept constant, and the particle mass reduced to, say, $\frac{1}{8}$, that is we think each large particle replaced by 8 smaller particles, then scattering curve 3 would result. From general principles we infer readily that $(I_0)_3 = \frac{1}{8}(I_0)_2$. Apart from this diminution of intensity, the second curve will differ from the first one by its extension over a larger range of angles, as the integral $\int_0^\infty Im^2 \cdot dm$ (the invariant) must be the same for both solutions. Assuming identical shape for both the large and the small particles—an assumption made only in the interest of easier illustration—the latter would have half the linear dimensions of the former, that is, curve 3 would be twice as broad as curve 2 (then one will understand at once that the invariant will remain unchanged in spite of the zero intensity being reduced to $\frac{1}{8}$). The relation between the solvent scattering (curve 1) and the scattering of the solution of large particles (curve 2) is chosen approximately as if we had protein molecules of molecular weight about 40,000 dissolved in water to a concentration of about 0.5 weight percent, while the corresponding small particles would then have a molecular weight of 5000. With the large particles, the inaccuracies are enlarged about five-fold by the subtraction of the experimental scattering curves of solution and solvent, when we confine ourselves to the zone from zero angle to 0.02 radians. This difficulty can be overcome by sufficient precision of both measurements. With the small particles, where a range up to 0.04 radians must be taken into account, the same procedure will bring about a multiplication of the error by a factor of 40 and more. With a further decrease of particle size, the scattering curve will extend up to the range of the beginning of the liquid maxima, which again increases the difficulty of taking the difference, since the background to be subtracted is no longer horizontal and cannot be determined by averaging.

With dye molecules of basic molecular weights between 300 and 1000, and their association products of low degree of association, we shall of course encounter the same difficulties to a high degree if, however, the electron density difference normally is a little higher. Further, for this group of materials in any event, but probably quite generally, the following difficulty is added. The interparticle interference effects, which bring about a falsification of the radius of gyration derived from a Guinier plot, and a reduction of the zero intensity obtained by extrapolation, and therefore also of the particle weight calculated from the absolute intensity, can for want of a generally applicable theory be eliminated only by studying a concentration series, and by extrapolating the above quantities to zero concentration. If this method is to promise success, one must not, with the most diluted solutions under investigation, be too far from the limiting case of pure particle scattering. Now for equal volume concentration and shape, the average distances between the particles are proportional to their linear dimension, that is, for small particles they are correspondingly smaller. This will bring about a stronger interaction, which will cause stronger interference effects. With dyes, it is true, the forces can be successfully shielded by salt addition, but then as a rule at the same time an association will be caused, that is an incipient salting-out. Therefore, no conclusion can be drawn from results obtained with salt-containing solutions which are also necessarily valid for saltfree ones. A variation of both concentration and salt content may here bring progress.

These remarks make clear that with dye solutions quite often only an extreme extension of the method will lead to success. Sometimes, however, the boundary of the normally possible will be surpassed.

In collaboration with Ciba AG,* we have studied a number of dyes, namely "All-4" and "All-3" copper phtalocyanine, several vat dyes, and the azo dye "Orange II."

(b) *Measurements With 9-Naphthol Orange Solutions ("Orange II")*†

This dye has the formula (31)

$$\text{OH-C}_6\text{H}_4\text{-N=N-C}_6\text{H}_4\text{-SO}_3^-\text{Na}^+$$

and possesses a molecular weight M of only 350. For this report, we have chosen the measurements with this dye, since here the association effects in water and salt solutions have already been investigated before by other methods, and therefore comparisons can be made. Milicevic and Eigenmann (31) have studied water solutions by means of equivalent conductance and viscosity. They have found that the dye is present as monomer first, but that the dimer will already prevail at a concentration of 10^{-2} mol/l. These results agree fully with our measurements. Investigations of "Orange II" by means of conductivity, viscosity and light-scattering were carried out by Frank (31), who found a strong increase of the degree of association on addition of salt, as is also in accordance with our findings. A concentration series in water is shown in Fig. 24. The measurements of the scattered intensity at the smallest concentration ($c = 0.212\%$) lie close to the limit of the method as is shown by the following consideration. Each point is the result of measuring 10^5 counts, so we may expect an average statistical error of 0.3%. The scattering curve of the dissolved dye is however the difference between dye solution and solvent and this is, for the above-mentioned most dilute solution, 3% of the experimental scattering intensity. Taking this differ-

* I want to express our very best thanks for the suggestion of these investigations, their support, and the preparation of the samples.

† Reference is made to the original papers in preparation.

ence, the average statistical error will therefore enlarge by about a factor 30 and mean errors of about 10% are to be expected. We have tried to keep this error as small as possible by smoothing the experimental scattering curve. But we are aware of the fact that with this measurement of intensity scattered by smallest particles ($M = 350$)

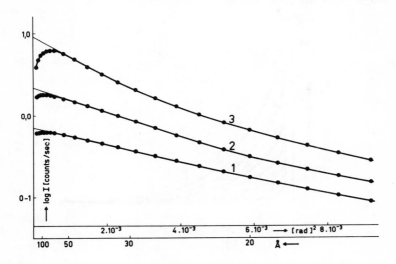

Fig. 24. Guinier plots for solutions of "Orange II" in water. The concentrations are:

1	0.212%
2	0.418%
3	1.015%

at lowest concentration, we have arrived at the limit of our technique. We therefore shall regard this scattering curve only as a supplement for the less problematical scattering curves at higher concentration. With the zero intensity obtained by extrapolation to (rad.)$^2 = 0$ and the measured primary intensity P_0, we obtain the absolute intensity I_0/P_0, which, together with the measured partial specific volume $\bar{v}_1 = 0.52$, yields the particle weights from known relationships (18, 24). They are given in curve 1 of Fig. 27 as multiples of the molecular weight. We note that the measurements with $c = 0.212\%$ fit well into the curve given by the other points. So we do regard it as a support for the extrapolation to $c = 0$ which we have made, in spite of the above-mentioned shortcomings. Therefore it appears very probable that in sufficiently dilute solution single molecules will be present, while with increasing concentration a considerable association will take place. We shall stress that the smallest particle weight determined from the absolute intensity of small-angle X-ray scattering is about 400. That corresponds to the single molecule of $M = 350$.

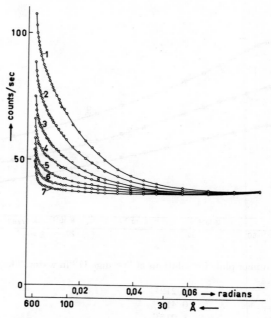

Fig. 25. Experimental scattering curves of "Orange II" in 0.3 m KCl solution with varying dye content. Concentrations:

1	1.09%,	2	0.74%
3	0.50%,	4	0.34%
5	0.22%,	6	0.11%,
7	0 (solvent scattering = curve of the 0.3 m KCl solution)		

The solutions with KCl addition shall now be discussed briefly. Here the association, and therefore also the intensity, are much higher, and taking the difference is much easier. We studied "Orange II" in 0.1, 0.2, 0.3 and 0.4 m KCl solution and found a strong rise in the degree of association. In Fig. 25 are shown several original scattering curves of "Orange II" in 0.3 m KCl solution. One can readily see even qualitatively, from the steep rise of the curves, that much larger aggregates than in water must be present. Figure 26 shows Guinier plots (after smoothing and collimation correction) for the solution with 0.1 m KCl and a dye content varying between 0.10 and 1.04%. The solution with 0.10% demonstrates a part of the problematic conditions setting the limits to our statement. Figure 27, curve 2 gives the degrees of association.

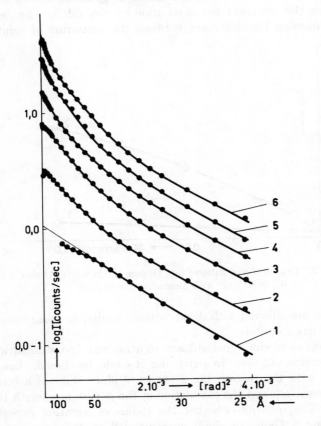

Fig. 26. Guinier plots for solutions of "Orange II" with 0.1 m KCl and varying dye content. Concentrations:

1 . . . . 0.10%,		2 . . . . 0.21%
3 . . . . 0.37%,		4 . . . . 0.51%
5 . . . . 0.76%,		6 . . . . 1.04%.

Dilute Solutions of Low Molecular Weight Cellulose Nitrate

(With H. Leopold and G. Puchwein*)

Solutions of chain molecules belong to the most difficult objects of small-angle research due to their poor scattering. If not only the central scattering range for the determination of the persistence length is to be measured, but the radius of gyration also is to be found, then the solution must in addition be very dilute. The conditions for forming the difference between the scattering of solution

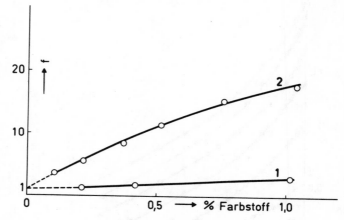

Fig. 27. Degree of association f for "Orange II" in water (1) and in 0.1 m KCl (2) with increasing dye content.

and solvent are afflicted with disadvantages similar to those encountered with dye solutions.

The behavior of chain molecules in solution was first studied with cellulose nitrate (32, 33). In particular it could be shown, that, in agreement with the theory, the $I \cdot \theta^2$ vs. θ plot exhibited a break, whose abscissa allowed the calculation of the persistence length (34) (Fig. 28). The question whether the radius of gyration expected by assuming a Gaussian chain corresponded to the experimental values remained open. Finally, from the point of view of the theory, the question had to be decided whether already for comparatively

* The measurements were carried out under the sponsorship of E. I. du Pont de Nemours and Company, Explosives Dept. We wish to express our sincere thanks for this support.

short chains a horizontal range in the $1/\theta^2$-plot would be found at the left-hand side of the break (Fig. 28), as expected from the analytical calculations made so far (33), or a range definitely decreasing towards smaller angles, as expected on the grounds of configurations set up statistically by means of random numbers [Monte Carlo method (35, 36, 37)] (Fig. 29). These questions are answered by new

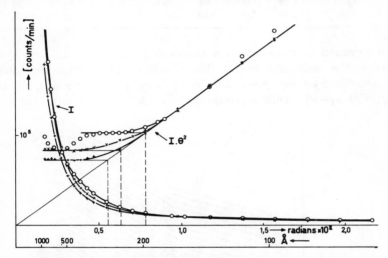

Fig. 28. High molecular weight cellulose nitrate in acetone solution ○ 6.65%, $a = 49.8$ Å; × 2.0%, $a = 62.3$ Å; + 1.0%, $a = 69.6$ Å.

measurements with four low molecular weight cellulose nitrates in acetone, containing 12 and 11% N. Here too the differences between scattering curve and background are very small. After subtraction and collimation correction, we obtain the Guinier plot in Fig. 30 and the $I\theta^2$ vs. 2θ plot in Fig. 31. We note that the type of the scattering curves corresponds exactly to the result of the statistical computation: namely a decreasing branch on the left-hand side of the break, and an increasing one on the right-hand side, which would extend exactly through the origin in its extrapolation on the left-hand side. The break can be determined very accurately, and leads to the persistence lengths a listed in Table II, between 49 and 59 Å. On the other hand, the limiting tangent (e.g. see Fig. 30) for the determination of radii of gyration can be fitted accurately. The values obtained are also given in Table II. To check whether the connection between persistence length, hydrodynamic length and radius

TABLE II
Low molecular weight cellulose nitrates in acetone.

Sample	%N	P	L(Å)	a(Å)	x = L/a	R_{calc}(Å)	R_{exp}(Å)
A	11.08	78	403	54.3	7.4	72.2	83.5
B	11.19	145	750	49.4	15.2	101.5	115.5
C	11.96	90	464	56.6	8.2	81.0	89
D	12.09	157	810	59.4	13.6	113.0	133

of gyration corresponds to a Gauss coil. (37) (Fig. 32), we must know the molecular weight. The absolute value of the zero-angle intensity does not yield sufficiently accurate information, as the partial specific volume could not be determined with sufficient

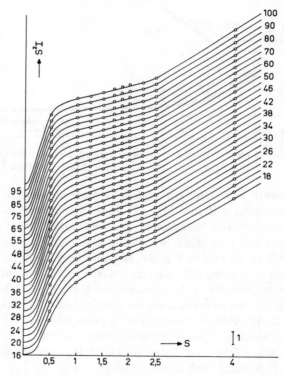

Fig. 29. Theoretical $Is^2 - s$ curves for coiled chains, calculated by the Monte Carlo method. Number of persistence lengths on the left and right hand side.

accuracy for the time being.* But it can be shown, that the comparison of both the intensity in the $1/\theta$ branch, which depends entirely on the mass per unit length of the chain, and the zero-angle intensity, which is proportional to the mass of the total coil, allows a determina-

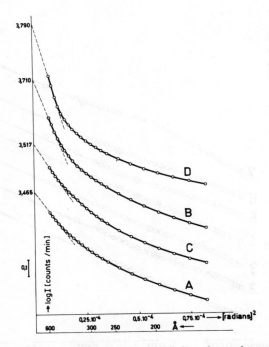

Fig. 30. Guinier plot for low molecular weight cellulose nitrates in acetone. Table II gives the characterization of the samples and the results of the evaluation.

tion of the molecular weight without knowing the partial specific volume. As can be immediately understood, the ratio of these quantities leads to the total length, to molecular weight and to the degree of polymerization, as mass and length of the monomer unit are known. We omit the derivation, and give only the result. We obtain:

$$\frac{M}{M_m} = P = 0.77 \frac{1}{l} \frac{I_0}{(I\theta)_0} \qquad 10.$$

* When using volatile solvents such as acetone, the determination of the density with extremely high accuracy meets with considerable difficulties.

Here M means the molecular weight, M_m the molecular weight of the basic (monomer) unit, l its length, and P the degree of polymerization. $(I\theta)_0$ is the value obtained from the linear, rising branch of the $I\theta^2$ plot. The degrees of polymerization obtained in this way, and the corresponding hydrodynamic lengths, are compiled in Tables

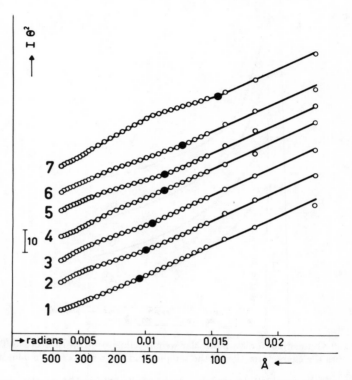

Fig. 31. $I\theta^2$-vs.-2θ-plot of low molecular weight cellulose nitrates. Curves 1, 2, 5 and 3 are measured with acetone solutions of the samples D, C, B and A; Curves 4, 6 and 7 with solutions in a poor solvent of the samples C, B and A— Tables II and III give the results of the evaluation.

II and III. Then we know at once the number of persistence lengths $x = L/a$ where L is the hydrodynamic length. From this we obtain the radius of gyration R_{calc} to be expected on the basis of a Gauss coil, using Fig. 32. Agreement with the experimental value R_{exp} is good. The fact that R_{exp} is 10–15% higher than R_{calc} is caused by the polydispersity.

TABLE III
Low molecular weight cellulose nitrates in a poor solvent.

Sample	%N	P	L(Å)	a(Å)	x = L/a	R_{calc}(Å)	R_{exp}(Å)
A	11.08	78	403	36.8	10.9	62.7	71
B	11.19	145	750	44.2	17.0	97.1	110.5
C	11.96	90	464	49.6	9.3	77.0	83

Measurements with the same sample in poor solvents yield significantly shorter persistence lengths (curves 4, 6 and 7 in Fig. 31 and Table III). Here part of the substance was present in the form of large clusters too.

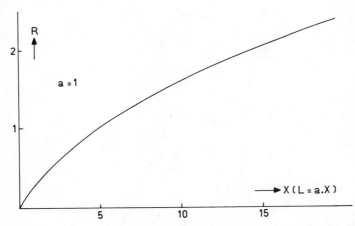

Fig. 32. Connection between the number of persistence lengths x and the radius of gyration (as a multiple of the persistence length "a") for a Gauss coil.

Measurements With Dilute Solutions of Ferritin and Apoferritin

Finally, we shall touch the almost "classical" field of small-angle X-ray measurements with proteins. We shall choose the example of the iron-containing protein ferritin, whose study has occupied us for several years. We refer to investigations carried through in cooperation with Bielig, Holasek, Wawra, Hauser, Steiner and Rohns (38, 39, 40, 41, 42). Since, as described later, very exact measurements with very dilute solutions were necessary, the investigations fit well into the scope of this report.

Ferritin consists of a protein shell surrounding an iron oxide hydrate micelle. The ratio of protein to iron may vary within rather wide limits. Among other things, we wanted to obtain more information about size and structure of this inclusion. This is done best by adding sugar to the solution until the electron density of the solvent will equal that of the protein. This condition is approximately met with a 66.65% solution of saccharose. The X-ray beam will then no

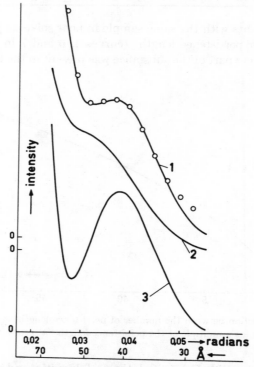

Fig. 33. Calculated scattering curves for three models of the iron III oxide hydrate micelle of ferritin. 1. four spheres, tetrahedral; 2. four spheres, square planar; 3. six spheres, octahedral. ○ measured intensity. $R = 26.7$ Å, $r = 18.5$ Å.

longer "see" the protein, and the scattering pattern is caused merely by the iron micelle. In a recent investigation we made use of a sample whose iron content had been increased to a maximum by "fattening" the horse donating the protein with an intramuscularly injected iron preparation (41). We hoped that the iron micelle would then acquire its optimal shaping. The scattering curve was obtained with such precision that statements about the fine structure of the micelle were possible. After subtraction of the solvent scattering and collimation correction we obtain the scattering curve, whose outer portion is shown in Fig. 33, curve 1. The step in the decrease of the main maximum is remarkable; its course suggests that it cannot be an effect caused by any simple shape (sphere, ellipsoid, etc.).

From electron microscopic investigations of G. W. Richter (43), the idea suggested itself that the iron III oxide hydrate micelle should consist of several subunits. In particular, a structure of six octahedrally arranged partial micelles of approximately spherical shape was conjectured. Now we have calculated the scattering curves for the following models of the iron III oxide hydrate micelle consisting of spherical subunits:

(1) 4 spheres in a square arrangement (2) 6 spheres in octahedral and (3) 4 spheres in tetrahedral arrangement.

If the absolute dimensions are chosen in agreement with the measured radius of gyration of $R = 26.7$ Å, then, as shown in Fig. 33, a side maximum is obtained for all three models, which agrees well with the actual one in its position. But only for the tetrahedral arrangement is there additional agreement with the shape of the curve in the neighborhood of the side maximum (41). In the other two cases we note large deviations. Because of this agreement, we come to the conclusion that the suggested model is the "scattering equivalent" of the real micelle. As long as there is no other contradictory experimental evidence, we therefore shall adopt it as a working model.

Following a procedure by Behrens and Taubert (44), it is possible for the isolation of the pure protein to remove the iron micelle from the protein shell. The apoferritin obtained in this way has been thoroughly investigated by us in solution (40, 42). The main maximum, obtained with solutions between 4.65% and 0.173%, is shown in Fig. 34. The directly measured values, after subtraction of the background scattering, are given. A radius of gyration of $R = 56$ Å is found. Figure 35 shows the first three of a total of 13 side maxima measured. If the main maximum at $\theta = 0$ is given the value 1, then the first side maximum corresponds to the intensity 0.0318, the third to 0.000549. From the fact, that between the maxima there are observed zero positions, the rather good spherical symmetry of the molecular shape can be inferred. With this, however, the prerequisites for a Fourier inversion are provided for the determination of the

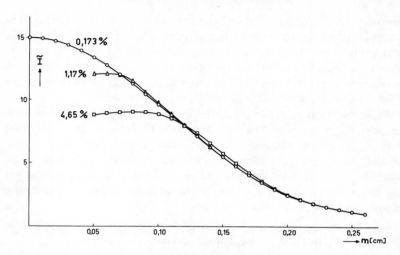

Fig. 34. Concentration series of apoferritin in water. $\tilde{I}$: measured intensity.

radial mass distribution. Figure 36 shows the result of this calculation. Because of the small number of undisturbed maxima caused by the radial mass distribution (the following 10 maxima are caused by the fine structure, that is the subunit structure of the molecule, a relatively large termination error occurs, which causes the fictitious density

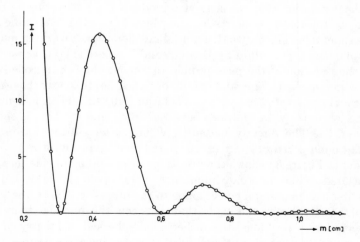

Fig. 35. Outer part of the scattering curve of apoferritin corrected for collimation error.

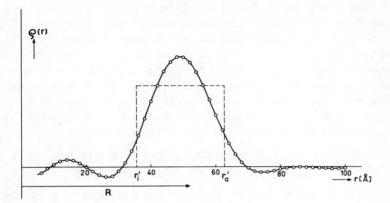

Fig. 36. Electron density distribution $\rho(r)$, obtained by means of Fourier transformation; R: calculated radius of gyration.

waves for small and large distances from the center. Apart from this we are led to the conviction that a spherical shell, replaced by a hollow sphere with a wall of homogeneous density, has an inner radius $r_i = 35.5$ Å and an outer radius $r_a = 62.7$ Å.

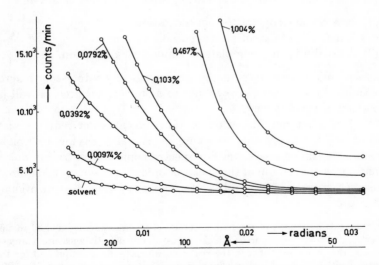

Fig. 37. Concentration series of ferritin, dissolved in 0.15 m aqueous NaCl solution; measured intensity.

Very similar results have been obtained in previous X-ray crystal structure investigations (45) as well as electron microscopic ones (46).

So far the measurements with ferritin agree to a first approximation, but not completely, with this total picture developed from iron core and shell (Fig. 37). As far as could be elucidated, in ferritin we have in part cores without, respectively with incomplete shells. Besides, complications due to association appear to arise. In any case, we are inclined to adopt the described "ideal structure" for this interesting protein, on the basis of the available results.

CONCLUDING REMARKS AS TO THE LIMITS AND ADVANTAGES OF THE METHOD

Some difficulties mentioned in the interpretation of small-angle scattering curves which were measured with high accuracy* direct our attention to the fact, that the restriction in the possibilities for evaluation is (for not too unfavorable experimental conditions—i.e. for molecules which are not too small, concentrations which are not too low, and electron density differences which are not too small) brought about not by the lack of accuracy of the measurements, but rather by the "imperfectness" of the samples. A maximum of information will be obtained only if (aside from the availability of exact scattering curves) the following three prerequisites are met:

(1) Exact knowledge of the concentration.
(2) Exact knowledge of partial specific volume.
(3) Availability of a completely homodisperse solution.

These prerequisites are, precisely for biologically interesting macromolecules, by no means always met to a sufficient degree, as will be explained in the following:

(1) The exact determination of the concentration requires very careful work, namely the removing of all salts, drying to weight constancy without losses etc. For spectroscopic concentration determinations, a calibration of the spectroscopic measurement is required. This necessitates a careful performance of the just-mentioned procedure.

---

* The best impression of the high accuracy of the measurements cited in this paper is given by those curves which represent the directly measured intensities (i.e. the difference between the scattered intensity for solution and for the solvent) without smoothing or elimination of the collimation error. We refer especially to Figs. 19, 25, 33, 34, 37.

(2) Still more difficult is a very exact determination of the partial specific volume. For dilute solutions, it requires a density determination to six places. This severe demand in accuracy is brought about because the square of the expression $(z_1 - \bar{v}_1\rho_2)$ enters the molecular weight determination. z_1 is the number of electrons per gram of solute, ρ_2 the electron density of the solvent. Both quantities are straightforward. But considerable errors are introduced by an incorrect partial specific volume of the solute, $\bar{v}_1$.

(3) The requirement of homodispersity is, in most cases, insufficiently met. At least a very considerable fraction of dimeric molecules will be present in most cases; often enough partial denaturation will take place leading to the formation of higher aggregates. It is also possible that the solution is alright at the beginning of the measurement, but after a measuring time of several hours changes may have occurred. Serological or immunological tests mean little with regard to morphological uniformity. For its determination, the ultracentrifuge is above all to be used. But the precision of the results obtainable in this regard is rather narrowly limited, as shown in a systematic investigation carried out recently (47).

It is clear that beside the old methods, such as recrystallization, newer purification methods such as gel filtration, deviation electrophoresis etc. are of highest importance.

It is known that he who works with small-angle X-ray scattering will not always find sufficient understanding among associates studying the respective materials from a biochemical viewpoint, where purity means a completely different thing. The often-heard advice—to proceed with the investigation, and then to see whether the sample is all right or not—is not always good. With small-angle scattering, one notes certainly the presence of large aggregates (however, they may escape observation if they exceed the studied particles by one order of magnitude or more in the linear dimension), but one will not note an addition of, say, double molecules, and necessarily very small particles will also escape. One must not forget that the interpretation of a small-angle picture, if a size, mass, and shape determination is attempted, is bound to the condition of homodispersity. If this is not given, but assumed for the evaluation, then a "scattering-equivalent system" is obtained. That is, one evaluates a homodisperse system, which has the same scattering as the actual one. The determination of several particle sizes side by side is possible only in the limiting case of very different particle sizes, but is excluded in general.

Thus attempts at a refinement of the small-angle method, and its adaptation to extreme experimental conditions may, as can be inferred from what has been said above, very often be fully effective only if certain prerequisites are fulfilled also on the part of the sample. Thus the foregoing remarks belong to the topic of this paper.

This restriction shall not make us forget that the small-angle method offers tremendous advantages for the study of macromolecules and colloid particles. Without entering into a closer discussion of this question, some advantages may be listed which it provides for the investigation of solutions of macromolecules of biological interest:

(1) Coarse impurities ("dust") offer practically no disturbance to the investigation, since their scattering lies at angles which are normally not included in the investigation. By contrast, with the light-scattering method, the large particles represent the major disturbance, since here a considerable angular range is covered by the disturbing scattering due to the about 4000-fold longer wavelength.

(2) The absolute intensity measurement, leading to mass determination, attains an accuracy of about 1%.

(3) While inner swelling is included in the determination of size and volume, outer solvation is not at all effective in small-angle scattering. The difficulty of the ultracentrifuge, which yields both swelling effects together with the anisotropy in the disymmetry factor, is thereby eliminated. The combination of both methods thus allows a determination of outer solvation as well.

(4) With chain molecules in solution, only the small-angle method can measure the coiling as a short-range effect, due to its use of X-rays of short wavelength as a probe. Light-scattering and viscosity, on the other hand, allow only an indirect calculation of the persistence length from molecular weight and radius of gyration.

The stressing of these advantages shall not be understood to mean that other pertinent methods do not also have their special advantages: for instance, the possibility for light-scattering to yield information about very much larger particles in very diluted solution also, the direct observation of electron microscopic results, the possibility of checking homodispersity in the ultracentrifuge, and others. But, in order to assign the small-angle method its proper place in the cooperative employment of the various methods, its possibilities, advantages and limitations must be known.

BIBLIOGRAPHY

1. Beeman, W. W., Kaesberg, P., Anderegg, J. W., and Webb, M. B.: "Size of Particles and Lattice Defects"; in: *Handbuch der Physik*, **32**, ed. by S. Flügge, Springer-Verlag, Berlin, 1957.
2. Kratky, O., Porod, G. and Skala, Z., Acta Phys. Austriaca, **13**, 76 (1960).
3. Ross, P. A., J. Opt. Soc. Am., **16**, 433 (1928).
4. Guinier, A., Compt. Rend., **223**, 31 (1946).
5. Johansson, J., Z. Physik, **82**, 587 (1933).
6. Jagodzinsky, H. and Wohlleben, K., Z. Elektrochem., **64**, 212 (1960).
7. Cauchois, Y., J. Phys. Radium, **6**, 89 (1945).
8. Damaschun, G., Naturwiss., **51**, 378 (1964); Exp. Technik der Physik, **13**, 224 (1965).
9. Ehrenberg, W. and Franks, A., Nature, **170**, 1076 (1952).
10. Franks, A., Proc. Phys. Soc., **B68**, 1054 (1955).
11. Bonse, U. and Hart, M., Conference on Small-Angle X-Ray Scattering, Syracuse, New York; June 1965.
12. Kratky, O., Z. Elektrochem., **58**, 49 (1954); **62**, 66 (1958); Kolloid-Z., **144**, 110 (1955); Kratky, O. and Skala, Z., Z. Elektrochem., **62**, 73 (1958).
13. Kratky, O. and Leopold, H., Makromol. Chem., **75**, 69 (1964).
14. Kratky, Ch. and Kratky, O., Z. Instrumentenk., **72**, 302 (1964).
15. Leopold, H., Elektronik, **14**, 359 (1965).
16. Jecny, J., Preprint P. 144 for the "International Symposium on Macromolecular Chemistry," Prague, C.S.R., 1965.
17. Porod, G., Kolloid-Z., **124**, 83 (1951); **125**, 51 (1951).
18. Kratky, O., Porod, G. and Kahovec, L., Z. Elektrochem., **55**, 53 (1951).
19. Kratky, O., Makromol. Chem., **35A**, 12 (1960).
20. Kratky, O. and Wawra, H., Monatsh. Chem., **94**, 981 (1963).
21. Luzzati, V., in: *X-Ray Optics and X-Ray Microanalysis*, p. 133; Acad. Press, New York, 1963.
22. Beeman, W. W., Conference on Small-Angle X-Ray Scattering, Syracuse, New York; June 1965.
23. Kratky, O., Z. analyt. Chem., **201**, 161 (1964).
24. Kratky, O., Progr. Biophys., **13**, 105 (1963).
25. Cleeman, J. C. and Kratky, O., Z. Naturforschg., **15b**, 525 (1960).
26. Damaschun, G., ibid., **20b**, 1274 (1965).
27. Wunderlich, W. and Kirste, G., Ber. Bunsenges. physik. Chem., **68**, 646 (1964).
28. Guinier, A. and Fournet, G., J. Phys. Radium, **8**, 345 (1947).
29. Heine, S. and Roppert, J., Acta Phys. Austriaca, **15**, 148 (1962); Heine, S., ibid. **16**, 144 (1963).
30. Kratky, O. and Leopold, H., J. Colloid Sci., in press.
31. Milicévic, B. and Eigenmann, G., Helv. Chim. Acta, **47**, 1039 (1964); Frank, H. P., J. Colloid Sci., **12**, 480 (1957).
32. Kratky, O., Sekora, A. and Treer, R., Z. Elektrochem., **48**, 587 (1942); Kratky, O. and Sembach, H., Makromol. Chem., **18–19**, 463 (1956); Kratky, O. and Breiner, R., ibid., **25**, 134 (1958).

33. Heine, S., Kratky, O., Porod, G., and Schmitz, P. J., Makromol. Chem., **44–46,** 682 (1961).
34. Kratky, O. and Porod, G., Rec. Trav. Chim. Pays-Bas, **68,** 1106 (1949); Porod, G., J. Polymer Sci., **10,** 157 (1953).
35. Peterlin, A., J. Polymer Sci., **47,** 403 (1960); Cleveland Meeting Amer. Chem. Soc. April 1960.
36. Heine, S., Kratky, O. and Roppert, J., Makromol. Chem., **56,** 150 (1962); Kratky, O., Kolloid-Z., **182,** 7 (1962).
37. Heine, S., Makromol. Chem., **71,** 86 (1964).
38. Kratky, O., ibid., **35A,** 12 (1960).
39. Bielig, H.-J., Kratky, O., Steiner, H. and Wawra, H., Monatsh. Chem., **94,** 989 (1963).
40. Bielig, H.-J., Kratky, O., Rohns, G. and Wawra, H., Tetrahedron Letters, Nr. **39,** 2701 (1964).
41. Hauser, H., Holasek, A., Kratky, O. and Wawra, H., Monatsh. Chem., **96,** 1103 (1965).
42. Bielig, H.-J., Kratky, O., Rohns, G. and Wawra, H., Biochim. Biophys. Acta, **112,** 110 (1966).
43. Richter, G. W., Lab. Invest., **12,** 1026 (1963).
44. Behrens, M. and Taubert, M., Hoppe-Seyler's Z. physiol. Chem., **290,** 156 (1952).
45. Fankuchen, I., J. Biol. Chem., **150,** 57 (1943); Hodgkin, D. C., Cold Spring Harbor Symp. Quant. Biol., **14,** 65 (1949); Harrison, P. M., J. Mol. Biol., **1,** 69 (1959); **6,** 404 (1963).
46. Richter, G. W., J. Biophys. Biochem. Cytol., **6,** 531 (1959).
47. Bodmann, O., Kranz, D., and Mutzbauer, H., Makromol. Chem., **87,** 282 (1965).
48. Kratky, O., Pilz, I. and Schmitz, P. J., J. Coll. Interf. Sci., **21,** 24 (1966).

A New Tool for Small-Angle X-Ray Scattering and X-Ray Spectroscopy: the Multiple Reflection Diffractometer

U. BONSE* AND M. HART†

Department of Materials Science and Engineering, Cornell University, Ithaca, New York

INTRODUCTION

Many experimental arrangements for small-angle X-ray scattering have been proposed and are outlined in several review articles (1, 2). The fundamental requirement of such cameras is that they should provide *monochromatic X-ray beams of narrow angular width*. In most cases the permissible degrees of collimation and monochromatization are dictated by intensity considerations. Often, such considerations, lead to the design of slit systems (rather than pinhole systems) providing angular collimation only in the direction normal to the effective slit.

Of the possible experimental arrangements, the highest angular resolution can be achieved with two crystals set in the parallel position (3–7). In addition the X-ray beam in that case is crystal-monochromatized and, although narrow in effective *angular* width, can be several millimeters in *spatial extent*. Thus, in spite of the high angular resolution, high intensity can also be obtained because a large X-ray source can be completely utilized. Since the beam is spatially extended, a large volume of scattering material can be used with thin samples, in contrast to slit systems. Thus problems due to multiple scattering can be avoided and at the same time many particles can be irradiated, so that meaningful average particle sizes can be measured.

Unfortunately the simple double-crystal diffractometer is not very suitable for small-angle scattering studies because the reflection

* Present address: Physikalisches Institut der Universität, Münster, Germany.
† Present address: H. H. Wills Physics Laboratory, Royal Fort, Bristol 2, England.

121

curves of perfect crystals have tails extending to high angles. Outside of the principal range of reflection the diffracted intensity decreases only with the square of the angle. Since scattered intensities only 10^{-4} or 10^{-5} of the incident intensity are typically encountered, the tails transmit too much of the primary beam (3).

Single crystals of germanium and silicon which are now freely available are sufficiently perfect so that the atomic planes remain flat and parallel to within a small fraction of a second of arc over distances of several centimeters [see, for example (8)]. In these crystals multiple Bragg reflections, alternately h and $\bar{h}$, can be obtained between the walls of a groove cut in the crystal. Using multiple reflections, we have been able to eliminate the tails of the single-crystal reflection curve. With a pair of *grooved* crystals in place of the usual pair of plane crystals in a double-crystal diffractometer the troublesome background of the rocking curve is eliminated and the advantages of the diffractometer mentioned before can be exploited.

With such a system we have measured the small-angle X-ray scattering of Dow polystyrene and polyvinyltoluene latex spheres with diameters ranging from 800 Å to 20,000 Å. Here we will give only a brief description of some of the experiments; complete details have been presented elsewhere (9). Some estimate of the performance of the multiple reflection diffractometer can be gained from the fact that, while the resolution is sufficient to measure the diffraction maxima of 20,000 Å-diameter particles with copper Kα radiation, there is also sufficient intensity to observe up to twenty orders.

APPARATUS

Multiple Reflection Diffractometer With a Large Effective Slit Height

In Fig. 1 a plan of the experimental arrangement is shown. Scattering curves are obtained by rotating the second grooved crystal about an axis normal to the plane of the diagram by means of a lever and tangent micrometer screw. We have used a highly stabilized PW1010 generator and a copper tube as X-ray source, operating at 40 kV and 24 mA. By pulse-height analysis of the output from the sealed-off, xenon-filled proportional counter, the effect of the harmonics of copper Kα was eliminated. For convenience, the intensities were measured with a three-decade logarithmic ratemeter connected to a strip-chart recorder which was coupled to the angle drive mechanism. The highest intensities were attenuated to within the range of the ratemeter by carefully calibrated aluminum foils. The crystals which we have used for making the grooves were all grown parallel to a ⟨111⟩ direction and, solely for convenience, the 220 Bragg reflection from planes containing the growth axis was always used.

If a crystal is cut obliquely so that the incident beam glancing angle is very small, the range of total reflection on the incident beam side of the crystal is increased (10-12) and higher diffracted intensity can be obtained since each point on the crystal can reflect X-rays from a larger area of the focus. The size of crystals was such that we could not cut oblique grooves; instead the separate asymmetric fore crystal is used as a "radiation collector." For a symmetric groove completely filled with m Bragg reflections, the crystal length is $(m + 1) w \cot \theta$, where w is the groove width. The beam width is

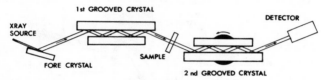

Fig. 1. Plan view of the multiple reflection diffractometer. Each grooved crystal contains five Bragg reflections and the fore-crystal acts as a radiation collector providing an enhanced diffracted intensity from the source.

then $2 w \cos \theta$. On the other hand the peak-to-background ratio of the reflection curve increases with m. Narrow grooves are difficult to prepare and restrict the working beam width. As a compromise the crystals are 6·5 cm in length and the grooves each contain five Bragg reflections. In this case the intensity in the tails of the reflection curve should decrease with the tenth power of the angular deviation from the parallel position. Thus the tails are eliminated (13).

For example, with no specimens in the beam and with grooved silicon crystals, the rocking curve half-width at 10^{-5} of peak intensity is only 9 seconds of arc and the measured intensity is down to counter background at only 30 seconds of arc from the parallel position. For the corresponding *double* crystal rocking curves the intensities are 7×10^{-2} and 6×10^{-3} of the peak at 9 seconds and 30 seconds of arc respectively. Of particular importance is the fact that the elimination of the tails has been accomplished with only a small loss in peak intensity, a factor of four. With the pair of grooved germanium crystals and the oblique fore-crystal, a fifteen-fold increase in peak intensity (compared with silicon crystals) is obtained at the expense of slightly broadening the beam. All of the experimental results to be described were measured with a set of germanium crystals.

Multiple Reflection Diffractometer With Zero Effective Slit Height

It is well known that the true scattering curve can be calculated from the experimental data obtained with an infinite slit height by means of an unsmearing process which is an integration involving the first derivative of the experimental scattering curve. However, at small angles this is a very uncertain process because the gradient

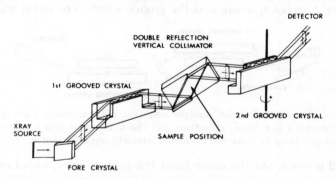

Fig. 2. A multiple reflection arrangement which provides collimation with zero effective slit height. The apparatus is the same as in Fig. 1 with the addition of a double reflection vertical collimator.

of the experimental curve is large and rapidly varying. Meaningful results can be obtained if the experimental curve is measured with zero effective slit height.

By using two double crystal arrangements with mutually orthogonal axes the effective slit height can be made negligible. However the stability problems associated with a system of separate crystals would be extremely severe. A simple solution is illustrated in Fig. 2.

The vertical collimator consists of a pair of Bragg reflections within a single crystal block which functions as a complete double-crystal diffractometer permanently set at the peak position. Since the focus subtends an angle of almost 2° at that crystal, the addition of the vertical collimator imposes no more stability problems than those already encountered in the basic double-crystal arrangement. Of course, the improved collimation is gained at the expense of intensity and a higher power X-ray source would be desirable.

THEORY OF SMALL ANGLE SCATTERING BY UNIFORM SPHERES

If the incident beam can be considered as a monochromatic plane wave, $(n - 1)a/\lambda$ is smaller than one and the attenuation and double scattering of the incident beam within a particle can be neglected, Rayleigh (14, 15) and Gans (16) have shown that the scattered intensity due to a uniform sphere of radius a is

$$i(h) = \Phi^2(ha) = \left[3 \frac{\sin ha - ha \cos ha}{h^3 a^3} \right]^2 = \frac{9\pi}{2} \left[\frac{J_{\frac{3}{2}}(ha)}{h^{\frac{3}{2}} a^{\frac{3}{2}}} \right]^2 \qquad 1.$$

where $h = 4\pi \sin \theta / \lambda$, θ is half of the scattering angle, λ the X-ray wavelength, and n is the refractive index of the sphere. The most restrictive of the premises is the assumption that multiple scattering within the particle can be neglected. For latex particles the upper limit of particle size for which Eq. 1 is valid should be close to a diameter of 20,000 Å. The function $i(h)$ has a maximum at the origin and a series of maxima at the positions where the Bessel function $J_{\frac{3}{2}}$ vanishes. For large values of the argument the spacing of the maxima converges to π while the intensity ratio of successive maxima tends to one. The first order maximum occurs near $ha = 5.765$ with an intensity only 0.00742 of the zero-order maximum. If the sample consists of N randomly arranged independent particles then the total scattered intensity is just the sum of the intensities scattered by each particle

$$i(h) = N\Phi^2(ha) \qquad 1a.$$

Examination of dried layers of latex spheres in the electron microscope showed considerable ordering of the particles into close-packed arrays.

By considering the sample as a fluid of mutually impenetrable spheres Fournet (17, 18) has derived Eq. 2 to account for the interparticle interference:

$$i_F(h) = N\Phi^2(ha) \left[1 + \frac{8v_0}{v_1} \epsilon \Phi(2ha) \right]^{-1} \qquad 2.$$

$$0 \leq \frac{8v_0}{v_1} \epsilon \leq 6$$

v_0 is the volume of a particle, v_1 is the mean volume available to each particle and ϵ is a constant which is close to one.

By comparing the particle packing with a close-packed array, Gingrich and Warren (19) obtained an equation of the form

$$i_G(h) = N\Phi^2(ha)\left[1 + P\left\{5\frac{\sin 2ha}{2ha} - 6\Phi(2ha)\right\}\right] \qquad 3.$$

The packing parameter P varies between one for a close-packed array and zero for an infinitely dilute solution. Since $i_G(h)$ is negative in part of its range for $P \geq 0.6$, values of P greater than 0.6 should not be considered.

Both i_F and i_G are equal to $i(ha)$ for large values of the argument $(ha \gtrsim 10)$ and when the dilution is large $(8v_0/v_1)\epsilon \simeq P \simeq 0$. So that the experimental curves measured with effectively infinite slit height could be compared with the theoretical equations, the observed scattered intensity $I(h)$ has been calculated from

$$I(ha) = \int_0^\infty i_0(y)i([h^2a^2 + y^2]^{\frac{1}{2}})\,dy \qquad 4.$$

using the experimentally measured particle size. Since the primary beam has constant power per unit height, $i_0(y)$ is a constant. The

TABLE I

Peak l	0	1	2	3	4	5
ha	0	5.30	8.62	11.86	15.00	18.24
$I(ha) \times 10^5$	10^5	1640	388	143	69.6	38.8
ha(Schmidt)	0	5.31	8.63	11.85	15.04	18.21
I(Schmidt)	10^5	1650	383	145	69.6	38.6
l	6	7	8	9	10	11
ha	21.33	24.60	27.70	30.88	34.00	37.12
$I(ha) \times 10^5$	23.5	15.4	10.8	7.61	5.59	4.25
l	12	13	14	15	16	17
ha	40.34	43.43	46.58	49.75	52.95	56.06
$I(ha) \times 10^5$	3.36	2.62	2.18	1.72	1.45	1.21
l	18	19	20	21	22	23
ha	59.25	62.24	65.48	68.67	71.74	74.86
$I(ha) \times 10^5$	1.02	0.872	0.750	0.654	0.571	0.499

Positions and intensities of the maxima of $I(h)$ for spherical particles of uniform electron density.

integral 4 has been evaluated for $0.5 \leq ha \leq 76$ at intervals of $ha = 0.5$ and the positions and intensities of the maxima (Table 1) were determined by graphical interpolation. For comparison Schmidt's (20) results are also included.

EXPERIMENTAL RESULTS

Thin samples for small-angle X-ray scattering experiments were prepared by drying a dilute suspension of latex particles under reduced pressure and spreading the dry powder on a "Mylar" foil.

Figure 3 shows the experimental scattering curve obtained from

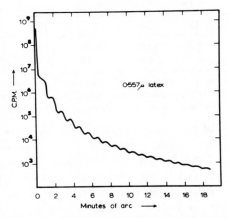

Fig. 3. Experimental small-angle scattering curve of 0.557 μm diameter Dow polystyrene latex spheres obtained with effectively infinite slit height. Copper $K\alpha$ radiation, specimen absorption 0.12, absorption + total scattering 0.21.

latex spheres of (nominally) 0.557 μm diameter, with the first experimental arrangement shown in Fig. 1. If the first plateau is assumed to be caused by interparticle interference we obtain good agreement between the measured and calculated positions and intensities of the maxima of $I(h)$ (see Table II). Since the first few maxima might be affected by interparticle interference the ordinates were normalized at the fourth maximum. Table II clearly shows a decrease in the intensity of the first two maxima which is greater than that calculated from $I_F(h)$ or $I_G(h)$ even with the largest permitted values for the packing parameters. The internal consistency of the peak positions is good and the calculated weighted mean particle size is

$$2a = 0.551 \pm 0.002 \; \mu m \qquad \qquad 5.$$

This is well within the stated standard deviation (± 0.011 μm) of the particle size distribution (Dow run No. LS-063-A).

TABLE II

Peak l	1	2	3	4	5	6
Position	18.60	18.57	18.46	18.40	18.31	18.45
Intensity	0.75	0.81	1.02	1.00	0.98	0.98
l	7	8	9	10	11	12
Position	18.19	18.27	18.26	18.35	18.32	18.31
Intensity	0.98	1.02	0.99	1.05	1.04	1.02
l	13	14	15	16	17	18
Position	18.35	18.41	18.40	18.26	18.35	18.52
Intensity	1.04	1.07	1.11	1.16	1.16	1.15

Measured positions and intensities of maxima of the small angle scattering from 0.557 μm diameter latex spheres. Position is measured (seconds of arc)/value of ha_{max} calculated. Intensity is measured/$I(ha)_{max}$ normalized for $l = 4$.

With particles of the same size, a scattering curve has also been measured using the second experimental arrangement (Fig. 2) so that the effect of slit height was negligible. Figure 4 shows the experimental curve with the theoretical curves $i_G(h)$ and $i_F(h)$ superimposed. The abscissae were normalized using the measured particle size while the ordinates were normalized at the first-order maximum. In spite of the large effect of slit width on the measured pattern it is quite clear that neither of the two theoretical curves provides an adequate description of the interparticle interference.

CONCLUSIONS

The application of multiple Bragg reflections to the basic double crystal diffractometer has been described. Using such arrangements small-angle X-ray cameras have been constructed which are superior to slit systems in resolution and at the same time provide high intensity crystal monochromatized X-ray beams. If we compare the germanium crystal apparatus described earlier with a slit system with an angular width of 4 minutes of arc (3), both using a similar X-ray source, the advantages of the multiple reflection system are obvious (Table III).

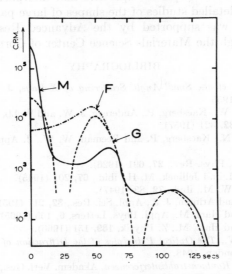

Fig. 4. Experimental small-angle scattering curve obtained with zero effective slit height. Dow polystyrene latex spheres 0.557 μm stated diameter. Copper Kα radiation, specimen absorption 0.40.

Experimental scattering curves of latex spheres have also been obtained with high angular resolution providing a precise verification of the Rayleigh scattering function. The high resolution should be extremely valuable in precise studies of the effects of interparticle

TABLE III

	Multiple reflection diffractometer	Four slit system (3)
Angular resolution	15 secs arc	240 secs arc
Monochromatization	Crystal	Ross filters or β filters
Peak intensity	10^9 c.p.m.	6×10^7 c.p.m.
Area under the rocking curve	1.5×10^{10} c.p.m. secs arc	1.4×10^{10} c.p.m. secs arc
Number of independent adjustable parts	3	4
Air scatter	Unimportant	Beam paths must be long so evacuation is desirable
Cross section of beam at the sample	4×17 mm²	0.25×10 mm²

interference and multiple intraparticle and interparticle scattering, as well as in detailed studies of the shapes of large particles.

This work was supported by the Advanced Research Projects Agency through the Materials Science Center of Cornell University.

BIBLIOGRAPHY

1. Guinier, A., et al., *Small-Angle Scattering of X-Rays*, J. Wiley and Sons, New York, 1955.
2. Beeman, W. W., Kaesberg, P., Anderegg, J. W. and Webb, M. B., *Handbuch der Physik*, **32**, 321 (1957).
3. Ritland, H. N., Kaesberg, P. and Beeman, W. W., J. Appl. Phys., **21**, 838 (1950).
4. Slack, C. M., Phys. Rev., **27**, 691 (1926).
5. Fankuchen, I. and Jellinek, M. H., ibid., **67**, 201 (1945).
6. Dumond, J. W. M., ibid., **72**, 83 (1947).
7. Daams, H. and Arlman, J. J., Appl. Sci. Res., **32**, 217 (1951).
8. Bonse, U. and Hart, M., Appl. Phys. Letters, **6**, 155 (1965).
9. Bonse, U. and Hart, M., Z. Physik, **189**, 151 (1966).
10. James, R. W., *The Optical Principles of the Diffraction of X-Rays*, G. Bell and Sons, London, 1948.
11. Laue, M. V., *Röntgenstrahlinterferenzen*, Akadem. Verl. Ges., Frankfurt, 1960.
12. Renninger, M., Z. Naturforschg., **16a,** 1110 (1961).
13. Bonse, U. and Hart, M., Appl. Phys. Letters, **7,** 238 (1965).
14. Lord Rayleigh, Proc. Roy. Soc. (London), **A84,** 25 (1911).
15. Lord Rayleigh, ibid., **A90,** 219 (1914).
16. Gans, R., Ann. Physik, **76,** 29 (1925).
17. Fournet, G., Compt. Rend., **228,** 1801 (1949).
18. Fournet, G., Acta Cryst., **4,** 289 (1951).
19. Gingrich, N. S. and Warren, B. E., Phys. Rev., **46,** 248 (1934).
20. Schmidt, P. W., Acta Cryst., **8,** 772 (1955).

Optical Analogs of Small-Angle X-Ray Diffraction from Drawn Fibers

PAUL PREDECKI AND W. O. STATTON

Carothers Research Laboratory, E. I. duPont de Nemours and Co.,
Wilmington, Delaware

INTRODUCTION

Several contributions have considerably increased our knowledge of small-angle X-ray patterns of polymeric fibers, and a recent review (1) has summarized the range of information that can be obtained from such studies. Despite this, however, their interpretation is not satisfactory in the light of current knowledge of polymers, and much uncertainty exists. Experimentally, the only certainty is that the observed long-period intensities can be attributed to a partially disordered arrangement of scattering centers; it is the nature and detailed arrangement of these scattering centers that is open to question. A firm interpretation in terms of a structural model would be of value since the patterns are variable and sensitive to thermal and mechanical treatment of the fibers, and consequently must also be related to their physical properties. It is the purpose of this paper to make the first step towards a firm interpretation by showing from diffraction principles and optical analogies the possible arrangements of scattering centers which could give rise to the X-ray observations. A later publication (2) will attempt to show in detail a model in terms of polymer molecules.

The use of optical methods as an aid in the determination of crystal structures is not new (3). More recently, optical diffraction patterns were prepared by Bonart (4) in his demonstration of Hosemann's paracrystalline model of polymer structure and also to support his paracrystalline model of the long periods in drawn polyethylene fibers (5). The same technique is used in the present paper except that our approach is to find possible arrangements of scattering centers in two-dimensional masks which give the same type of small-angle diffraction patterns with light as are obtained from drawn nylon fibers with X-rays. An assessment of the results, including those of other investigators, is attempted in the discussion, and a most likely model is selected.

A disadvantage of the optical method is its inability to provide quantitative judgment of intensities. However, the qualitative information which it furnishes can be obtained rapidly and conveniently. The method is therefore valuable at the outset of an investigation in order to narrow down quickly the number of possible structure models.

EXPERIMENTAL

The optical diffractometer instrument is not described in detail here; it is commercially manufactured by the R. B. Pullin Company

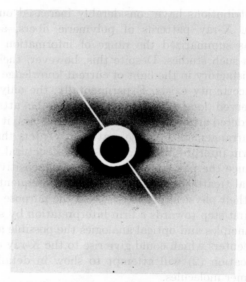

Fig. 1. Small-angle X-ray diffraction pattern from drawn 66 Nylon fibers.
Fiber axis is vertical.

of London. A camera attachment with a "Polaroid" back was added, allowing diffraction patterns first to be enlarged with a microscope and then photographically recorded.

The region of interest in typical small-angle X-ray patterns obtained from nylon fibers with copper $K\alpha$ radiation corresponds to scattering angles of less than three degrees (Figs. 1 and 2). It is desirable for the analogous optical experiment to have scattering

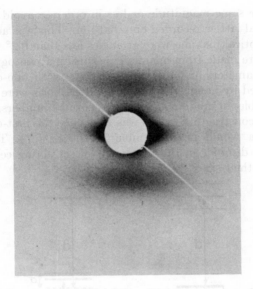

Fig. 2 Small-angle X-ray diffraction pattern from drawn and annealed 66 Nylon fibers. Fiber axis is vertical.

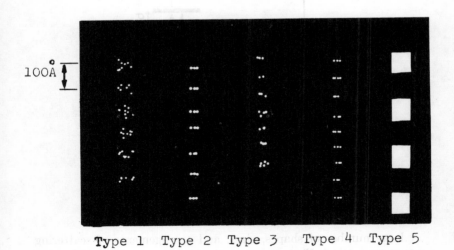

Type 1   Type 2   Type 3   Type 4   Type 5

Fig. 3. Possible arrangements of scattering centers.

angles of the same magnitude or less in order to avoid distortion of the reciprocal lattice recorded on a flat film. This was achieved in all cases; the optical maxima were at angles less than 0.2°.

Masks were made from opaque paper using a pantographic punch with hole diameters of $\frac{3}{4}$ and 1 mm. Masks with non-circular holes were prepared by hand. The positions of the holes were plotted with the aid of tables of random and random normal numbers. The guiding principle in constructing masks was that only the first-order diffraction maxima would be discernible in the pattern. This required considerable disorder in the placement of scattering centers and/or variation of their size.

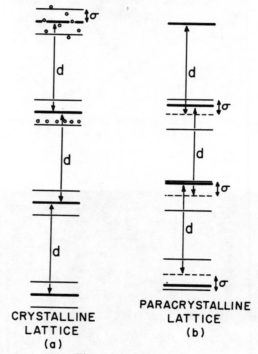

Fig. 4. Disorder types.

RESULTS

A large number of shapes, sizes, and arrangements of scattering centers were tried. Five representative types emerged which gave small-angle optical diffraction patterns similar to Fig. 2; these five types are summarized in Fig. 3. Each arrangement was characterized by an average spacing d in the vertical (fiber) direction, usually taken as 100 Å. The other dimensions of the fibril or scattering center arrangement were then described in terms of this spacing. For example, a fibril width of about $d/2$ gave a dimension to the diffraction maxima in the direction parallel to the equator which corresponds to the width of the streak in the X-ray pattern.

The five types could further be divided into two groups.

Group 1: Small Scattering Centers

Traditionally, discrete small-angle maxima from fibers have been attributed to large, bulky, scattering centers probably 50 Å or more in extent (1). A new approach taken in this work was to substitute for these large units many scattering centers having considerably smaller dimensions. To demonstrate this approach with the smallest feasible structural entity, a hole diameter of about $d/15$ was chosen to approximate an interchain spacing in 66 nylon. This might roughly simulate the lattice void created by a chain fold. The small scattering centers were arranged into clusters, and these clusters were then arranged into a fibrillar line lattice which was aligned parallel to the fiber axis. There were several aspects of this clustering which were varied and which gave rise to Types 1 through 4. The manner in which these arrangements were generated is most easily described with the aid of Fig. 4.

Type 1: Gaussian Distribution of Small Scattering Centers on a Crystalline Lattice

On each of the crystalline lattice lines shown in Fig. 4a, holes were clustered with a Gaussian distribution centered around the line as shown in the top part of the figure. The Gaussian displacements (which may be considered analogous to pronounced thermal motion) were normal to the lattice lines; i.e., in the fiber direction. The lateral positions of the holes were chosen at random in order to prevent statistical correlation in the direction normal to the fiber axis. The standard deviation σ of the Gaussian distribution was found to be $d/7.5$ to produce first-order diffraction only. The criticality of this disorder is shown in Fig. 5 where increasing the deviation to $d/5$ results in only the zeroth order occurring in the diffraction pattern, whereas decreasing the deviation to $d/10$ results in a second order.

Type 2: Gaussian Distribution of Rows of Scattering Centers on a Crystalline Lattice

The arrangement of holes was similar to Type 1, but in this case the holes were placed in straight rows normal to the fiber axis. These rows were distributed (one per lattice line) with a Gaussian distribution centered on the lines of the crystalline lattice as shown in the second line of Fig. 4a. The required standard deviation was again found to be $d/7.5$.

Type 3: Gaussian Distribution of Scattering Centers on a Paracrystalline Lattice

In a one-dimensional paracrystalline lattice, the position of a particular line is determined by the positions of its nearest neighbors, as shown in Fig. 4b.

Thus in Type 3, the reference lattice has paracrystalline disorder, and the positions of holes on the lattice have thermal disorder as in Type 1. It was found that the standard deviation of the lattice disorder could be continuously varied from zero to $d/6$ and still yield only the first-order diffracted maxima by suitably adjusting the thermal disorder of the holes. In the particular arrangement shown in Fig. 3, the standard deviation of the lattice disorder was $d/6.5$ (where d is the paracrystalline lattice spacing) and that of the hole disorder was $d/13$.

Type 4: Paracrystalline Lattice

In Type 4, all the disorder was accounted for by the paracrystalline disorder of the lattice, the holes being positioned exactly on the paracrystalline lattice lines. The standard deviation of the lattice disorder was found to be about $d/6$ for this arrangement.

Group 2: Large Scattering Centers

Following the traditional concept of fiber structure, which proposes alternating regions of varying electron density, Type 5 was obtained to represent the fringe micelle model. In this case, rectangular holes were placed on the reference lattice of Type 1. The dimensions of these large holes markedly affected the number and width of the diffracted maxima. For example, slightly elongating the holes in the fiber direction resulted in only the zeroth order remaining in the diffraction pattern, while shortening them caused two or more orders

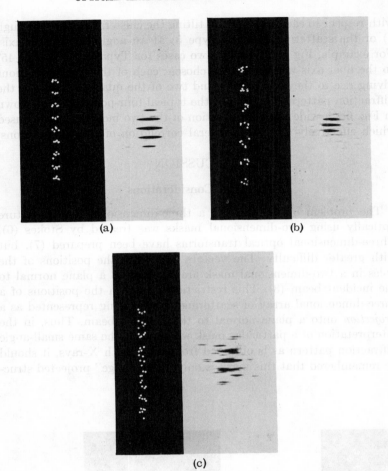

Fig. 5. Effect of varying the standard deviation σ of the disorder in Type 1. (a) $\sigma = d/10$ showing two orders (b) $\sigma = d/7.5$ showing one order and (c) $\sigma = d/5$ showing the zeroth order only.

to appear. The width of the holes was inversely proportional to the length of the diffracted maxima parallel to the equator.

Four-Point Diagrams

Diffraction patterns similar to the four-point pattern of Fig. 1 could be obtained with all five types by suitably staggering two fibril rows with respect to each other, or by tilting the clusters (Types 1 through 4) or the scattering centers (Type 5) at an angle to the fiber axis. For example, Fig. 6 shows the two cases for Type 3. A tilt of ±45° to the fiber axis was arbitrarily chosen; each of the two inclinations giving rise to the zeroth order and two of the quadrant spots in the diffraction pattern. To produce the typical four-point pattern shown in Fig. 6b, a wide lateral separation of the two inclinations was used which effectively eliminated lateral correlation of the tilt positions.

DISCUSSION

General Considerations

The problem of representing a three-dimensional fiber structure optically using two-dimensional masks was treated by Stokes (6). Three-dimensional optical transforms have been prepared (7), but with greater difficulty. The vectors describing the positions of the holes in a two-dimensional mask are confined to a plane normal to the incident beam (3). This restriction results in the positions of a three-dimensional array of scattering centers being represented as a *projection* onto a plane normal to the incident beam. Thus, in the interpretation of a particular mask which gives the same small-angle diffraction pattern as is obtained from fibers with X-rays, it should be remembered that this mask is only an "average" projected struc-

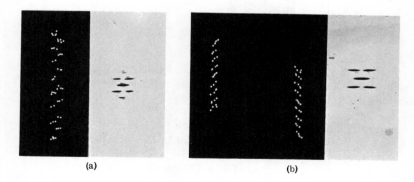

Fig. 6. Four-point optical diffraction patterns obtained from the Type 3 arrangement in the (a) staggered, (b) tilted configurations.

ture of the three-dimensional arrangement of scattering centers in the fiber.

Each of the arrangements in Types 1 to 4 was represented on a mask by a single row of from 6 to 13 clusters or by large holes in Type 5. Increasing the length of a row or fibril resulted only in an intensity increase in the first-order diffracted maxima. Decreasing the length to less than about 5 repeats produced a broadening of the first and zeroth orders as well as a decrease in intensity. The relative insensitivity to short fibrils is due primarily to the large amount of disorder present within the lattice. With all five types, the hole arrangements can be extended laterally by adding fibrils which are all aligned parallel to the fiber axis. The diffraction patterns will not be different from those due to one fibril provided there is no statistical correlation between the positions of the fibrils in the direction normal to the fiber axis, or provided the width of the fiber units varies with a sufficiently large standard deviation to give only the zeroth order in the lateral direction. The first-order reflections can be broadened by slightly varying the spacing d of the reference lattice in the fibrils. Thus, each of the five types can be "tailored" to give the desired diffraction pattern.

Furthermore, varying the size, shape, and number of small holes in the masks of the group 1 arrangements produced only changes in *intensity* of the diffracted maxima; no change was produced in the shape or in the number of discernible orders, provided the variations were not so large as to markedly affect the disorder. In the Type 1 and 2 arrangements, the number of discernible orders was the result of the amount of disorder of the hole arrangements in the fiber direction. This is not expected in normal diffraction, since thermal motion of a lattice does not generally broaden the maxima, but progressively reduces their intensity. However, if the thermal displacements are sufficiently large as in our case, higher orders may be submerged into the background, as discussed in the appendix. This can adequately explain the presence of only one order in the corresponding diffraction patterns.

Paracrystallinity, on the other hand, progressively broadens and reduces the intensity of higher order maxima, explaining the absence of all but the first order in Types 3 and 4. An expression for the diffracted intensity from essentially a Type 4 arrangement is given by Guinier (8) as discussed in the appendix. It also applies approximately to Type 3 for small thermal displacements of the scattering centers.

The experimentally obtained values of $d/6.5$ and $d/6$ for the standard deviation σ of the disorder in Types 3 and 4 agree well with the range $d/4 > \sigma > d/8$ within which only the zeroth and first-order maxima should theoretically be discernible.

Type 3 is actually the most general arrangement, and Types 1, 2, and 4 are special cases of Type 3.

In contrast to Types 1 through 4, Type 5 had no lattice or scattering center disorder. Here the size of the holes was the controlling factor in the number of orders and also the width of the diffraction maxima. The patterns in this case are simply a product of the transform of the delta function describing the lattice with the transform of one of the rectangular holes.

Relation to Fiber Structure

In order to prove if any one of these arrangements is responsible for the small-angle diffraction from fibers, additional experimental evidence is necessary. However, these general results can guide in the construction of a model, or conversely, one can construct masks of proposed models and check their suitability in actual diffraction.

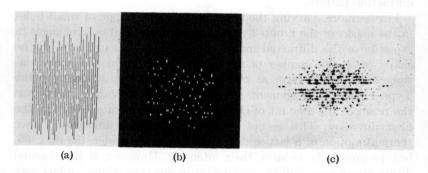

Fig. 7. A chain-folded model (9) (a), its corresponding scattering center mask (b), and optical diffraction pattern (c).

Existing evidence would seem to us to rule out the Type 5 arrangement which involves large regions of uniformly low and high electron density. Beresford and Bevan (9) have recently shown that tension causes the small-angle patterns from drawn nylon fibers to change from four-point to two-point. The necessary movement of scattering centers required for this change is difficult to explain in terms of the bulk motion of large regions of different electron density, and is more

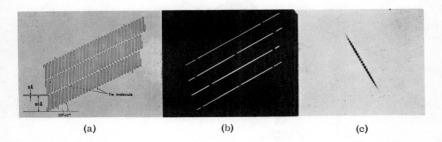

Fig. 8. A chain-folded model of a fiber after annealing (9) (a), its corresponding scattering center mask (b), and optical diffraction pattern (c).

likely to be due to the movement of individual small scattering centers. For this reason, both the older fringe micelle model and the paracrystalline layer model proposed for polyethylene (5) seem unlikely to us.

Furthermore, the chain-folded model recently suggested (9) for as-drawn nylon fibers also appears unlikely. It consists of randomly placed pairs of folds lying in the fiber direction; this model results in a small-angle optical pattern in which the many orders of diffracted maxima are excessively elongated parallel to the equator, as shown in Fig. 7. The proposed model of an annealed fiber which is shown in Fig. 8 is much too perfect and yields too many orders in its small-angle optical pattern.

From the above arguments and on the basis of our present results we are led to conclude that one of the arrangements in Types 1 through 4 will be the most representative of the actual scattering centers in fibers. This is compatible with the assumption that the scattering centers are chain folds (9, 10, 11) and/or chain ends (2). It would appear that these defects are neither isolated nor arranged in an extensive, well-defined lamellar array but exist in clusters of limited lateral extent (10). Additional experimental evidence is necessary in order to assess the validity of this model, and is being sought.

SUMMARY

The five types of scattering center arrangements described in this work which give the desired kind of small-angle pattern are special

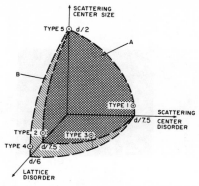

Fig. 9. Interaction of scattering center variables.

cases of a continuous range of mutually varying hole and lattice disorders and hole sizes. The interplay of these three variables is shown schematically in Fig. 9 which defines a likely volume enclosed by the surfaces A and B which would encompass the actual structural possibilities in a fiber.

A new result found in this work is that many small scattering centers can be substituted for the conventional large units usually envisioned as the source of small-angle discrete diffraction. Based on the evidence available to us at this time, these small scattering centers appear most likely to represent the actual fiber structure.

APPENDIX

The intensity $I(s)$ of a unit cell in a fiber unit is given by (8)

$$I(s) = \frac{|\Sigma(s)|^2}{V} \star Y(s) \qquad (\star = \text{convolution})$$

where V is the volume of the fiber unit and $\Sigma(s)$ the transform of its shape function. The function $Y(s)$ is equal to the square of the structure factor F of the thermally vibrating cell. F is given by

$$F = f \exp(-M)$$

where f is the structure factor of the nonvibrating cell and M is the Debye-Waller factor and is given by

$$M = \frac{8\pi^2 \sin^2 \theta \Delta X^2}{3\lambda^2}$$

Here θ is the Bragg angle, λ the wavelength and ΔX^2 the quadratic average of the square of the displacement of the scattering centers. In Types 1 and 2, the function $Y(s)$ is wide compared with $|\Sigma(s)|^2$, and since the volume under the $|\Sigma(s)|^2$ peak is equal to V, the convolution can be evaluated and $I(s)$ is proportional to $Y(s)$. Thus at low angles, log $I(s)$ is proportional to $-\theta^2$; i.e., higher orders are less intense. In addition, most of the temperature diffuse scattering is to be found near the intensity maxima, further reducing the peak-to-background ratio at higher orders.

The diffracted intensity $I(s)$ from a paracrystalline one-dimensional lattice is given by

$$I(s) = \frac{Z(s)F^2}{N} \star |\Sigma(s)|^2$$

where $Z(s)$ is the transform of the distribution of scattering centers and F is the structure factor of the unit cell.

In Types 3 and 4 the width of the form factor $\Sigma(s)$ is negligible compared with $Z(s)$, and the intensity is proportional to $Z(s)$ even for the first order reflection. Using Hosemann's criterion that maxima are not resolved when the modulation depth (Z_{max}/Z_{min}) is smaller than 4, maxima will only be visible when the condition $n\sigma/d \leqslant 0.25$ given by Guinier (8) is satisfied. Here n is the order of the reflection, σ the standard deviation of the disorder and d the lattice spacing. Thus, if only zeroth and first-order maxima are to be discernible, then $d/4 \geqslant \sigma \geqslant d/8$, which agrees well with the values of $d/6.5$ and $d/6$ experimentally obtained for σ with Types 3 and 4.

BIBLIOGRAPHY

1. Statton, W. O., Chapter in *"Newer Methods of Polymer Characterization,"* ed. B. Ke, Interscience, New York, 1963.
2. Predecki, Paul and Statton, W. O., to be published in J. Appl. Phys.
3. See for example Taylor, C. A. and Lipson, H., *"Optical Transforms,"* G. Bell and Sons, London, 1964.
4. Bonart, R., Z. Krist., **109**, 309 (1957).
5. Bonart, R., Kolloid-Z., **194**, 97 (1964).
6. Stokes, A. R., Acta Cryst., **8**, 27 (1955).
7. Harburn, G., and Taylor, C. A., Proc. Roy. Soc., **A264**, 339 (1961).
8. Guinier, A., *"X-Ray Diffraction,"* W. H. Freeman and Co., San Francisco, 1963.
9. Beresford, D. R., and Bevan, H., Polymer, **5**, 247 (1964).
10. Dismore, P. F., and Statton, W. O., J. Polymer Sci., **13C**, 133 (1966).
11. Hay, I. L., and Keller, A., Nature, **204**, 862 (1964).
12. Statton, W. O., J. Polymer Sci., **25**, 423 (1958).

Small-Angle X-Ray Scattering by Crystalline Polymers

A. PETERLIN

Camille Dreyfus Laboratory, Research Triangle Institute, Durham, N.C.

INTRODUCTION

The folded chain lamellae with a rather uniform thickness L of between one and a few hundred Å and much larger lateral extension are the main structural unit of the polymer solid. They form spherulites originating from primary crystallization nuclei which, by secondary nucleation and to a large extent by non-crystallographic branching, grow first into a bundle-like and later into a radial arrangement of ribbon-like, more or less twisted, lamellae. The size and perfection of spherulitic structure depend on crystallization conditions, on temperature, density of nuclei, impurity concentration, and so on. The parallel stacking of lamellae creates a nearly ideal periodic lattice which, due to the large period L, gives rise to a small-angle X-ray scattering. The intensity of scattering depends on the square of the electron density difference between the crystal lattice and the region between the adjacent crystals, which contains chain folds, tie molecules and free ends of chains incorporated only partially in the crystalline lattice. The thickness of the density-deficient layer is the second factor affecting the scattering intensity. Small-angle X-ray scattering, however, does not yield any information about the fine structure of the intercrystal layers, particularly whether they are amorphous or not, how many and how long the folds are, and so on.

A combination with wide-angle X-ray or electron scattering, which yield the crystallographic orientation of crystalline regions, and with electron microscopy, which reveals the morphology of the sample, turns out to be an extremely rewarding approach for the study of fine structure of crystalline polymers and the modifications induced by different mechanical and thermal treatment. The results of such investigations of polymer solids will be reported here. Polyethylene was primarily studied because it has the simplest molecular configuration with no complication due to incomplete steric and positional tacticity. Undeformed and plastically deformed single crystals and bulk material isothermally crystallized or quenched from melt were investigated with particular emphasis on annealing effects.

SINGLE CRYSTAL MATS

Particularly simple systems are single-crystal mats obtained by sedimentation of solution-grown individual crystals. By rapid filtration one obtains well-oriented mats suitable for investigation in a

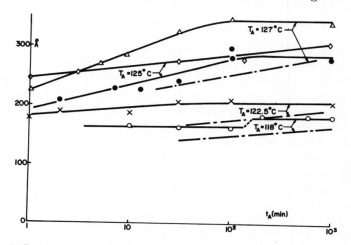

Fig. 1. Long period L of PE single crystal mats as function of $\log t_A$. Most samples were wrapped in aluminum foil for ensuring better thermal contact (full line). Note the less steep increase and smaller L values if the sample was in vacuum without aluminum foil (broken line). The influence of sample preparation shows up in the difference in the 125°C curves for crystals grown from xylene solution at 70° (●) and 80°C (◇).

linear slit camera (for instance a Kratky camera). Due to the uniformity of crystal thickness and the absence of crystal bending one obtains also higher orders at Bragg angles. Annealing of polyethylene single-crystal mats above 90°C increases the long period nearly linearly with the logarithm of annealing time (1, 2). The slope of L vs. $\log t_A$ plots increases with annealing temperature T_A. Recent investigations have shown, however, that as a rule the plots are not linear, that they depend on heating conditions and that the long period seems to approach a limiting value which is influenced by molecular weight, heating conditions and annealing temperature (3, 4) (Fig. 1). The faster the heating and the better the contact of crystals with the heating systems the faster the initial growth of crystal thickness.

An isolated single crystal grows in thickness in a rather irregular manner: some areas thicken at the expense of the surrounding sections of the same crystal where holes develop (5). In a multilayer crystal the bulges of one crystal penetrate into the holes of the adjacent crystals so that in spite of large growth of L the bulk sample (i.e. mat) does not change shape (Fig. 2). One usually assumes that by creeping

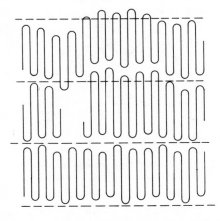

Fig. 2. Interpenetration of crystals in a multilayer sample during long period growth.

longitudinal motion of the chains every single molecule brings more chain elements into the surface areas containing the folds. By such a transport the free ends of the molecule came closer to the central section of the chain, thus creating new holes and shifting or enlarging the already existing holes in the lattice. Surface nucleation, hole formation at the free ends of macromolecules and the transport mechanism determine the growth rate (6, 7). Particularly important seem to be submicroscopic discontinuities, cracks and defects by which the crystal gets subdivided into a mosaic pattern. The existence of the latter was suggested by Hosemann (8) on the basis of the width of small-angle X-ray diffraction spots. According to his data, the mosaic crystal has lateral dimensions of about 300 Å. An indication of such a structure is also given by the short-period fluctuation of the Moiré pattern which, with polymer single crystals, never exhibits smooth lines (9). Such a mosaic crystal can indeed grow in thickness more easily than a larger crystal with no defects because it must not create any new lateral surface. Rapid heating favors the growth of submicroscopic discontinuities and hence creates the possibility for a more rapid thickening. On the other hand such a single block of the mosaic structure cannot grow indefinitely because soon the surface energy contribution of the lateral faces $(S_l \cdot \sigma_l)$ becomes comparable with that of the fold-containing surfaces $(S_e \cdot \sigma_e)$. As a consequence of this effect the volume of the block determines the maximum possible thickness L_∞ which depends on the ratio of σ_e/σ_l but not on tem-

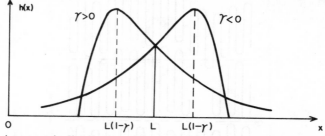

Fig. 3. Asymmetric distribution of long periods with positive and negative γ.

perature (10). According to this picture the annealing temperature can influence L_∞ only by yielding mosaic blocks of different size.

UNDEFORMED POLYMER SOLID

Bulk samples crystallized from melt exhibit a similar behavior. But there is a rather general observation that with annealed samples the first and second diffraction maxima do not occur at the proper Bragg angles. The first maximum appears at too small and the second one at too large an angle (11, 12). A possible explanation of this effect may be the asymmetry of the distribution function for the crystal thickness (11). In Fig. 3a distribution with a short shoulder on the side of smaller L and a long tail on the side of larger L is shown (γ is positive). Such a distribution was derived by Weeks and Hoffman from their melting experiments with bulk PE. It leads to a scattering pattern with a shift of maxima as experimentally observed (Fig. 4). In order to explain the rather large shifts one has to assume a very

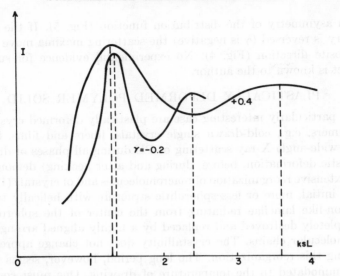

Fig. 4. First- and second-order diffraction maximum due to the asymmetric long period distribution from Fig. 3 for $\gamma = -0.2$ and $+0.4$.

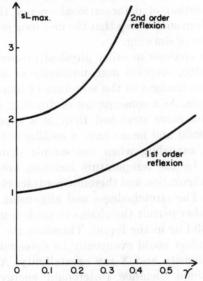

Fig. 5. Position of first and second maxima as function of dissymmetry parameter $\gamma > 0$.

high asymmetry of the distribution function (Fig. 5). If the asymmetry is reversed (γ is negative) the scattering maxima move in the opposite direction (Fig. 4). No experimental evidence for such an effect is known to the author.

PLASTICALLY DEFORMED POLYMER SOLID

A particularly interesting case are plastically deformed crystalline polymers, e.g., cold-drawn single crystals, fibers and films. Small- and wide-angle X-ray scattering of PE during all phases of drawing (plastic deformation, before, during and after necking) demonstrate an extensive reorganization of macromolecules and of crystals (13–15). The initial, more or less spherulitic structure with helically twisted ribbon-like lamellae radiating from the center of the spherulite is completely destroyed and replaced by a highly aligned arrangement of molecular chains. The crystallinity does not change appreciably during this reorganization. The long period, however, seems to get accommodated to the temperature of drawing. One must conclude that during deformation the original crystals are broken into smaller blocks and these blocks incorporated in the new structure. Due to the high concentration of deformational energy these blocks may rearrange the folds in such a way that the new long period corresponds to the temperature of drawing.

But the drastic changes in other physical properties, particularly in heat content (16), sorption and diffusivity of solvent molecules (17, 18), are due to changes in the structure of non-crystalline parts of the drawn sample. As a consequence of drawing, the tie molecules are certainly much more stretched than in a supercooled melt or even in the bulk solid and hence have a smaller entropy. This effect shows up during annealing when the sample shrinks because the chains, mobilized by the temperature increase, are able to reach a more probable configuration and therefore contract close to the equilibrium dimensions. The stretchedness and alignment of chains in the drawn sample further permit the chains to pack more closely than in the amorphous solid or in the liquid. Therefore the density must be also higher. The effect could eventually be detected by very precise measurement of density and X-ray crystallinity. As a consequence of smaller interchain distance (interchain energy) and of more stretched form favoring the trans conformation which has a lower energy content than the gauche conformation (intrachain energy), the amorphous component of the drawn sample has a smaller heat content than in undeformed material. Closer packing, lower inter- and intramolecular energy level and high strain drastically reduce the permeability of drawn sample. Solvent molecules simply cannot be accommodated in the amorphous regions of highly drawn PE.

We think (19) that the highly drawn crystalline polymer represents a very intricate blend of folded chain lamellae and fringed micelle models. A great many polymer chains run through more than one crystal. The chain sections between two consecutive crystals are rather strained, aligned and more closely packed than in a completely relaxed amorphous liquid. By this interconnection of crystals the drawn sample obtains a higher mechanical strength in the draw direction. The surface of every crystal contains a sufficient number of folds so that there is provided space for the lower density of non-crystalline regions separating the crystals. But in freshly drawn

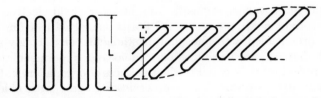

Fig. 6. Tilting, slip and crack formation in strained single crystals.

material the folds exhibit many irregularities and the crystal surface is so coarse that the meridional small-angle diffraction peak has a substantial lateral width.

In order to obtain a better understanding of the deformation mechanism, we studied the plastic deformation of polyethylene, polyoxymethylene and nylon-6 single crystals (20–24). Besides initial phase transformation and twinning, chain tilting and slip seem to be the main mechanism by which the crystal changes its shape. Both effects also reduce the thickness of the crystal and soon lead to crack formation (Fig. 6). Fibrils are pulled out, bridging the gap between the receding crystal boundaries. The fibrils show nearly complete orientation of chain, i.e., of the c-axis, in the fiber direction. But no long period is detectable in the electron micrographs. Small-angle electron scattering cannot be applied because there is not enough scattering power in the small amount of material contained in the fibrils. Annealing, however, produces extremely regular striations perpendicular to the fiber axis with a long period of about 250 Å, which is markedly higher than the thickness of original crystals. Different explanations of the appearance of such a striation mainly started with Kobayashi's picture of fibril drawing which assumes complete unfolding as the chains are pulled out of the crystals (25). The extended chains fold again (26) during annealing or there is epitaxial growth of crystals of low molecular weight components (27) yielding the observed periodicity. Since it is difficult to imagine why extended chains ought to prefer the folded conformation with higher free energy, we suspected that the folds must already exist in the freshly drawn fibrils, but due to insufficient regularity of fold arrangement and to a high amount of defects in the crystalline regions as a consequence of the drastic deformation, cannot be detected by the usual techniques of electron microscopy. With this in mind, we tried some less common methods for detecting a discontinuity and eventually also a periodicity in the structure of such fibrils which could be associated with folded chain crystals. Our experiments (24), i.e., dark field electron micrographs, bright field pictures of PE fibrils loaded with iodine and of POM fibrils loaded with phosphotungstic acid, convincingly demonstrated that an alternation of permeable and impermeable regions with a period roughly equal to the thickness of original crystals exists in freshly drawn fibrils. These observations agree with older data of Fischer and Schmidt (2) who shadowed PE fibrils with Au/Pt, and of Hess and Mahl (28), who loaded cellulose fibers with iodine. We, therefore, believe that during drawing the chains get only partially unfolded and that as a rule whole blocks are pulled out of the highly tilted crystal and incorporated in the fibril. By covering the original PE single crystal with a thin Pt layer we were able to observe the breaking of this layer into smaller pieces of nearly 200–400 Å extension in the fiber direction, suggesting that the blocks pulled out from the crystal may have such a width. It may be more than a fortuitous coincidence that this dimension roughly agrees with the diameter of mosaic blocks as determined by Hosemann. The deformed crystal breaks and disintegrates into blocks just at places of maximum slip, which primarily occur at lines of high concentration of lattice defects. Exactly the same conditions determine the boundary of the mosaic block.

Annealing enormously changes the drawn sample. It mobilizes the densely packed, strained and aligned chains so that they can nearly completely relax and assume a more probable conformation. The sample shrinks without a significant disorientation of crystals, which still have the chain axis oriented in the fiber axis. Due to strain relaxation and increased mobility they can be partially incorporated in adjacent crystals so that the crystallinity of the sample markedly increases. But the starting point for all these changes is the increased mobility of polymer chains in the crystal lattice. Due to the increased temperature they can slip along their axis and so permit relaxation of tie molecules. At the same time the crystal thickness starts growing. The initial growth is extremely rapid in agreement with the logarithmic dependence on time (29). In spite of the relatively small slope (Fig. 7) the whole transformation from

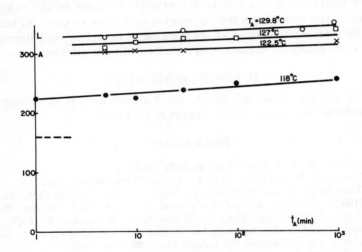

Fig. 7. Long period growth of drawn PE as function of annealing temperature
The broken line marks the value of long period before annealing.

the original $L = 160$ Å to the nearly constant value of 210 Å is achieved in less than 10 sec and may be in even less than 1 sec. The experimental technique is mainly limited by the finite heat conductivity and does not permit going to shorter times. The extrapolation of the straight line in Fig. 7 to smaller times demands a transformation from $L = 170$ Å to 240 Å in less than 10^{-6} sec. The relaxation of amorphous regions increases the entropy, the heat content and the permeability for solvent molecules to values which are very close to the values of supercooled liquid, indeed closer than they were in the original, rapidly quenched sample, where as a consequence of rapid crystallization a great many tie molecules are under high strain. Annealing also reduces the density of amorphous regions, smoothes the interface between them and the crystals, and heals some of the crystal defects. As a consequence of increased density difference and of sharper boundary between crystalline and amorphous regions, the intensity and sharpness of small-angle scattering very much increase so that soon higher orders or at least the second-order diffraction maximum can also be observed.

Drawing at higher temperature combines the effect of annealing with that of drawing. Particularly the long period corresponds to the draw temperature and not to the crystal thickness of the original sample. The amorphous regions exhibit very little or even no traces of the high strain and of closer packing obtained in cold drawing. They are nearly as relaxed as in undeformed bulk samples.

ACKNOWLEDGEMENT

The author wants to thank the Camille and Henry Dreyfus Foundation for the generous support of this work.

BIBLIOGRAPHY

1. Statton, W. O., J. Appl. Phys., **32,** 2332 (1961).
2. Fischer, E. W. and Schmidt, G. F., Angew. Chem., **74,** 551 (1962).
3. Takayanagi, M. and Nagatoshi, F., Mem. Fac. Eng. Kyushu Univ., **24,** 33 (1965); Nagatoshi, F. and Takayanagi, M., Rep. Prog. Polymer Phys. Japan, **7,** 77 (1964); Hirai, N., Tokumori, T., Katayama, T., Fujita, S. and Yamashita, Y., Rep. Res. Lab. Surface Sci., Okayama U., **2,** 91 (1963).
4. Peterlin, A. and Meinel, G., J. Appl. Phys., **36,** 3028 (1965).
5. Statton, W. O. and Geil, P. H., J. Appl. Polymer Sci., **3,** 357 (1960).
6. Peterlin, A. J. Polymer Sci. **B1,** 279 (1963); Polymer, **6,** 25 (1965).
7. Hirai, N., Yamashita, Y., Matsuhata, T. and Tamura, Y., Rep. Res. Lab. Surface Sci., Okayama U., **2,** 1 (1961); Hirai, N., and Yamashita, Y., Chem. High Polymers (Japan), **21,** 173 (1964).
8. Hosemann, R., (private communication).
9. Holland, V. F., J. Appl. Phys., **35,** 1351 (1964); Fischer, E. W., Kolloid-Z., **189,** 97 (1963).
10. Geil, P. H., *Polymer Single Crystals*, J. Wiley and Sons, New York, 1963.
11. Reinhold, Chr., Fischer, E. W. and Peterlin, A., J. Appl. Phys., **35,** 71 (1964).
12. Geil, P. H., Paper presented at Amer. Chem. Soc. Meeting, Detroit, April 1965.
13. Aggarwal, S. L., Tilley, G. P. and Sweeting, O. J., J. Polymer Sci., **51,** 551 (1961).
14. Kasai, N. and Kakudo, M., ibid., **A2,** 1955 (1964).
15. Hendus, H., (private communication). See also p. 428 in Ref. 10.

16. Peterlin, A. and Meinel, G., J. Polymer Sci., **B3,** 783 (1965); J. Appl. Phys., **36,** 3028 (1965).
17. Peterlin, A. and Olf, H. G., J. Polymer Sci., A2, **4,** 587 (1966).
18. Williams, H., Stannett, V. T., Peterlin, A. and Olf, H. G., (to be published).
19. Peterlin, A., J. Polymer Sci., **C9,** 61 (1965).
20. Geil, P. H., **A2,** 3813, 3835, 3857 (1964); Geil, P. H., Kiho, H. and Peterlin, A., ibid., **B2,** 71 (1964).
21. Peterlin, A., Kiho, H. and Geil, P. H., ibid., **B3,** 151 (1964).
22. Kiho, H., Peterlin, A. and Geil, P. H., J. Appl. Phys., **35,** 1599 (1964); J. Polymer Sci., **B3,** 157, 257, 263 (1965).
23. Ingram, P. and Peterlin, A., **B2,** 739 (1964).
24. Peterlin, A., Ingram, P. and Kiho, H., Makromol. Chem., **86,** 294 (1965).
25. Kobayashi, K., see Ref. 10; Hirai, N., Kiso, H. and Yasui, T., J. Polymer Sci., **61,** S1 (1962).
26. Dismore, P. F. and Statton, W. O., ibid., **B2,** 1116 (1964).
27. Anderson, F. R., Paper presented at Amer. Phys. Soc. Meeting, Philadelphia, March, 1964.
28. Hess, K. and Mahl, H., Naturwiss., **41,** 86 (1954).
29. Corneliussen, R., (private communication).

Small-Angle Equatorial Diffraction in Fibrous Proteins

S. KRIMM

Harrison M. Randall Laboratory of Physics and Biophysics Research Division, Institute of Science and Technology, University of Michigan, Ann Arbor, Mich.

INTRODUCTION

The X-ray diffraction patterns of many fibrous proteins, for example, α-keratin, feather keratin, collagen and muscle, exhibit discrete equatorial small-angle diffraction maxima, often corresponding to first-order spacings of about 100 Å and higher. In several cases efforts have been made to relate these maxima to interference arising from a specific arrangement of structural units. Thus, for muscle such reflections have been satisfactorily correlated (1, 2, 3) with the hexagonal array of actin and myosin filaments seen in electron micrographs of cross-sectioned muscle (4). However, the small-angle reflections observed for collagen (5, 6) have not as yet found any observable electron-micrographic counterpart. As a result attempts have been concentrated on determining a possible arrangement which would account for the observed spacings. These have included a monoclinic unit cell (5), a cylindrical lattice (7, 8), and a limited hexagonal lattice (9). As for the keratins, small-angle spacings of over 100 Å were observed about thirty years ago (10), but after a suggestion (11) that they were spurious in origin no attention was given to them by most authors. Recent careful small-angle X-ray studies on feather keratin (12, 13) and on α-keratin (14) have verified the genuine nature of these diffractions and established more reliable values for the spacings observed. As a consequence the studies on interpreting the small-angle pattern of α-keratin (15) are in need of revision, and a model is now required to explain the reflections found in feather keratin.

The study of such small-angle equatorial reflections in fibrous proteins is of importance for two main reasons. First, it can provide information on how the small-scale elements (often polypeptide chains or small aggregates of such chains) are organized into the larger-scale units which are present and which may be intimately related to the functioning components of the system—for the small-scale fibrous elements apparently do not simply fill space in some pseudo-hexagonal packing but have a specific structural arrangement in relation to their neighbors. Second, in some cases it may be possible to obtain information about the internal structure of the element associated with the small-angle scattering, since the transform of the element may determine some features of this scattering. This is obviously of importance in cases where the structure of the polypeptide chain is unknown, since knowledge even of the lateral dimension of this structure is a significant boundary condition on any proposed chain conformation.

In this paper we wish to comment on the nature of the information obtainable from discrete small-angle equatorial reflections, and the problem of its uniqueness. A new type of packing of fibrous elements will also be proposed which may be relevant in accounting for the diffraction pattern of feather keratin and also perhaps of other fibrous structures.

STRUCTURAL ORIGIN OF SMALL-ANGLE DIFFRACTION

Discrete small-angle equatorial diffraction maxima can originate from fibrous structures in two general ways. In the first case scattering arises from inter-unit interference, with the distance between centers of units being large enough to place the scattering in the small-angle region. In other words, the first-order diffraction occurs at a value of $2 \sin \theta/\lambda$ of the order of $1/r$, where r is the distance between centers of units (2θ is the scattering angle and λ is the wavelength of the radiation). In the second case small-angle diffraction results from intra-unit interference in a unit containing a small number of fibrous elements. In this instance the units are considered to scatter independently, and the maxima at small angles arise from the lack of extended lateral order. We shall first consider these two extreme cases separately.

Inter-Unit Interference

The first type of scattering can be considered from the point of view of a general or a specific structural model. In general, as is well known [i.e. reference (16)], the scattered intensity from a system of non-independent scatterers is determined by the radial distribution function for the system. For a two-dimensional array of fibrous units in which all orientations about the fiber axis are equally probable the scattered intensity is given by (17)

$$i(h) = F^2(h)\left\{1 - \nu \int_0^\infty 2\pi r[1 - g(r)]J_0(hr)\, dr\right\} \qquad 1.$$

In this equation $h = (4\pi/\lambda)\sin\theta$, $F(h)$ is the scattered amplitude for an individual unit, ν is the density of unit centers in the plane perpendicular to the fiber axis, $g(r)$ is the radial distribution function in this plane, and $J_0(hr)$ is the zero-order Bessel function. By means of the Fourier-Bessel theorem this integral can be inverted to give

$$2\pi[g(r) - 1] = \int_0^\infty h\left[\frac{i(h)}{F^2(h)} - 1\right] J_0(hr)\, dh \qquad 2.$$

so that from a measurement of the normalized intensity it is possible to evaluate the integral in 2 and thus obtain $g(r)$. This technique has been used for some fibrous systems (18) in order to derive the radial distribution function of their microcrystalline components. While such an approach has the advantage that $g(r)$ is obtained directly from the experimental data, with no intervening assumptions, it suffers from the shortcoming that $g(r)$ is not very informative: a radial distribution function is difficult to interpret in terms of specific structural groupings which may be present. A similar problem arises if the concept of a paracrystalline lattice (19) is invoked.

In order to determine possible specific structural origins of the small-angle scattering it is necessary to make assumptions about the nature of the structure and then to calculate the expected scattering. Agreement with the observed pattern may then provide presumptive evidence for the existence of the proposed structure. In the case of fibrous proteins various structural types have been considered: layer lattices (20), hexagonal lattices, cylindrical lattices (8), and limited hexagonal arrays (9, 21, 22). In each of these cases the scattering is of a predictable form; for example, for the layer and cylindrical lattices diffraction maxima occur at values of h in the ratio of $1:2:3:\ldots$, whereas for limited hexagonal arrays of less than about 12 units the diffraction maxima occur near h-values corresponding to the maxima of $J_0(hr)$. Distinguishing uniquely between the various possibilities, however, is not easy since the characteristics of the scattering are similar for the several structures, particularly at higher h- values. A possible diagnostic, however, exists in the ratios of the first few orders of diffraction. As can be seen in Table I for the case of inter-unit interferences, the ratio of h_2 to h_1 is significantly different, from the experimental point of view, for the several model structures mentioned (the J_0-dependence applies strictly to the case of a two-unit or three-unit structure (22); a structure consisting of one unit surrounded in hexagonal packing by six units, $1 + 6$, is seen to be significantly different). Of course, the models listed in Table I do not exhaust the structural possibilities, and herein lies

TABLE I
Ratios of successive orders of diffraction for various structures.

Structure	$h_2:h_1$	$h_3:h_2$	$h_4:h_3$	$h_5:h_4$	$h_6:h_5$
Inter-Unit Interference					
Layer lattice	2.00	1.50	1.33	1.25	1.20
Cylindrical lattice	2.00	1.50	1.33	1.25	1.20
Limited hexagonal: J_0	1.90	1.47	1.32	1.24	1.20
Limited hexagonal: $1 + 6$	1.84	1.44	1.35	1.26	1.17
Hexagonal lattice	1.73	1.15	1.32	1.14	1.15
Intra-Unit Interference					
Circular lattice: J_0^2	1.83	1.45	1.31	1.24	1.19
Slit stencil	1.72	1.41	1.29	1.23	1.18
Solid cylinder: $[J_1(x)/x]^2$	1.64	1.38	1.27	1.21	1.18
Cylindrical lattice A[1]*	1.64	1.33	1.31	1.20	1.17
Cylindrical lattice B[2]*	1.44	1.25	1.22	1.20	1.13

[1] Lattice of Fig. 2a of Sasisekharan and Ramachandran (8).
[2] Lattice of Fig. 2b of Sasisekharan and Ramachandran (8).
* Based on optical diffraction patterns by author.

the weakness of this approach. It may serve, however, to eliminate proposed structures from consideration. (It must be remembered that the ratios in Table I are associated only with diffraction maxima produced by the lattice. The positions of these could be shifted by the special characteristics of the transform of the scattering unit which must multiply the lattice transform.)

Intra-Unit Interference

It is well known that when an infinite periodic structure is reduced to one of finite extent by multiplication with a stencil function, the transform acquires subsidiary maxima associated with the characteristics of the stencil function. In particular, maxima in the transform are expected to appear between the origin and the first-order diffraction peak associated with the inter-element distance. This has naturally suggested the possibility that small-angle diffractions in fibrous proteins arise from limited arrays of scattering elements (7, 9). The relative positions of these small-angle subsidiary maxima are a function of the characteristics of the limited array in question, and again the ratios of successive orders of diffraction might be used to differentiate between structures. Such ratios arising from the internal characteristics of the scattering unit are listed in Table I for several structures which have been suggested in the literature. Again we see that the ratios of $h_2:h_1$ and $h_3:h_2$ are reasonably characteristic of the structure, and might at least be used to eliminate a proposed model. Also, the ratios of higher orders are so similar that we would not expect an experimental distinction between structures to be feasible by their use.

Fibrous Proteins

The above discussion presents two extreme cases for the origin of small-angle equatorial diffraction in fibrous systems. In actual systems it is likely that the scattering cannot be understood as arising entirely from one of these factors alone. Thus, inter-unit scattering would be superimposed on the transform associated with a limited array if the units do not scatter independently. In fact, such an assumption is at the basis of the calculation of scattering from several fibrous proteins (9). In other instances it may be possible, however, that relatively independent unit scattering occurs.

The distinction between these extreme cases is nevertheless an important one with respect to the physical nature of the structure. Furthermore an experimental differentiation between the two may be possible. Thus, if we are able to swell the system with a solvent, different and possibly distinguishable behaviors could result. If the elements within a unit were closely bound together and were unaffected by the solvent, then only the units would move apart on swelling. As a consequence the inter-element spacing would be substantially unchanged while the small-angle spacings would increase. At the other extreme, if the elements of a limited array move apart isotropically on swelling and the array maintains its shape and composition then there would be an affine change in the entire diffraction pattern, medium- and small-angle spacings changing in the same ratio. These two examples are by no means exhaustive, and other kinds of swelling behavior are possible. Unfortunately, little has been done to study small-angle scattering from fibrous proteins as a function of swelling, so these possibilities have as yet not been tested in detail.

Recently, however, a swelling study of feather keratin was undertaken (12, 13). The following important characteristics of the reversible swelling of feather in water were observed: (1) the small-angle spacings (first order of about 125 Å) increase ~40% in water whereas the medium-angle spacings (based on a first order of ~32 Å) increase by only ~10%; (2) the macroscopic lateral dimension of the sample increases only by ~7% on swelling; (3) the small-angle diffraction maxima are narrower in the wet state and more orders are seen; (4) the ratio of first- to second-order spacings is greater than 1.85 in the wet state. Points (1) and (2) make it unlikely that the small-angle scattering arises from inter-unit interference between large units which move apart on swelling. Rather these observations as well as point (3) suggest that the change in the diffraction pattern is due to an enlargement in the size of a finite domain from which the scattering may be considered coherent. Point (4) indicates (by comparison with Table I) that the limited structures discussed thus far do not satisfactorily explain the data, certainly not a cylindrical lattice as suggested by Sasisekharan and Ramachandran (8). These observations therefore prompt a search for other general types of lattices which could account for the observed $h_2:h_1$ ratio. One such is described below.

PARALLEL-ARC LATTICES

The kind of structure we are seeking is one in which the small-angle diffractions can change independently of those at medium angles, yet consistent with a meaningful physical mechanism. Finite domains of the slit stencil and cylindrical lattice types satisfy the qualitative requirement, in that the size of the independently scattering domains could increase without comparable changes in the inter-element distance; however, as we can see from Table I, their $h_2:h_1$ ratio appears to be significantly smaller than values which can be observed. It is therefore desirable to seek other structural types.

One system which satisfies the above requirements is based on an array of elements whose locus is the arc of a circle. The transform of such a structure has been determined (23, 24, 25), and is given by

$$T(h,\Phi) = M \sum_{p=-\infty}^{+\infty} J_p(hr) \exp\left[ip\left(\Phi + \frac{\pi}{2}\right)\right]\left[\frac{\sin(\pi pM/N)}{M \sin(\pi p/N)}\right] \quad 3.$$

where M is the number of evenly spaced elements on the arc, N is the number of such elements on a complete circle, and r is the radius of this circle. For a complete circle of elements the above equation becomes

$$T(h,\Phi) = N \sum_{n=-\infty}^{+\infty} J_{nN}(hr) \exp\left[inN\left(\Phi + \frac{\pi}{2}\right)\right] \qquad 4.$$

where $n = \pm 1, \pm 2, \ldots$. In the latter case we see that the diffraction pattern is dominated in the region near the origin, (i.e., in the small-angle region) by $J_0^2(hr)$ and in the medium-angle region by $J_N^2(hr)$, $J_{2N}^2(hr), \ldots$. Since the maxima of $J_N^2(hr)$, $J_{2N}^2(hr), \ldots$ are very

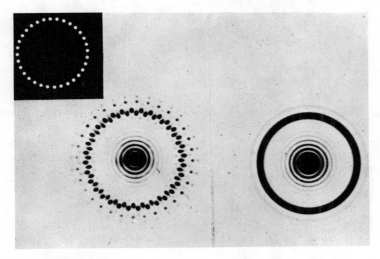

Fig. 1. Optical diffraction pattern of a circular array of elements (inset), unrotated and rotated.

close to the maxima of $J_0(h \cdot 2\pi r/N)$ (the ratios of successive maxima are indistinguishable), we see that the small-angle diffraction is determined by the radius of the circle while the medium-angle diffraction is governed by the distance between elements. This is qualitatively true also for the elements on an arc, and optical diffraction patterns show that this characteristic persists down to a semicircle of elements in random orientations, as can be seen in Figs. 1 and 2.

As soon as the number of elements occupies less than a semicircle, however, the small-angle diffractions begin to smear out and distinct maxima are no longer evident. Furthermore the ratio of $h_2:h_1$ falls

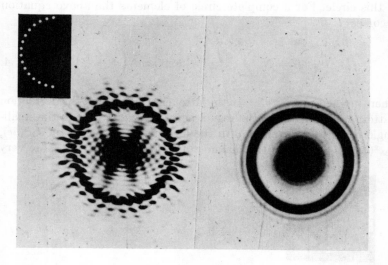

Fig. 2. Optical diffraction pattern of a semicircular array of elements (inset), unrotated and rotated.

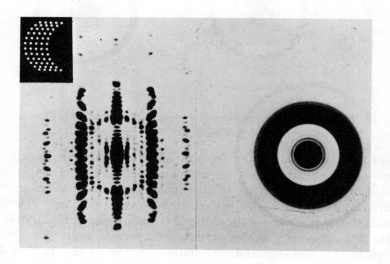

Fig. 3 Optical diffraction pattern of a parallel-arc lattice, 5 arcs and 12 elements/arc (inset), unrotated and rotated.

below its value of 1.83 for a full circle (for a semicircle of elements $h_2:h_1$ was measured to be about 1.6). Single arc arrays therefore are not satisfactory. Both of these difficulties are overcome by placing several arcs "parallel" to each other, as illustrated for example in Figs. 3 and 4. The interference function associated with the parallel array of arcs eliminates much of the diffraction in the small-angle region, giving rise to more distinct maxima in the rotated pattern. The influence of the arc arrangement of elements nevertheless mani-

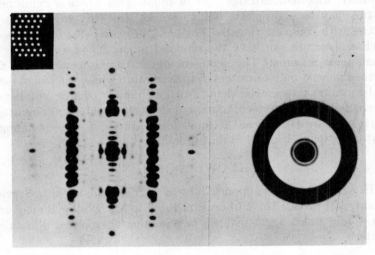

Fig. 4. Optical diffraction pattern of a parallel-arc lattice, 5 arcs and 8 elements/arc (inset), unrotated and rotated.

fests itself in the ratio of orders. For a series of models containing 10 parallel arcs with 8 elements per arc and varying radius of the arc (the separation between elements being kept constant), the ratio of $h_2:h_1$ was found to be about 1.90 for the set (probably with an error no larger than ±0.05). This is of course significantly higher than the value of about 1.7 which is found for a rectangular lattice of finite extent.

The above model thus provides a structural scheme which could account for the characteristics of the small-angle scattering from feather keratin. This of course does not prove that this model is actually the basis for the small-angle diffraction from feather, but analysis of electron micrograph cross-sections (13) indicates that such structures may be present. A parallel-arc lattice is physically more reasonable than a cylindrical lattice since neighboring units along an arc "interact" with each other in the same way throughout the structure, whereas this is not true for a cylindrical lattice since each ring of elements has a different radius of curvature. Although fragments of a cylindrical lattice (such as sectors, portions of several concentric rings, etc.) will also give rise to small-angle diffractions, none of these seem to give values of $h_2:h_1$ greater than about 1.75. (Incidentally, this demonstrates that in those cases where a cylindrical lattice may be appropriate it may not be necessary to postulate a lattice with complete rings, as originally proposed (7); fragments of such a structure can have the required properties of giving rise to small-angle maxima.) The details of the scattering from parallel-arc lattices depend on parameters of the structure whose influences need to be examined in greater detail (such as number of arcs, number of elements per arc, radius of arcs), but it seems apparent at this stage that such structures can account for features of the small-angle scattering found in some fibrous proteins.

ACKNOWLEDGEMENT

This research was supported by a U.S. Public Health Service Grant, AM 02830. The author is indebted to Mr. Walter Bigney and Mr. Craig Baker for assistance in obtaining the optical diffraction patterns.

BIBLIOGRAPHY

1. Huxley, H. E., Proc. Roy. Soc., **B141**, 59 (1953).
2. Worthington, C. R., J. Mol. Biol., **3**, 618 (1961).
3. Elliott, G. F., Lowy, J. and Worthington, C. R., ibid., **6**, 295 (1963).
4. Huxley, H. E., J. Biophys. Biochem. Cytol., **3**, 631 (1957).
5. North, A. C. T., Cowan, P. M. and Randall, J. T., Nature, **174**, 1142 (1954).
6. Cowan, P. M., North, A. C. T. and Randall, J. T., Proc. Symp. Soc. Exp. Biol., **9**, 115 (1955).
7. Ramachandran, G. N. and Sasisekharan, V., Arch. Biochem. Biophys., **63**, 255 (1956).
8. Sasisekharan, V. and Ramachandran, G. N., Proc. Ind. Acad. Sci., **45**, 363 (1957).
9. Burge, R. E., J. Mol. Biol., **7**, 213 (1963).
10. Corey, R. B. and Wyckoff, R. W. G., J. Biol. Chem., **114**, 407 (1936).
11. Bear, R. S., J. Am. Chem. Soc., **66**, 2043 (1944).
12. Krimm, S., Acta Cryst., **16**, A82 (1963).
13. Krimm, S., to be published.

14. Krimm, S., Abstracts Amer. Chem. Soc., Detroit, March 1965.
15. Fraser, R. D. B., MacRae, T. P. and Miller, A., J. Mol. Biol., **10,** 147 (1964).
16. Guinier, A., *et al.*, *Small-Angle Scattering of X-Rays*, J. Wiley and Sons, New York, 1955.
17. Oster, G. and Riley, D. P., Acta Cryst., **5,** 272 (1952).
18. Heyn, A. N. J., J. Appl. Phys., **26,** 519, 1113 (1955).
19. Hosemann, R. and Bagchi, S. N., *Direct Analysis of Diffraction by Matter*, North-Holland Publishing Co., Amsterdam, 1962.
20. Fraser, R. D. B. and MacRae, T. P., Biochim. Biophys. Acta, **29,** 229 (1958).
21. Burge, R. E., Acta Cryst., **12,** 285 (1959).
22. Burge, R. E., Proc. Roy. Soc., **A260,** 558 (1961).
23. Whittaker, E. J. W., Acta Cryst., **7,** 827 (1954).
24. Whittaker, E. J. W., ibid., **8,** 261, 265, 726 (1955).
25. Waser, J., ibid., **8,** 142 (1955).

Combined Analysis of Small-Angle Scattering of Reflections (000) and (hkl)

ROLF HOSEMANN

Fritz-Haber-Institut der Max-Planck-Gesellschaft, Berlin
Faradayweg 4-6, Germany

INTRODUCTION

It is well known that small-angle scattering in the conventional sense is understood as diffraction and scattering of X-rays, electrons, etc. with diffraction angles smaller than about 1°. If $\mathbf{s}$, $\mathbf{s}_0$ are unit vectors parallel to the diffracted and primary beams and λ is the wavelength, the "reciprocal vector"

$$\mathbf{b} = \frac{\mathbf{s} - \mathbf{s}_0}{\lambda}; \qquad |\mathbf{b}| = \frac{2 \sin \theta}{\lambda} \qquad 1.$$

expands a three-dimensional Fourier space. 2θ is the scattering angle. If $\rho(\mathbf{x})$ is the electron density distribution of the structure under consideration in the physical space, expanded by the vector $\mathbf{x}$, then the amplitude of elastically scattered waves is proportional to

$$R(\mathbf{b}) = \mathfrak{F}(\rho); \qquad \mathfrak{F} = \int e^{-2\pi i (\mathbf{bx})} dv_x \qquad 2.$$

($\mathfrak{F}$ = symbol of Fourier transform, dv_x volume element of physical space).

Introducing a shape function

$$s(b) = \begin{matrix} 1 & \text{if} & b < b_0 \\ 0 & \text{if} & b > b_0 \end{matrix} \qquad 3.$$

where $b_0 = 1°/\lambda = (1/57\lambda)$, $R \cdot s$ is the function which can be studied by small-angle scattering. If

$$s(\mathbf{b}) = \mathfrak{F} S(\mathbf{x}) \qquad 4.$$

$S(\mathbf{x})$ is a function with a maximum at $\mathbf{x} = 0$, which has a width of about 57 λ. According to the convolution theorem of Fourier-transformation one obtains from Rs by an inverse Fourier transform not

ρ, but the convolution product (symbolized by $\frown$)*between ρ and S:

$$\rho \frown S = \mathfrak{F}^{-1}(Rs) \qquad 5.$$

with

$$\rho \frown S = \int \rho(\mathbf{y}) S(\mathbf{x} - \mathbf{y}) \, dv_y \qquad 6.$$

This means that the unknown real structure $\rho(\mathbf{x})$ is smeared out by $S(\mathbf{x})$ over domains of diameter 57 λ. Using wavelengths of the order of 1 Å all atomic details are lost and information is available only about details larger than 57 λ.

From Eq. 5 we learn, that the analysis of the small-angle scattering gives information only of the smeared, non-atomistic structures. In the section dealing with the scattering from linear polyethylene an example is given, which shows, that following this line wrong solutions with regard to the conformation of chain molecules can be discarded. On the other hand, taking into account certain information from wide-angle diffraction, the results of the small-angle pattern can be understood much better. This is shown in the section following the one mentioned above by combining the results of the profiles both of the reflection (000) (= small-angle scattering) and the wide-angle reflections ($hh0$) of linear polyethylene. In other words: small-angle studies not only around (000) but also around (hkl) can give us important information.

In the next section, continuous small-angle scattering is discussed briefly. It demonstrates that quite new diffraction phenomena arise, unknown on the atomic scale, since now the particles can vary much more in shape and size. If, moreover, they build up macrolattices, a series of further new phenomena occur. To understand them well, the subsequent sections deal with the concept of paracrystals, with some results for macrolattices from small-angle scattering, and with results for macrolattices and atomic lattices by the combined analysis of small- and wide-angle scattering.

More details of the two interference theories and the relevant proofs of equations dealt with in this paper can be found in the book of Hosemann and Bagchi (1).

CONTINUOUS SMALL-ANGLE SCATTERING

Most of the small-angle studies are concerned with a continuous intensity distribution $i(\mathbf{b})$, which, without pronounced interference

* The symbol ★ is used elsewhere in this volume to denote the convolution product. The notation in this paper, however, is retained to conform to Ref. (1), where much of the theoretical material is treated (Ed.).

maxima, drops to zero with increasing **b**. This type of scattering was first discussed by Lord Rayleigh (2). It means that we have the addition of the scattered intensities i_n of single "particles" which practically do not interfere. Rayleigh's formula is of the kind

$$\overline{RR^*} = \sum_{n=1}^{N} i_n \qquad 7.$$

The bar indicates the average or expectation value of the scattered squared amplitude. Rayleigh was able to prove that 7 holds exactly, if the phase between the amplitudes R_n, R_m scattered by the particles n, m has the same *a priori* probability between 0 and 2π.

The particles need not move; they can be fixed in a frozen structure. But their positions must be quite irregular. For wide-angle scattering a relation similar to 7 was proposed by Debye (3) for molecules or atoms in an ideal gas. Since in an ideal gas the particles have diameters negligibly small relative to their mean separations, Rayleigh's condition of vanishing phase-relations is realized. It is not in a real gas near the critical point. Eisenstein and Gingrich (4), for instance, found that argon gas gives rise to a continuous small-angle scattering of the type of Eq. 7, with a halfwidth of the order of $\delta b \sim 0.03$ Å^{-1}. According to 5, there must exist statistical inhomogeneities of diameters of the order of 15 Å, which were explained by Vineyard (5) as "droplets," e.g. clusters of argon atoms.

The experimenter who does not investigate the small-angle scattering, in this case gets no information about these inhomogeneities. The wide-angle intensity function really is of the type of Eq. 7, where i_n is the intensity scattered by a single argon atom. Hence with respect to this domain of information the atoms behave like totally randomly distributed particles, producing a real Rayleigh scattering.

In Fig. 1a a model of statistically homogeneously distributed points is given. Its Fraunhofer pattern (Fig. 1b) shows a small-angle scattering, which depends on the volume of the whole model (Debye's volume scattering). Contrary to Fig. 1, in Fig. 2a the point atoms cluster together in "droplets," hence the Fraunhofer pattern of Fig. 2c shows a small-angle scattering depending on the size and size distribution of the clusters. Since the packing density of the atoms in both figures is relatively high, it was not possible to distribute them quite randomly as in the gaseous state. Hence both Fraunhofer patterns show at larger angles a diffuse ring, as observed in the X-ray scattering of real liquids. Figure 2b is a quite schematic representation of the smeared-out function of Eq. 5. Its Fraunhofer pattern would offer the same small-angle scattering as in Fig. 2c, but no liquid-like halo.

The first observation of a continuous small-angle scattering was made on amorphous graphite by Raman and Krishnamurti (6) and explained by interferences between widely separated conduction electrons. Following Debye's concept of the gaseous state, it seemed to be impossible that single scattering particles in a solid body could

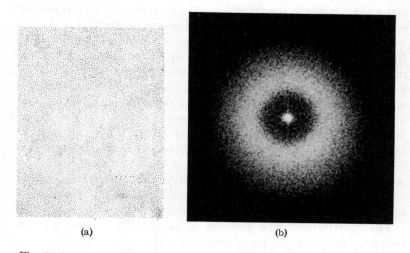

Fig. 1. A statistically homogeneous lattice and its Fraunhofer pattern. At small angles Debye's volume scattering occurs. At wide angles a fluid ring exists whose diameter is reciprocal to the mean distance between the atoms.

have such an irregular arrangement that a Rayleigh-type scattering occurs. Hence Kratky (7, 8) for instance explained such scattering by the simultaneous presence of many different interplanar spacings, which produce a set of Bragg-reflections, each displaced from the other sufficiently so that a continuous small-angle scattering arises. Kruyt (9) emphasizes "that this could only hold for particles which all have practically the same dimensions. In that case the scattering at small angles will show a pronounced maximum, which is quite comparable with the amorphous ring which occurs in liquids and amorphous solid substances."

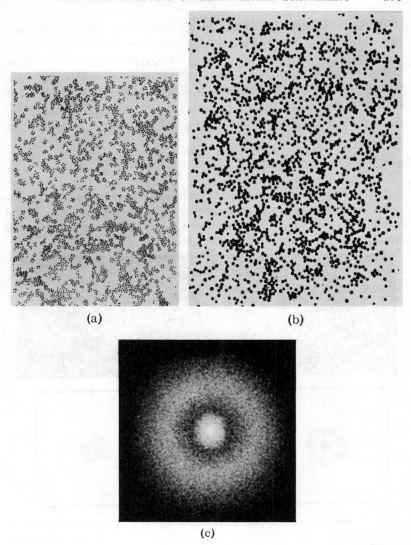

Fig. 2. A statistically inhomogeneous lattice with the same mean distance between the atoms within the clusters (a). The smeared-out colloidal structure (schematically) according to Eq. 5 (b). Fraunhofer pattern of (a) with the continuous small-angle scattering of the polydisperse clusters (c).

Taking into account not only the positions of the different particles but also their different shapes one comes to quite new intensity formulae (10). In the theory of polydisperse particles [for details see Ref. (1)], it is proved that a continuous small-angle scattering of the Rayleigh type exists, if the polydispersity g_y is larger than the packing density ϵ

$$g_y \geq \epsilon \qquad 8.$$

with

$$g_y = \left[\left(\frac{\overline{y^2} - \bar{y}^2}{\bar{y}^2}\right)\right]^{\frac{1}{2}}; \qquad \epsilon = \frac{\Omega}{V} \qquad 9.$$

$\bar{y}$ is the weight-averaged radius (resp. radius of gyration) of the particles, $\overline{y^2}$ the weight average of its square, Ω the volume occupied by all particles, V the total volume. Clustering effects, macrolattice formations etc. of the particles are excluded.

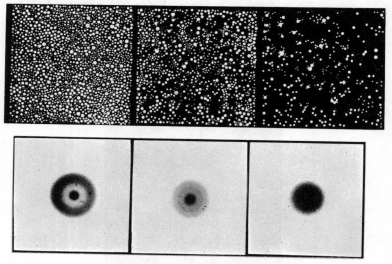

Fig. 3. Experimental results of two-dimensional polydispersed and amorphous substances with polydispersity $g_y = 0.3$ and packing densities $\epsilon = 0.5$, 0.3, 0.1.

If inequality 8 holds the interference effects between the particles are so small, that they can be neglected in a b^2i vs. b plot. Figure 3 shows three models with $g_y = 0.3$ and the packing densities 0.5, 0.3, 0.1, and their Fraunhofer patterns. An interference halo arises only for $\epsilon = 0.5$. In the b^2i-b-plot (Fig. 4) the curves for $\epsilon = 0.1$ and 0.3 are nearly the same. Using the values b_m and b_0 of the position of the maximum of $b^2 i$ and the intersection of its tangent through the inflection point P with the abscissa, one gets for this two-dimensional case

$$\bar{\bar{y}} = \frac{4.6}{2\pi b_m} \cdot \frac{1}{\sqrt{\dfrac{b_0}{b_m} + 3.3}}; \qquad g_y = 0.43 \sqrt{\frac{b_0}{b_m} - 2.1} \qquad 10.$$

Table I gives the results obtained from the Fraunhofer pattern with the help of this formula. They agree quite satisfactorily with the real statistical parameters of the model and demonstrate the validity of inequality 8.

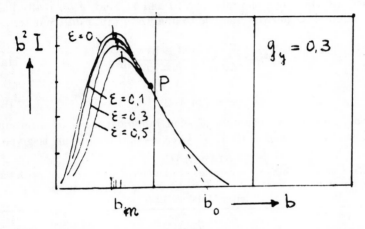

Fig. 4. $b^2 i$-microphotometer curves of Fig. 3. The curve $\epsilon = 0$ is calculated.

TABLE I
Analysis of two-dimensional models of polydispersed systems [after Joerchel (11)]

Fig.	From the model			From the diffraction	
	ϵ	g_y	$\bar{\bar{y}}$ (mm)	g_y	$\bar{\bar{y}}$ (mm)
3a	0.5	0.3	0.155	0.3 ± 0.06	0.142 ± 0.02
3b	0.3	0.3	0.155	0.27 ± 0.06	0.145 ± 0.02
3c	0.1	0.3	0.155	0.26 ± 0.06	0.137 ± 0.02

The three-dimensional case is treated in Table II. Motzkus (12) analyzed the continuous small-angle X-ray scattering of an amorphous carbon black as a function of packing density ϵ. Packing densities of $\epsilon = 0.08$ are powders, loosely poured between two sheets of cellophane. The samples with $\epsilon = 0.44$ were pressed together hydraulically at 11,500 kg cm^{-2}. Only in this case is the g_y-value affected by interference effects, while $\bar{\bar{y}}$ remains constant for $\epsilon \lesssim 0.28$. For this three-dimensional case the following equations hold:

$$\bar{\bar{y}} = \frac{0.96}{2\pi b_m} \frac{9.6 - 2\frac{b_0}{b_m}}{\sqrt{7.5 - \frac{b_0}{b_m}}}; \quad g_y = \sqrt{\frac{b_0 - 2.1 b_m}{9.6 b_m - 2 b_0}} \quad 11.$$

A correction formula depending on ϵ gives the corrected values of the last two columns. The relations 10 and 11 hold, if $b^2 i$ really has one maximum. Otherwise one has to analyze $b^2 i$ into different terms

$$b^2 i = \sum_{m=1}^{N} \frac{A_m \cdot b^2}{[1 + (bB_m)^2]^{(n+4)/2}} \quad 11a.$$

with

$$g_{yn} = (2n + 2)^{-\frac{1}{2}} \quad 11b.$$

In a later section an example of such inhomogeneous distribution functions with $N = 3$ and $A_m = \frac{1}{3}$, $g_{yn} \sim 0$, $B_m = B \cdot m$ is given.

TABLE II

Analysis of small-angle X-ray scattering of carbon black ELF 5 (Cabot) with the $b^2 i$-b-method and corrected for different values of the packing density ϵ [after Motzkus (12)].

			Corrected	
ϵ	g_y	$\bar{\bar{y}}$ (Å)	g_y	y (Å)
0.08	0.45	187	0.47	190
0.18	0.46	187	0.49	196
0.25	0.46	182	0.50	196
0.28	0.47	182	0.51	197
0.33	0.44	159	0.50	174
0.35	0.45	156	0.51	173
0.45	0.40	140	0.49	157

Inequality 8 in every case is a sufficient but not necessary condition for continuous small-angle scattering. It is interesting to learn that even very densely packed systems behave like gases with respect to interference phenomena, if 8 holds. Such phenomena in the conventional theories, dealing with wide-angle patterns, are totally unknown.

Hence, studying the small-angle scattering, one has to build up entirely new diffraction theories. In the following chapters discontinuous small-angle scattering, showing real interference maxima, is treated.

CONCEPT OF PARACRYSTALS

The classical concept of M. v. Laue (13) of a crystal starts from the assumption that each lattice cell, averaged over time, has the same shape. Hence, if d is the lattice constant, one finds the 50,000th neighbor at a distance $D = 50{,}000\, d \pm \Delta$, where $\Delta \ll d$ takes into account thermal vibrations etc. The long-range order is established *usque ad infinitum*.

In small-angle diffraction we have to deal with macrolattices, if small-angle reflections appear. Now it is quite possible that each cell of this macrolattice does not contain the same number of atoms, since the constitution of the macromolecules or monomeric units from cell to cell may differ. Even for a constant constitution the configuration or conformation within each macrocell may be different. It follows that each cell then has another shape.

In the classical concept of liquids formulated by Debye (14) on the other hand no lattices exist and each atom resp. molecule has the same *a priori* statistical distance distribution to the other atoms or molecules. As a direct consequence the long-range order is lost. Moreover the concept of a lattice is lost too, since this density distribution is spherically symmetric.

TABLE III
Basic properties of the theories.

	Crystal (v. Laue)	Paracrystal	Liquid (Debye)
Long-range order	yes	no	no
Lattice	yes	yes	no
The same *a priori* distance distribution	no	yes	yes
Spherically symmetric correlation function	no	no	yes

As shown in Table III, the concept of paracrystals stands in between these two theories. If in a crystalline lattice a certain atom n oscillates thermally around its mean (ideal) position and at a certain time is shifted by $\delta \mathbf{x}_n$ from this ideal position, all other atoms with respect to its momentary position are shifted by $-\delta \mathbf{x}_n$. Hence *a priori* it has another distance statistic to all other atoms than the atom m, which is shifted by $\delta \mathbf{x}_m$.

In a paracrystalline lattice the ideal positions of the atoms (or molecules) are defined in this way: If $\mathbf{a}_{11}$ is a vector from the first ideal position to the next, it has an *a priori* frequency $H_1(\mathbf{x})$, with $\mathbf{x} = \mathbf{a}_{11}$. This density distribution function is normalized

$$\int H_1(\mathbf{x})\, dv_x = 1 \qquad 12.$$

and has its center of gravity at

$$\mathbf{a}_1 = \int \mathbf{x} H_1(\mathbf{x})\, dv_x \qquad 13.$$

Now regarding the distance frequency distribution from this neighbor to the next neighbor, its frequency distribution is given by the same *a priori* function H_1, which is not affected by the special value a_{11}. Hence the frequency of finding a special combination

$$\mathbf{a}_{11} + \mathbf{a}_{12} = \mathbf{x} \qquad 14.$$

is given by

$$H_1(\mathbf{a}_{11})\, H_1(\mathbf{a}_{12}) \qquad 15.$$

Hence the distance vector $\mathbf{x}$ between next nearest neighbors has a frequency $H_{20}(\mathbf{x})$ which can be obtained from 14 and 15 by integration over all position $\mathbf{a}_{11}$ of the next neighbor

$$H_{20}(\mathbf{x}) = \int H_1(\mathbf{a}_{11}) H_1(\mathbf{x} - \mathbf{a}_{11})\, dv_a \qquad 16.$$

dv_a is a three-dimensional volume element in the space spanned by the vector $\mathbf{a}_{11}$. 16 is the convolution product of H_1 with H_1 (cf. 6). In the same way the distance statistics $H_{30}(\mathbf{x})$ to the third neighbor are given by

$$H_{30}(\mathbf{x}) = \widehat{H_1\, H_1}\, \widehat{H_1} \qquad 17.$$

In Fig. 5 the mean types of point lattices are drawn. On the upper left side a crystalline lattice is given, the next is an ideal paracrystalline lattice. If $H_{10}(\mathbf{x})$ is the so-called coordination statistic in the direction $(h0)$, $H_{01}(\mathbf{x})$ in the direction $(0h)$, the statistic to the paracrystalline lattice point (p,q) is given by

$$H_{p,q}(\mathbf{x}) = P(x - 0) \overbrace{\widehat{H_{10}\, H_{10}} \cdots \widehat{H_{10}}}^{p\text{-times}} \overbrace{\widehat{H_{01}\, H_{01}} \cdots \widehat{H_{01}}}^{q\text{-times}} \qquad 18.$$

$P(x - 0)$ is a Dirac point-function. All lattice cells are parallelopipeds. The summation over H_{pq} [in the three-dimensional case over all $H_{pqr}(\mathbf{x})$] gives a convolution polynomial, which depends only on the three coordination statistics. Figure 6a gives a special two-dimensional model of Schoknecht-Kast (15) where both coordination statistics H_{10} and H_{01} are line functions parallel to the abscissa. Its correlation function (Fig. 6b) is obtained by folding two models in

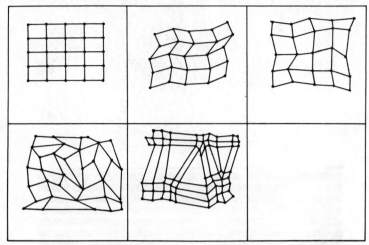

Fig. 5. Different types of point lattices (a) crystalline, (b) ideal paracrystalline, (c) paracrystalline with correlation corrections, (d) amorphous, but statistically homogeneous, (e) statistically inhomogeneous. The clusters build up a macrolattice and consist of lattices on the atomic scale.

a light-optical "folding machine." The maxima of this correlation function in the vertical direction never touch each other. Hence long-range order is established in this direction. In the horizontal direction the zone of non-touching maxima is quite limited and has the shape of an oblique ellipsoid.

Figure 7 gives all distance statistics, calculated from the model, as black strips. The statistics H_{pq}, calculated from the coordination statistics H_{10}, H_{01} and H_{11}, are drawn as white strips. They agree with the black ones within the statistical errors and demonstrate the advantage of the paracrystalline theory: The manifold of distance statistics can be reduced to a folding polynomial of only a few so-called coordination statistics.

Since according to 5 the Fourier transform of a convolution product of two functions is given by the product of the Fourier-transforms of the two functions, the transform of 18 is nothing but a geometric series:

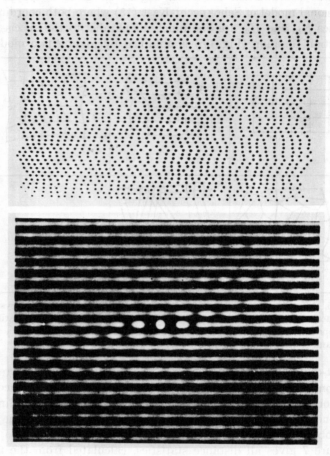

Fig. 6. A two-dimensional paracrystalline lattice and its correlation function, produced in an optical folding machine.

$$\mathfrak{F} \sum_{p,q} H_{pq} = (1 + F_1 + F_1^2 + \cdots F_1^* + F_1^{*2} + \cdots)$$

$$\times (1 + F_2 + F_2^2 + \cdots F_2^* + F_2^{*2} + \cdots) \quad 19.$$

$$F_1 = \mathfrak{F}(H_{10}); \quad F_2 = \mathfrak{F}(H_{01}); \quad 20.$$

For the three-dimensional case one has to add all H_{pqr} and obtains the paracrystalline lattice factor $Z(\mathbf{b})$

$$Z(\mathbf{b}) = \mathfrak{F} \sum_{pqr} H_{pqr} = \frac{1}{\bar{v}_r} P(\mathbf{b} - 0) + \prod_{k=1}^{3} \operatorname{Re} \frac{1 + F_k}{1 - F_k} \qquad 21.$$

$\bar{v}_r$ is the averaged volume of a lattice cell of the paracrystal, $P(\mathbf{b} - 0)$ a point function at $\mathbf{b} = 0$. This point function produces the small-angle scattering, since the second summand of 21 has maxima around

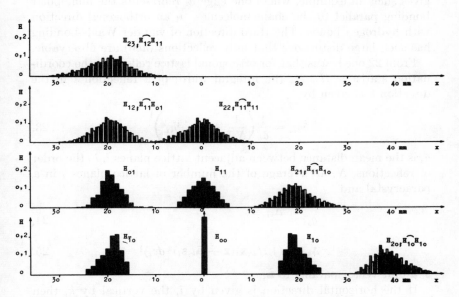

Fig. 7. Comparison between the distance statistics, calculated from the model in Fig. 6 (black strips) and from the folding polynomial of Eq. 16 (white strips).

the reciprocal lattice points (hkl), but not at (000). For a bounded paracrystal we have to multiply 18 with the shape function $s(\mathbf{x})$ of this lattice and then obtain for the intensity function

$$i \cong \frac{1}{v_r} f^2 \widehat{Z} |S|^2; \qquad S(\mathbf{b}) = \mathfrak{F}(s) \qquad 22.$$

f^2 stands for the averaged structure factor. If the lattice cells are no longer parallelopipeds (right upper corner of Fig. 5), certain correction terms must be introduced in 21, which take into account statistical correlations between adjacent edges of a paracrystalline lattice cell. In the model of Fig. 6 these correlations cause 18 to hold only in the two quadrants $p \cdot q \geq 0$. The octant between $p + q = 0$ and $q = 0$ must be developed as a folding polynomial of H_{11} and H_{01}, the octant between $p + q = 0$ and $p = 0$ by H_{11} and H_{10}.

Contrary to classical crystallography, the cell edges of a paracrystal cannot be chosen arbitrarily by any set of three nonplanar periodicities, but have a real physical meaning: they define the vectors to adjacent atoms, which are chemically bonded. Figure 9 below gives such an example, where one edge a_2 represents the homopolar bonding parallel to the chain molecules, a_1 an orthogonal direction with hydrogen bonds. The third direction of van der Waals-bonding has such large distortions that only reflections $(hk0)$ are observable.

From 22 one learns that for orthogonal lattice cells with the coordination statistics H_i, H_k the integral width of a reflection i in the direction k is given by

$$\delta b_{ik} = \frac{1}{\bar{a}_k}\left(\frac{1}{N_k^2} + \pi^4 g_{ki}^4 h_i^4\right)^{\frac{1}{2}} \qquad 23.$$

$\bar{a}_i$ is the mean distance between adjacent lattice planes i, h_i the order of reflections, N_i an average of the number of lattice planes j in a paracrystal and

$$g_{ki} = \frac{\Delta_{ki}}{\bar{a}_k} \qquad 24.$$

$$\Delta_{ki} = (\int H_k(\mathbf{x})(\mathbf{x} - \bar{\mathbf{a}}_k, \mathbf{s}_i)^2 \, dv_x)^{\frac{1}{2}} \qquad 25.$$

$\mathbf{s}_i$ a unit vector parallel to $\bar{\mathbf{a}}_i$.

If the horizontal direction is given by i, the vertical by k, then in the model of Fig. 6:

$$g_{ki} = g_{ii} = 0.2; \qquad g_{ik} = g_{kk} = 0 \qquad 26.$$

Hence $\delta b_{kk} = \delta b_{ki} = 0$ and all reflections $(0k0)$ along the meridian are crystalline-like. This is shown in the Fraunhofer pattern of the model (Fig. 8). The reflections $(h00)$ on the other hand have widths δb_{ii}, δb_{ik} increasing in a parabolic way with h. A quite similar diffraction pattern is given in Fig. 9. The fiber axis lies in the vertical direction parallel to k, the horizontal direction is given by i, which corresponds to the backbone linkage (hydrogen bonds) in β-keratin, $\bar{a}_i = 34$ Å, $\bar{a}_k = 183$ Å.

$$g_{ii} = 0.06; \qquad g_{ik} = 0.05; \qquad g_{ki} = 0.02; \qquad g_{kk} = 0.007 \qquad 27.$$

The coordination statistic H_k shows, that the 183 Å macrolattice edge nearly does not change its length, but mainly its direction

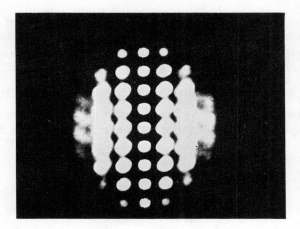

Fig. 8. Fraunhofer pattern of the model in Fig. 6. Only the reflections (0k) are crystalline, all others more or less diffuse according to Eq. 22 and 23.

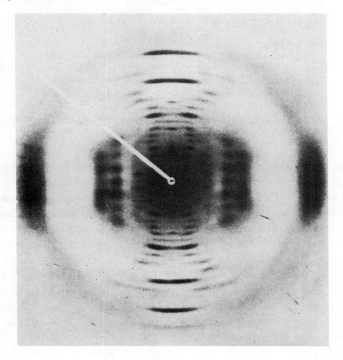

Fig. 9. Discontinuous small-angle X-ray pattern of β-keratin (dry) of the quill of a sea gull [Ref. (16)].

($g_{kk} \ll g_{ki}$), while the backbone linkage changes length and direction with larger relative fluctuation ($g_{ii} \sim g_{ik} \gg g_{kk}$). This seems to be the atomistic geometric explanation why β-keratin behaves like a fiber on the macroscopic scale.

SMALL-ANGLE SCATTERING OF LINEAR POLYETHYLENE

Hess and Kiessig (17, 18) first observed discontinuous small-angle scattering in stretched synthetic polyamids and polyesters. Figure 10

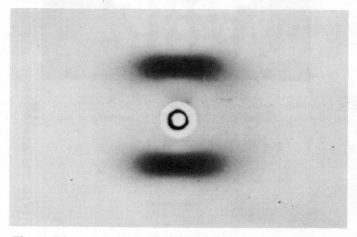

Fig. 10. Discontinuous small-angle X-ray pattern of hot-stretched linear polyethylene.

gives a similar pattern for hot-stretched linear polyethylene; the direction of the chains is again vertical. The paracrystalline layer structure of Fig. 11 produces a similar Fraunhofer pattern. Shortening of these layers (Fig. 12) gives another pattern. The horizontal width δb_{ki} of the meridional small-angle reflection according to 23 depends both on the length $a_i N_i$ of the lamellae and their statistical curvature, which is proportional to g_{ik}. In Fig. 11 g_{ik} dominates, in Fig. 12 $a_i N_i$ (particle size effect). A careful analysis of the X-ray intensity of Fig. 10 by microphotometer curves proved that we have to do with a case similar to Fig. 12. According to Hosemann and Wilke (19) hot-stretched material gives particle sizes of $\bar{a}_i \bar{N}_i \sim 91$ Å, while Bonart and Hosemann (20) found 100 Å for cold-stretched linear polyethylene. From the position of the meridional reflection one

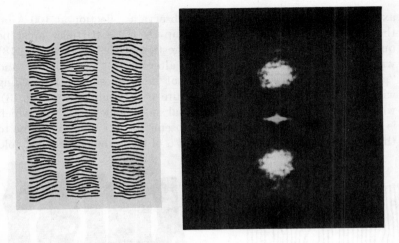

Fig. 11. Paracrystalline layer structure and its Fraunhofer pattern.

obtains a "periodicity" in direction of the chains of $\bar{a}_k = 150$ Å (cold-stretched) and 200 Å and more (hot-stretched).

The question arises, what is the nature of the particles with a periodicity of 150 Å and more and a thickness orthogonal to the fiber

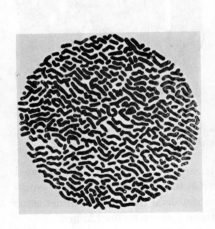

Fig. 12. A model similar to Fig. 11 but with short lamellae, and its Fraunhofer pattern.

axis of 90 or 100 Å. Since we don't observe a reflection at 100 Å on the equator, the macrolattice built up by these particles must be quite irregular (cf. Fig. 12). The suggestion of Hess-Kiessig (17, 18) with bundles of chain molecules, alternating "crystalline" and "amorphous" regions (Fig. 13) cannot be true, since after folding with $S(\mathbf{x})$ (cf. Eq. 5) a colloidal structure arises (right side of Fig. 13), which must produce a strong equatorial small-angle scattering, which is not observed in Fig. 10. Bonart-Hosemann (21) therefore came to the conclusion that a certain amount α of chain molecules must fold

Fig. 13. Structure of chain molecules in linear polyethylene proposed by Hess-Kiessig (17, 18) and its smeared-out colloidal structure.

back on the surface of the crystalline regions (Fig. 14). Cold-stretched material, on the other hand, gives rise to a strong continuous equatorial scattering (Fig. 15). Now the material in colloidal dimensions splits similar to Fig. 13, but back-folding remains unchanged.

This equatorial small-angle scattering is observable in hot-stretched material too, though very weak, if one stretches more than to 12-fold length. The intensity distribution is given in Fig. 16 and compared with that of small-angle meridional reflections (layer line) parallel to the equator. The equator runs quite parallel to it for $b > 0.007$ Å^{-1}. Extrapolating this linear branch to $b = 0$ (dashed line) one finds that about 34 weight percent of chains build up ultrafibrils of 91 Å diameter quite uniformly (polydispersity $g_y \ll 0.2$).

If one subtracts this dashed line, the residue again can be analyzed in the same way. One finds another particle size of 192 Å, again with an unobservably small polydispersity g_y, now with 38 weight percent

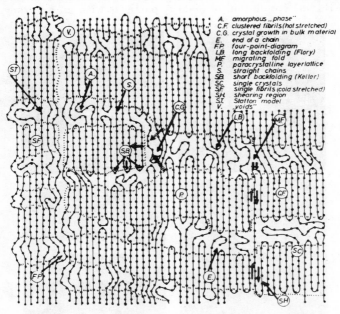

Fig. 14. Model of linear polyethylene. On the left side cold-stretched material with an isolated ultrafibril. On the bottom of the right side a single crystal. Between them intermediate paracrystalline states with partially smectic and nematic features.

and finally a third component with 309 Å and 28 weight percent. We now have to discuss a system of polydisperse particles with an inhomogeneous distribution statistic consisting of three sharp maxima at 91 Å, 2 × 96 Å, 3 × 103 Å, corresponding to an integral polydispersity

$$g_y \sim 0.4$$

The question arises, what is the physical nature of these three sorts of particles and how can they build up a macrolattice, which does not satisfy inequality 8 but nevertheless produces a continuous small-angle scattering. To find an answer, one has to investigate the line profiles of wide-angle reflections.

COMBINATION WITH LINE PROFILES OF WIDE-ANGLE REFLECTIONS

A high resolution diffraction camera was used (manufactured by AEG-Berlin) (resolution of 3000 Å particles with an accuracy of 10

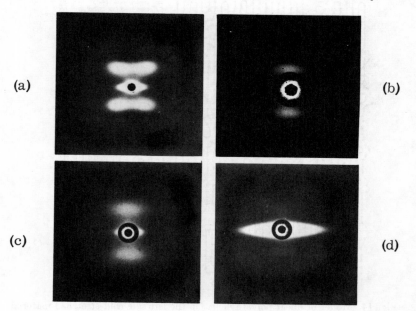

Fig. 15. Small-angle pattern of linear polyethylene, fiber axis vertical; (d) cold stretched, (c) then laterally pressed, primary beam parallel to the direction of press, (a) ditto but orthogonally to the direction of press, (b) then annealed.

percent is possible). It is a double cylinder unit of the Guinier-Jagodzinski type with a bent Johansson monochromator. $CuK\alpha_1$ radiation was used.

In Fig. 17 the same analysis of the profile of the reflection is carried out for (110), (220), (330) wide-angle reflections as in Fig. 16 for (000). The component i_1 of 91 Å ultrafibrils can be found in (110) and (220), while i_2 (with 192 Å) is detectable in (000), (110), (220) and (330). In both cases the slope of the curves decreases with increasing h. Drawing the squared integral widths $\delta\beta_1^2$, $\delta\beta_2^2$ of these components against h^4 (Fig. 18), one finds the parabolic law of Eq. 23. From the slope one obtains g-values (cf. 24) of 3.15% (for i_1) and 2.2% (for i_2).

We conclude, that the 91 Å particles build a paracrystalline lattice with a g-value of 3.15%; 34 weight percent of these ultrafibrils exist isolated from the other chains by splits (cf. Fig. 13 and the left hand part of Fig. 14). 38 weight percent of these ultrafibrils build up thicker fibrils, consisting of 4 ultrafibrils each. The boundary

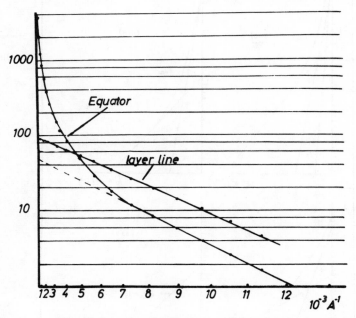

Fig. 16. Intensity distribution parallel to the equator of hot-stretched polyethylene.

between them is much less disturbed than that of the "isolated" ultrafibrils. Otherwise one could not observe i_2 at the (000) reflection. The lattice planes of the single 4 ultrafibrils are so well adjusted, that they scatter more or less coherently (otherwise i_2 could not appear in $(hh0)$. The g-value of 2.2% of this macrolattice, consisting of 4 bricks, can be understood in this way: While inside of the ultrafibrils relative statistical distance fluctuations of the chain molecules exist of $g = 3.15\%$, at the boundary neighboring chain molecules, belonging to different ultrafibrils, have g-values of 42%.

Then the centers of the ultrafibrils fluctuate with a g-value of 2.2%. If the boundary consists of more than one layer of chain molecules, the g-value inside the boundary is smaller.

Particles consisting of 9 ultrafibrils have so much distortion in the boundaries, that the component i_3 cannot be scattered coherently, hence does not appear in the wide-angle reflections (Fig. 17). In

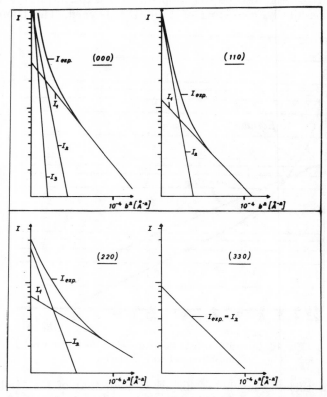

Fig. 17. Gaussian analysis of the line profiles of $(hh0)$ reflections of hot-stretched polyethylene.

Fig. 19 a cross-section with four coherent ultrafibrils is drawn schematically. The paracrystalline distortions are distributed over all chains, but larger in the boundaries. Figure 20 shows a Fraunhofer pattern of this model, which corresponds to the equatorial wide-angle reflections of linear polyethylene. Figure 19 can explain why the small-angle scattering is continuous, though $g_y < \epsilon$ (cf. Eq. 8). The single ultrafibrils are packed together so closely, having common highly distorted boundaries, that their Fourier transform is zero just at the points where the macrolattice factor 21 has a peak.

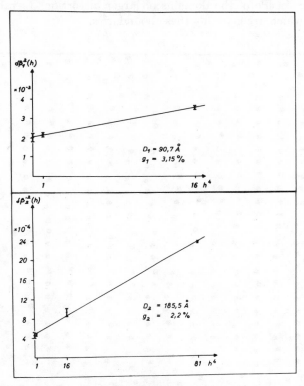

Fig. 18. Squared integral widths $\delta\beta_1^2$ and $\delta\beta_2^2$ against h^4 of the components i_1, and i_2 of Fig. 17.

This model of Fig. 20 gives an interesting explanation for the shearing effect we had observed in stretched polyethylene (Fig. 21). We called this a shearing of the second kind (right side of Fig. 22).

The vertical dashed lines represent the direction of the chains at the two ends of the stretched material. The paracrystalline macro layers (Fig. 11) inside of the sample remain orthogonal to them over the entire volume of the sample. Hence the vector $\bar{a}_k$ parallel to the chains is no longer orthogonal to the vector $\bar{a}_i$ building up the lateral edge of the macrolattice. This is proved by the small-angle scattering (Fig. 21), which both for cold- and hot-stretched material has a monoclinic structure. Taking into account the boundaries between the ultrafibrils, it is quite probable that gliding of the second kind takes place inside of the domains of larger distortions (g-values up to 45%), which are given by these boundaries.

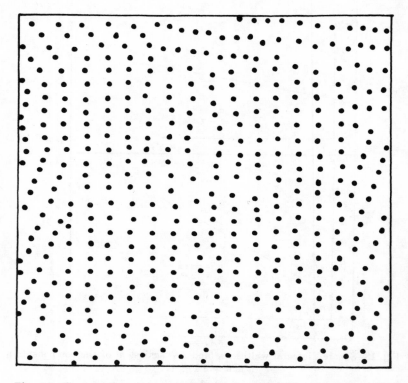

Fig. 19. Cross-section (schematically) of hot-stretched polyethylene. Four coherent scattering ultrafibrils, connected by boundaries of higher paracrystalline distortions.

Another proof for the high (liquid-like) mobility within the boundaries is given by the small-angle diagrams of Fig. 15. In Fig. 15a a cold-stretched material is pressed together from the left- and right-hand side of the figure. Now the laterally more extended amorphous regions along the single isolated ultrafibrils (see left side of Fig. 14) try to come into contact with the crystalline regions of the neighboring ultrafibrils to produce a higher packing density. The meridional small-angle reflections now consist of 4 peaks (Fig. 15a). The colloidal structure must now show paracrystalline hexagonal lattices (Fig. 23b). If the primary beam is directed parallel to the direction of press, only 2 points appear (Fig. 15c). Moreover, by pressing, the intensity of equatorial small-angle scattering is weakened, indicating that many of the splits between the fibrils become smaller or disappear.

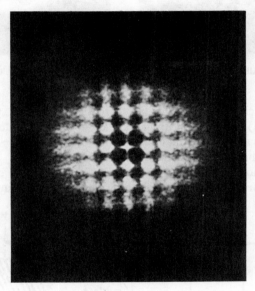

Fig. 20. Fraunhofer pattern of Fig. 19.

If then the pressed material is annealed, the 4-point diagram loses its character and the horizontal width of the meridional small-angle reflections becomes smaller (Fig. 15b). The ultrafibrils now tend to build up more or less undistorted paracrystalline layer lattices (Fig. 23c). Even in the case of parallel-oriented bundles of single crystals (Fig. 23d) we could find a continuous equatorial small-angle scattering, which proves the existence of about 300 Å-"particles" in the direction normal to the chains. Its shape is elongated parallel to the meridian (Fig. 24) and indicates that the particles are smaller in the direction of the chain than orthogonal to it. In Fig. 24 one can see the second order of the meridional reflections, corresponding to

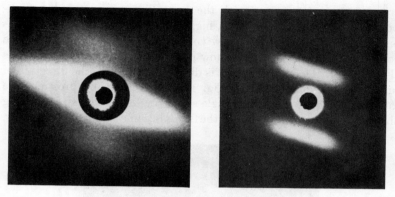

Fig. 21. Small-angle scattering of sheared polyethylene, left: cold-stretched, right: hot-stretched.

the "periodicity" in the vertical direction inside of the bundles (cf. Fig. 23d). Since four orders are observable it follows from 22 that the g_{kk}-value of this macrolattice is very small ($g_{kk} \sim 0.05$) compared with that in stretched bulk material ($g_{kk} \sim 0.2$).

The wide-angle reflections ($hh0$) of bundles of single crystals give results similar to Fig. 17 but only with one component indicating the

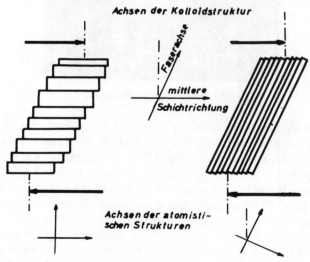

Fig. 22. Schematic representation of shearing in metals (left) and in linear high polymers (right).

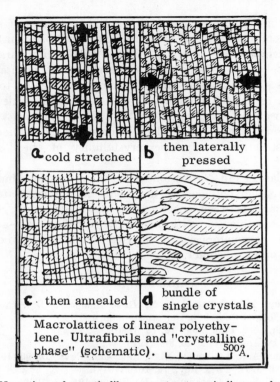

Fig. 23. Nematic- and smectic-like macrostructures in linear polyethylene.

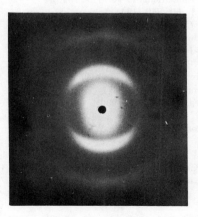

Fig. 24. Small-angle scattering of a parallel bundle of single crystals of polyethylene.

thickness of 300 Å mentioned above and a g-value of about 2%. Because here practically all chain molecules fold back on the surface of the single crystals ($\alpha = 1$), we have to do with paracrystalline mosaic blocks. In the bulk material a certain number of chain molecules passes through the amorphous region directly to the next crystalline region (cf. Fig. 14). Since $\alpha < 1$, the particles of 100 Å thickness we now better call ultrafibrils. Their g-value is remarkably higher than in the mosaic block, indicating the influence of the micellar structure on the paracrystallinity of the "crystalline" domains.

From Figs. 14 and 23 it becomes clear that in the solid bulk material of a linear high polymer we meet with all intermediate states of colloidal order between smectic and nematic structures. Small-angle analysis in combination with line profile studies of wide-angle reflections gives parameters which describe quantitatively the degree of statistical order—disorder.

BIBLIOGRAPHY

1. Hosemann, R. and Bagchi, S. N., *Direct Analysis of Diffraction by Matter*, North-Holland Publishing Co., Amsterdam, 1962.
2. Rayleigh, Lord, Phil. Mag., **10**, 73 (1880).
3. Debye, P., Ann. Physik, **46**, 809 (1915).
4. Eisenstein, A. and Gingrich, N. S., Rev. Mod. Phys., **15**, 90 (1943).
5. Vineyard, G. H., Phys. Rev., **74**, 1076 (1948).
6. Raman, C. V. and Krishnamurti, P., Nature, **124**, 53 (1929).
7. Kratky, O., Z. Elektrochem., **46**, 535 (1940).
8. Kratky, O., ibid., **50**, 249 (1944).
9. Kruyt, H. R., *Colloid Science I. Irreversible Systems. Hydrophobic Colloids*, Elsevier Publishing Co., Amsterdam, 1952.
10. Hosemann, R., Kolloid-Z., **117**, 13 (1950).
11. Joerchel, D., Z. Natforschg., **125a**, 123, 200 (1957).
12. Motzkus, F., Acta Cryst., **12**, 773 (1959).
13. Laue, M. v., Ann. Physik, **41**, 371 (1913).
14. Debye, P., Phys. Z., **31**, 797 (1930).
15. Schoknecht, G. and Kast, W., Forschgs. Ber. Nr. 173, Wirtschafts-u. Verkehrsministerium Nordrhein-Westfalen; Westdeutscher Verlag Köln-Opladen, 1956.
16. Bear, R. S. and Rugo, H. J., Ann. N.Y. Acad. Sci., **53**, 627 (1951).
17. Hess, K. and Kiessig, H., Z. physik. Chem., **193**, 16 (1944).
18. Hess, K. and Kiessig, H., ibid., **193**, 196 (1944).
19. Hosemann, R. and Wilke, W., Faserforschung und Textiltechn., **15**, 522 (1964).
20. Bonart, R. and Hosemann, R., Makromol. Chem., **34**, 105 (1960).
21. Bonart, R. and Hosemann, R., Kolloid-Z., **186**, 16 (1962).

Some Experimental Procedures and Recent Results on Liquids, Solutions and Globular Proteins

W. W. BEEMAN

Biophysics Laboratory, University of Wisconsin, Madison, Wisc. 53706

We discuss here some of the X-ray equipment and procedures in use in the Biophysics Laboratory at Wisconsin and a few recent results on pure liquids, solutions and globular proteins, in particular myoglobin. In another contribution to this volume Professor Anderegg discusses the work on certain viruses.

REMARKS ON EQUIPMENT

We continue to use as an X-ray source a simple and rugged rotating anode tube developed in our laboratory about fifteen years ago. These tubes were first described by Leonard (1). In succeeding years a number of engineering improvements have been made but no major changes in design. The anode is a copper cylinder about 10 cm in diameter and 3.5 cm high. It is attached to the top of a vertical hollow steel shaft through which cooling water is circulated to the anode. The shaft and anode are usually run at between 1200 and 1800 RPM. This speed gives good vacuum seal lifetime and permits a continuous focal line loading of 6000 watts without serious anode deterioration.

The electron beam is focussed on the outside vertical surface of the anode. The focal line is approximately 1 cm high, 1 mm wide, and, of course, vertical. As viewed along the axis of the slit system the focal line appears narrowed to 0.1 or 0.2 mm.

A high intensity source is almost a necessity for most of our work with biological macromolecules and we have found our present tubes to be convenient and reliable. However there are some disadvantages, compared to commercial, stationary-target, sealed-off tubes, which stem from a slight positional instability of the focal line. This instability has a high frequency component due to vibration of the anode and other parts of the X-ray tube and a slow drift which we believe reflects slight changes in the position of filament and focussing cup.

The focal line motions are of the order of 10^{-2} mm and are troublesome only when attempting precise measurements with the narrowest slits.

Next we describe briefly the symmetrical four-slit scattering geometry which has been used in nearly all our experiments. Figure 1 shows schematically a top view of the source and slit geometry. Two stationary slits collimate a beam from the focal line, which illuminates the scattering sample. A third and fourth slit view the sample and define the scattered beam reaching the counter. These slits and the counter are fixed to an arm which rotates about a vertical axis through the sample. A parallelogram-shaped region about the sample is simul-

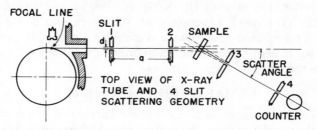

Fig. 1. Schematic view from above of the X-ray tube and four-slit scattering geometry.

taneously illuminated by the incident beam and seen by the counter. This crossover volume is the source of parasitic scattering from sample holder windows, air, and at sufficiently small angles, the edges of the second and third slits. Successive slits are the same distance apart and the sample is half-way between the second and third slits. The slits themselves are of tungsten or tantalum a little less than a millimeter thick.

Some of the advantages of the four slit geometry are the ease and accuracy with which the center of the scattering pattern, or zero scattering angle, may be determined by taking data on both sides of center. The symmetry of the scattered intensity is also an important check on alignment. Alignment is relatively straightforward. In addition the calibrated bilateral slits permit a rapid change in slit width without affecting alignment or centering. Thus the best compromise between intensity and angular resolution can be quickly achieved. Background scattering is very low. At scattering angles greater than $4d/a$, where d is the slit width and a the separation of successive slits, the second and third slits are no longer within the parallelogram referred to and therefore slit edge scattering is eliminated at least in principle. In practice background scattering from sources other than the sample and sample holder remains appreciable until $5d/a$ at which angle we usually start taking data.

We have in use two separate scattering assemblies both based upon the principles just described. In the first, successive slits are 10 cm apart and all four slits and the sample may be placed in an evacuated chamber or in helium. This device is useful when very weak scattering is to be measured and one must therefore eliminate the parasitic scattering from air near the sample. It is not convenient at the smallest angles although with 0.1 mm wide slits (which are used routinely) one can begin taking data at a scattering angle of a little less than one third of a degree. Using CuKα radiation this is equivalent to a Bragg spacing of 300 Å. The large-angle limit on scattering angle is about 105°.

The second assembly, and the one customarily used for our experiments on biological macromolecules, has successive slits 50 cm apart. With 0.1 mm slits we begin taking data at an equivalent Bragg spacing of 1500 Å. It is seldom necessary to use narrower slits although it is feasible to go to about 6000 Å, equivalent spacing. The large-angle limit of this second assembly is 15°. With X-ray paths of over a meter and a half, air absorption is a problem and the collimating slits and, separately, the analyzing slits are placed in long tubes which can be evacuated or filled with helium. More recently we have in addition placed a helium-filled jacket around the sample.

Perhaps the most informative data we can give are some counting rates under various conditions of operation of the two slit assemblies. We assume X-ray tube operation at 6000 watts although the 10 cm assembly is usually run at 4200 watts because of power supply limitations. The detector is a xenon-filled proportional counter with a counting efficiency of 70 or 80%; pulse height discrimination is used. A nickel filter (0.0012 cm thick) is used to reduce the Kβ intensity.

Under the conditions just described actual counting rates are 0.1 to 0.2 cts/sec with the X-ray tube operating but without a scattering sample and with the scattering chamber evacuated. More than half of this rate is cosmic rays or local radioactivity, the remainder is multiply scattered X-rays or electrical noise. While this irreducible background rate is relatively insensitive to slit settings other rates are not and we now refer to the 10 cm machine adjusted to begin taking data at an equivalent Bragg spacing of 75 or 80 Å. The slits are 6 mm high. With such an adjustment the scattering from the windows of the sample cell can easily be measured. Two crystalline quartz windows, each 0.0025 cm thick, scatter about one count per second and transmit two-thirds of the incident beam. Crystalline quartz is the best window material we have discovered. Good quality mica of comparable transmission scatters more than ten times as much. However at small angles, less than 5×10^{-3} radians, the quartz scattering rises rapidly and becomes greater than that of mica.

Now the rate just referred to is for quartz windows in vacuum. If the scattering chamber is filled with helium the total scattering, helium plus quartz, rises to 8 or 9 cts/sec at 1° scattering angle and 2 or 3 cts/sec at 5°. Air in the scattering chamber leads to much higher and, in general, unusably high counting rates.

Next, in Table I, we give a few rates for a system consisting of the helium-filled scattering chamber, and a quartz window sample holder approximately 1.0 mm thick filled with various scattering samples.

TABLE I
Observed counting rates from 1 mm-thick samples of various scatterers in a geometry which will begin taking data at an equivalent Bragg spacing of 75 or 80 Å. All samples measured at atmospheric pressure. These data are from the 10 cm machine.

Scatterer	Counting rate
Water	65 cts/sec
Ethyl alcohol	190 cts/sec
Xylene	150 cts/sec
Octofluorocyclobutane (C_4F_8)	160 cts/sec
Sulfur hexafluoride (SF_6)	90 cts/sec
Dichlorodifluoromethane (CCl_2F_2)	60 cts/sec

The 50 cm assembly is used when greater angular resolution is desired and of course in such use gives less intensity. A few representative counting rates follow. With the slits adjusted for a maximum equivalent spacing of 170 Å and with a quartz window sample holder 1 mm thick in a helium atmosphere one gets roughly 1.5 cts/sec from the quartz and helium. Water in the sample holder furnishes a total of 10 cts/sec and C_4F_8 about 20 cts/sec. A 0.3% by weight solution of myoglobin in water also gives about 20 cts/sec (in the radius of gyration region). These data are taken with slits 7.5 mm high.

With a given scattering geometry and sample and the only variable the width d of each slit the scattered intensity varies as the square of the slit width, providing that the slits are always wider than the projected width of the focal line (0.1 or 0.2 mm). Put another way the scattered intensity varies as the inverse square of the maximum equivalent Bragg spacing which can be measured. If the slits are narrower than the projected focal line the scattered intensity varies as the cube of d.

However the larger molecules, whose measurement demands the narrower slits, are the better scatterers on a weight concentration basis. We find experimentally in our work with solutions of macromolecules that counting rates in the radius of gyration region are, for a fixed weight concentration of solute, relatively independent of molecular weight. The minimum weight concentration that can be used is about 0.1%. Obviously these generalizations depend also upon the shape and average electron density of the molecule. The results given apply to the globular proteins and nucleoproteins.

ABSOLUTE INTENSITY MEASUREMENTS

A number of different methods of calibration of incident beam intensity are in use (2, 3, 4). The one which seems most natural with our scattering geometry and which permits a high degree of accuracy is the intercomparison in identical geometries of the scattering from the unknown and from a gas of accurately known composition and equation of state. The gas, ideally, should contain only atoms of low atomic number, to avoid anomalous dispersion problems, and should exhibit close to perfect gas behavior at one atmosphere and normal room temperatures. It is well known that such gases are very common. However if scattered intensities are to be conveniently high, rather more than fifteen or so electrons per molecule must be available.

Our first efforts are due to Katz (5). He was able to use air and sulfur hexafluoride as standards but only in samples about 1.0 cm thick. Our regular liquid or solution sample holder is 0.1 cm thick and there are troublesome geometric corrections in comparing thick and thin sample scattering.

More recently Shaffer (6), using a more intense incident X-ray beam and aided, in particular, by the very low scattering from crystalline quartz sample holder windows has been able to work with gas scattering samples only 0.1 cm thick. The three most useful standards appear to be C_4F_8, SF_6 and $CClF_2$. Counting rates in one geometry are shown in Table I and, as mentioned earlier, even a geometry permitting radius of gyration measurements on myoglobin ($R_g \simeq 16.0$ Å) will give 20 cts/sec from 0.1 cm of C_4F_8. The two fluorocarbons are commercially available at purities of about 99.9%. The equations of state are well known. The deviations from perfect gas behavior are small, not more than two percent in density and compressibility.

Details of the calibration procedure are available in the thesis of Dr. Shaffer (6). Here we will only remind the reader that in the experiments we discuss, an observed counting rate depends on the intensity of the incident beam, on the mass per unit area of the sample and on the structure of the sample (molecular form factors, pair distribution functions etc.). Everything else is geometry, in which term, however, we include window transmission, counter efficiency and various parasitic scatterings. The comparison in identical geometries of the scattering rates from a sample of known structure and from the unknown eliminates almost all of these variables. Masses per unit area must still be determined and account must be taken of differences in transmission of standard and unknown. It appears that most of the individual measurements can be made with an accuracy of one per cent or better and that the actual determination of an absolute scattering cross-section should be good to two per cent. To be more precise it is the ratio of the scattering cross-section of the unknown to that of a gas used as a primary standard which is determined. The cross-section of the gas is assumed to be calculable. An extended series of intercomparisons of the three gases mentioned indicated that $\pm 2.0\%$ was achieved. There was no indication that any of the scattering ratios differed from theory by more than this amount.

However, other comparisons were not as good and we are thus led to the subject of slit height corrections. In our experiments scattered intensities are plotted at a number of angles and the intensity extrapolated to zero angle on a log intensity versus angle squared plot. Such plots are excellent straight lines. However, the extrapolated zero-angle intensities from two different samples (perhaps a standard and an unknown) are not in the same ratio as their actual forward scatterings per unit solid angle unless the two scattering curves have the same shape over the angular region which the slit heights integrate.

The slit height correction problem can be put in best perspective if we use a Gaussian representation of the scattering functions and the instrumental weighting function. We have

$$I_1(h) = \int_0^\infty W(z) i_1(\sqrt{h^2 + z^2})\, dz \qquad 1.$$

$I_1(h)$ is the observed scattered intensity from sample 1 as a function of $h = 4\pi\lambda^{-1} \sin\theta$, $W(z)$ is the instrumental weighting function and $i_1(h)$ is the desired scattering function in pinhole geometry. Let us assume that $i_1(h) = A_1 e^{-k_1^2 h^2}$ and $W(z) = e^{-\omega^2 z^2}$. The value of k_1 can be gotten from the observed curve, $I_1(h)$, since in the Gaussian approximation the pinhole and slit-smeared curves have the same shape. A short calculation leads to the following result

$$\frac{A_1}{A_2} = \frac{I_1(0)}{I_2(0)} \cdot \left[\frac{\omega^2 + k_1^2}{\omega^2 + k_2^2}\right]^{\frac{1}{2}} \qquad 2.$$

where the subscripts refer to two different samples, one of them, in general, a gas used as a primary standard. A_1/A_2 is the desired ratio of forward scattered intensities in pinhole geometry. $I_1(0)/I_2(0)$ is the observed ratio of slit-smeared forward intensities and ω, k_1 and k_2 are known. If $\omega \gg k_1$ and k_2, the correction factor is close to unity and insensitive to ω. This is the case in intercomparing gases, liquids and solutions of small molecules where the angular dependence is slight. If $k_1 \gg \omega$ and $k_2 \gg \omega$ again the correction factor is insensitive to ω although not necessarily close to unity. This case is important, for instance, in determining relative molecular weights of large molecules. The most unfavorable situation is $k_1 > \omega > k_2$ which arises if one wishes to make a molecular weight determination on a large molecule (k_1) by direct comparison with the scattering from a gas (k_2). For such a comparison the shape of the weighting function must be accurately known.

The discussion just given also reminds us that slit height corrections must be done much more carefully if absolute intensities are involved than when just the shape of the scattering curve is important.

To summarize we find that slit height corrections appear to be the biggest remaining problem in our method of absolute intensity determination. When all the samples have scattering curves much broader than the window passed by the weighting function accuracies of $\pm 2.0\%$ are easy but it is difficult to compare the scattering of a large molecule directly with a gas standard. Shaffer (6) has calculated improved weighting functions for both the 10 cm and 50 cm geometries and Sztankay (7) has experimented with shorter slits and approximations using several Gaussians to the scattering functions.

We are able to determine molecular weights in the range of 10^4 to 10^5 to perhaps 5% and should be able to improve this.

Before leaving the subject of absolute intensity determinations we will mention two applications to low molecular weight systems. One confirms our belief in the reliability of our procedures. The other does not. Of course we consider the latter puzzling rather than disheartening. The successful application is shown in Fig. 2. The observed forward scattering from solutions of sucrose in water is plotted against the weight concentration. The solid line represents a calculation from thermodynamic data of the theoretical forward scattering. Quantities involved in the calculation are the number of electrons per sucrose

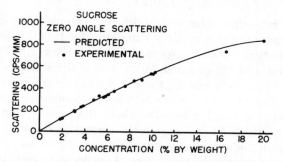

Fig. 2. The forward X-ray scattering of sucrose solutions as measured (dots) and as calculated (solid line) from thermodynamic data. There are no adjustable parameters in this comparison.

molecule, the partial specific volume and the osmotic pressure. An excellent treatment is given by Kirkwood and Buff (8), who express the thermodynamic quantities in terms of pair distribution functions which can then easily be related to the forward scattering cross-section per molecule. As just stated what is actually calculated is a forward scattering cross-section per molecule, or its equivalent. This is translated to a counting rate in a given geometry by calibrating that geometry with a gas as has been described.

We would emphasize that there are no adjustable parameters in the comparison shown in Fig. 2. The observed and predicted scattering rates are in excellent agreement up to weight concentrations of 20% at which concentration deviations from ideal solution behavior are already quite obvious.

The experiment where the expected result fails to materialize is the measurement of the forward scattering of pure water. Careful measurements were made by Shaffer (6) between one and six degrees scattering angle in which range the angular dependence is very slight. Slit height corrections on the extrapolated zero-angle scattering should be reliable. Background and parasitic scattering rates were low.

The theoretical forward scattering of a pure liquid is given by the isothermal compressibility times kT. Our experimental result is about ten per cent higher than the theoretical prediction. At the moment we are unable to offer an explanation.

Our result confirms the work of Weinberg (4) who measured a forward scattering 9% above theory. Weinberg and Shaffer used quite similar four-slit scattering geometries but different methods of intensity calibration. More work needs to be done if one is to be quite certain that systematic errors have been avoided.

It is abundantly clear that absolute intensity measurements can be very important in many applications of small-angle X-ray scattering. Such experiments would be facilitated if the several laboratories active in the field could agree on primary intensity standards or perhaps each calibrate and intercompare some convenient secondary standard.

THE SCATTERING FROM DILUTE SOLUTIONS OF SPERM WHALE METMYOGLOBIN

Since the myoglobin (9) and hemoglobin molecules are now known in detail, as they occur in the wet crystal, it is of interest to compare the molecular configuration in the crystal with that observed in dilute solution using small-angle X-ray scattering. At the moment we have reliable data only on myoglobin.

Sperm whale metmyoglobin was obtained from two commercial sources (Mann and Serevac) and also from Prof. Frank Gurd of the University of Indiana Medical School. The samples after dialysis or gel filtration were run in a phosphate buffer at a pH near 7.3. In some cases sodium chloride was also present. The weight concentration of total salt varied from 1.0% to 3.0% in different runs. The weight concentration of myoglobin varied from 0.3% to 20%. The lowest myoglobin concentrations used were necessary to avoid interparticle interference effects in the radius of gyration region from $h = 0.02$ to 0.05 Å^{-1}. Figure 3 shows a plot of radius of gyration against myoglobin concentration and the customary extrapolation to zero concentration. There is also a rough determination of the radius of gyration of a myoglobin dimer. Of this more will be said in a moment.

At larger angles the scattering is less sensitive to interparticle interference effects and concentrations up to a maximum of 20% by weight were used in order to gain intensity. The extended experimental scattering curves shown in Figs. 4 and 5 are composites run at different myoglobin concentrations in different angular ranges but with large

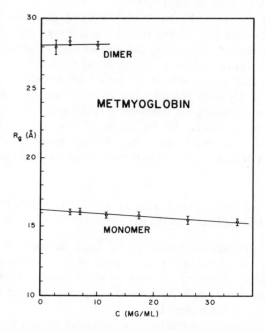

Fig. 3. The radii of gyration of metmyoglobin monomer and dimer as a function of concentration of the protein. The values used in the discussion are those extrapolated to zero concentration.

angular regions of overlap. The curves have been corrected for slit height smearing. The shapes are not a function of salt concentration in the range mentioned nor of pH variations between 7.0 and 7.4. We believe the data accurately represent the spherically averaged scattering of a solvated myoglobin molecule.

In the course of our measurements a difficulty became obvious which limits the usefulness of radius of gyration (R_g) measurements unless careful precautions are taken. Many of our earlier determinations gave an R_g of 18 Å and more. Internal consistency indicated the accuracy was 0.2 or 0.3 Å. However it was clear that R_g for the crystallographic molecule was probably less than 16.0 Å. We decided it would be necessary to have a better check on the homogeneity of our preparations than can be got from the sedimentation pattern in the analytic ultracentrifuge. Gel filtration using Sephadex G-50 and G-75 was tried and well-resolved subsidiary peaks were found representing a few per cent dimer and higher aggregates. As is well known

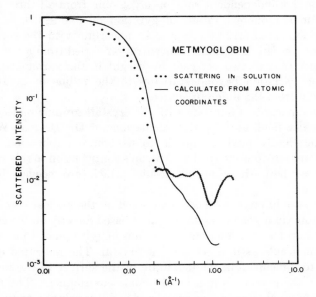

Fig. 4. The experimental and calculated scattering curves of metmyoglobin. The two curves have been set equal at zero angle.

this gives an average radius of gyration weighted in favor of the larger particles. In addition, if the amount of dimer is only a few per cent, one may not be able to detect any curvature of the Guinier plot.

In our later measurements a Sephadex column was used as a preparative as well as an analytic tool. The column fraction corresponding to the monomer was separated, reconcentrated, and used in the X-ray experiments. Enough dimer was also separated to furnish the radius of gyration determination shown in Fig. 3. Both the separated monomer and dimer when stored in solution approach an equilibrium mixture. We did not investigate the dependence of this equilibrium on the various relevant parameters except to note that under most of the solution conditions we used, the equilibrium dimer concentration was more than 5% by weight. Such concentrations of dimer easily explain our earlier high values of R_g. Fortunately the approach to equilibrium is quite slow (a day or two) so that runs could be made before the dimer concentration had built up too greatly. Dimer concentrations were measured before and after each run and corrections made to the observed R_g using the rough dimer R_g of Fig. 3. Based on five independent runs on myoglobin from all three sources we obtain a monomer radius of gyration of 16.0 ± 0.4 Å. About half the indicated error represents an uncertainty in the correction for dimer content. The dimer concentration varied from a few tenths of one per cent to two per cent by weight in the various runs. We found no variation beyond statistics in the radius of gyration of myoglobin from the different sources.

The comparison of our data with the crystallographic molecule has been greatly facilitated by the calculations of Dr. Herman Watson, which he kindly made available to us before publication. These calculations are described in Dr. Watson's contribution to this volume. We repeat here those of his results which are needed for our comparison.

Using two different methods, one based on the nuclear coordinates of the atoms in myoglobin and the other based directly on the electron density map Dr. Watson obtains a radius of gyration of 15.5 ± 0.5 Å for the molecule as it is seen in the crystal. The indicated error is related to difficulties in the definition of the surface of the molecule which, in the crystal, is in a highly saline environment. The radii of gyration of myoglobin in the crystal and in dilute solution agree within 3% which is well within the uncertainties of the two determinations. It should be noted that the radius of gyration calculated by Dr. Watson is for the molecule in vacuum, that measured is of a molecule in water.

More interesting than just the radius of gyration is the comparison of the extended scattering curves out to values of h where shape and internal periodicities become important. Dr. Watson has evaluated the Debye formula using again the known nuclear coordinates for myoglobin and has made available the spherically averaged scattering to $h = 1.2$ Å^{-1}. This is plotted as the solid line in Fig. 4 and also (out to $h = 0.3$ Å^{-1}) as the triangular data points of Fig. 5. The extended curve also is calculated for the molecule in vacuum.

Let us consider in more detail the comparison of Fig. 4. Out to values of h of about 0.2 or 0.3 Å^{-1} the experimental and calculated scattering curves have very similar shapes but are displaced relative to one another along the (logarithmic) h axis. Beyond $h = 0.3$ Å^{-1} (an equivalent Bragg spacing of about 20 Å) the calculated curve falls rapidly below the experimental. This latter behavior is, of course, expected since the calculated curve is for the molecule in vacuum and the experimental curve is for the molecule in water. Only at

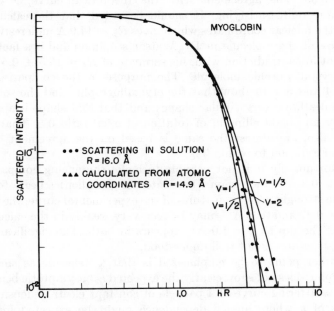

Fig. 5. The experimental and calculated scattering curves plotted against hR. The intensities have been made equal at zero angle. The solid curves are the theoretical scattering curves for ellipsoids of revolution of constant electron density and of axial ratio v as indicated.

values of h so small that internal fluctuations of electron density do not contribute to the scattering does one expect the shape of the curve to be independent of the surrounding medium.

The near identity of the curves at the smaller values of h, except for a translation on the logarithmic axis, implies that the crystallographic and solvated molecule have the same shape but different sizes. In fact the experimental curve gives $R_g = 16.0$ Å and the calculated curve $R_g = 14.9$ Å. In Fig. 5 we plot the same curves (for $h < 0.3$ Å^{-1}) against hR_g using the above values of R_g. The two curves are in excellent agreement and each agrees with the theoretical curve for an oblate ellipsoid of revolution of axial ratio $\frac{1}{2}$ and constant electron density. We include for comparison the curves for a sphere, $v = 1$, for an oblate ellipsoid, $v = \frac{1}{3}$, and a prolate ellipsoid, $v = 2$. The overall shape of the crystallographic molecule is best described (10) as an oblate ellipsoid of revolution of axial ratio 0.55. The agreement with the theoretical curve, $v = \frac{1}{2}$, is therefore quite satisfying. It should be noted that the calculation from the nuclear coordinates which gives $R_g = 14.9$ Å underestimates the size of the molecule in Dr. Watson's opinion and is a factor in, but not in contradiction with, his estimate of $R_g = 15.5 \pm 0.5$ Å for the crystallographic molecule. The purpose of the comparisons of Figs. 4 and 5 is to show that the crystallographic and the solvated molecules have very similar shapes and that that shape is approximately an oblate ellipsoid of rotation of axial ratio 0.5. That their sizes, also, are almost the same is based on the agreement of the radii of gyration to within 3%.

Unfortunately it is not possible to make convincing comparisons for $h > 0.3$ Å^{-1}, the region where internal periodicities begin to contribute strongly. Certain features of the experimental curve, the peaks at $h = 0.47$ and 0.68 Å^{-1}, may be seen very weakly in the calculated curve. The dip at $h = 1.0$ Å^{-1} appears in both. The significance of these structures is not well understood.

Another point to be emphasized is that a molecule as small as myoglobin does not show clearly the expected h^{-4} asymptotic behavior of the scattering curve of a particle of constant electron density. At values of h where an h^{-4} dependence might be expected internal periodicities are already dominating the scattering. We find in our experimental curve a short region between $h = 0.1$ and 0.2 Å^{-1} where the dependence changes from about h^{-3} to h^{-5}. From $h = 0.6$ Å^{-1} to 4.2 Å^{-1} (where we have some data not shown in Fig. 4) the experimental scattering curve oscillates around that calculated on an independent atom basis. Obviously in this angular range we are learning very little of the overall size and shape of the molecule.

We may say, in conclusion, that to the extent that the spherically averaged scattering defines the molecule and to an experimental accuracy of 3 or 4% we see the same myoglobin molecule in dilute solution that is seen in the wet crystal. In some respects the agreement is a little better than we can understand. The outside of the molecule essentially consists of amino acid side chains whose electron density is about the same as that of water. It can be argued that these would not be seen in the experimental work on the solvated molecules but would be seen in the calculations from the crystallographic molecule. Thus the solvated molecule should appear perhaps 1 Å smaller (in R_g), rather than 0.5 Å bigger, than the crystallographic molecule. Additional measurements, particularly in solutions of much lower salt concentration, may help clear this up.

At present our work on hemoglobin is quite preliminary and entirely dependent on getting monodisperse preparations. We measure radii of gyration between 28 and 30 Å. For the crystallographic molecule $R_g \simeq 22$ Å. The difficulty is almost certainly one of aggregation and is at present unsolved in our laboratory.

The work reported on myoglobin is from the thesis of Dr. Zoltan Sztankay (7) where additional detail may be found. All of the work reported was supported by research and training grants of the U.S. Public Health Service.

BIBLIOGRAPHY

1. Leonard, B. R., Ph.D. Thesis, Univ. of Wisc. (1952).
2. Kratky, O., Z. analyt. Chem., **201,** 161 (1964).
3. Luzzati, V., Acta Cryst., **13,** 939 (1960).
4. Weinberg, D. L., Rev. Sci. Instr., **34,** 691 (1963).
5. Katz, L., Ph.D. Thesis, Univ. of Wisc. (1958).
6. Shaffer, L., Ph.D. Thesis, Univ. of Wisc. (1963).
7. Sztankay, Z., Ph.D. Thesis, Univ. of Wisc. (1965).
8. Kirkwood, J. G. and Buff, F. P., J. Chem. Phys., **19,** 774 (1951).
9. Kendrew, J. C., *et al.*, Nature, **185,** 422 (1960).
10. Watson, H., (personal communication).

A Low-Angle X-Ray Diffraction Study of γ-Globulin

W. R. KRIGBAUM AND R. T. BRIERRE, JR.*

Department of Chemistry, Duke University, Durham, N.C.

INTRODUCTION

γ-Globulins form a group of blood proteins which exhibit some variability in peptide sequence, and a considerable range of antibody specificities. Despite these differences, present evidence suggests that γ-globulins from various sources all have the same gross molecular structure and shape.

Molecular weights reported for γ-globulin fall in the range $150,000 \pm 10,000$, and each molecule is composed of four peptide chains (1–11). Two of these (designated A chains) are of molecular weight $55,000 \pm 5,000$, while the remaining pair of smaller chains (designated B) each have a molecular weight of 20,000–25,000. Some workers refer to these as H and L chains, respectively.

Porter (12) showed that γ-globulin could be split in a unique manner into three fragments of approximately equal molecular weight by the action of a proteolytic enzyme, papain, followed by the reduction of a disulfide bond. Two of these, Fragments I and II, are composed of one-half an A (or H) chain and one B (or L) chain, while Fragment III consists of two identical halves of the A (or H) chain. It is now believed that fragments I and II are identical (13). Nelson (14) recently demonstrated that cleavage of the two A chains proceeds in a stepwise manner, with one Fragment I being the first product. These results lead to the proposed chain structure for γ-globulin shown schematically in Fig. 1.

Noelken, Nelson, Buckley, and Tanford (15) have performed viscosity and sedimentation measurements upon the intact molecule and the two types of fragments. These data indicate that Fragments I and III are globular and are not highly hydrated, while the γ-globulin molecule is either of a more irregular shape, or is highly hydrated.

* Present address: Chemstrand Research Corporation, Research Triangle Park, Durham, N.C.

Optical rotatory dispersion measurements (15–17) show that neither the intact molecule nor the fragments have an appreciable α-helix content.

A more detailed knowledge of the molecular structure and conformation of γ-globulin would obviously be of assistance in understanding the mechanism of the binding to antigens. The electron microscope studies (18, 19) suggest a strand-shaped molecule. Almeida, Cinader, and Howatson (19) estimated the length of the

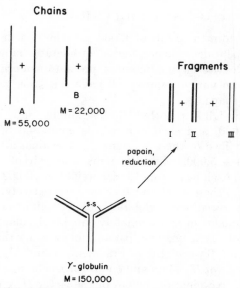

Fig. 1. Proposed chain structure for γ-globulin.

molecule by a study of the distance between antibody-linked virus particles. This averaged 150 Å, with a maximum in the vicinity of 270–290 Å. They indicated the maximum width of the strand to be 35 ± 5 Å. Some of their plates suggest a beaded appearance of the molecule. Also, γ-globulin must have some flexibility, since both studies cited revealed cases in which the antibody was bent into a loop, with both sites attached to the same virus particle. Further evidence for flexibility is provided by the low value calculated for the rotational relaxation time from fluorescence depolarization (20, 21).

A considerable body of knowledge concerning the molecular conformation of γ-globulin has come from the extensive low-angle X-ray studies of Kratky and coworkers (22–24). The scattering curve is rod-like in form, and yields a radius of gyration of 72 Å. The radius of gyration of the cross-section is approximately 23 Å. If one assumes a rigid rod, these values correspond to a rod length of 240 Å, which stands in fair agreement with the electron microscope results. The cross-section appears to be rather complicated. If one assumes a uniform cross-section along the length, it turns out to be elliptical with a radius ratio of 2.5:1. According to their most recent work (24), the cross-section scattering curve, $\log [hI(h)]$ vs. h^2, shows an inner region of higher slope. This behavior might arise from a mixture of particles differing in size, or from a single type of particle having either an asymmetric shape, or two rather distinct regions of electron density. The first possibility could be eliminated since the dimer, which was the only known impurity, was not present in sufficient amount to account for the observed behavior. No satisfactory asymmetric model was found, so the authors adopted the third possibility and treated a rigid cylindrical model composed of a concentric sheath and core of equal length. By factoring the amplitudes of the cross-section curve they deduced cross-sectional radii of gyration of 13.6 and 38.5 Å for the core and sheath, respectively, and concluded that the core comprised 76% of the molecule. Thus, their final model is a rigid cylinder of length 240 Å having an elliptical cross-section with major axes 50 Å and 22 Å and a non-uniform electron density when viewed in cross-section.

Holasek, Kratky, Mittelbach and Wawra (25) have also reported measurements of the Bence-Jones protein. This is composed of B (or L) chains, and since the molecular weight was found to be 43,000 the particle under investigation must have been a dimer. They found the radius of gyration to be 26.3 Å, and stated that the cross-section curve could be explained by an ellipsoid having major axes 48.3 and 21 Å. Although γ-globulin does not contain a $B–B$ pair, it is of interest that the latter dimensions are rather close to those deduced for γ-globulin.

Edelman and Gally (26) have proposed one hypothetical structure for γ-globulin, shown in Fig. 2a, in which the two A and two B chains are combined to form a rigid cylinder having the dimensions deduced by Kratky and coworkers (22). On the other hand, Noelken, Nelson, Buckley and Tanford (15) favor the model depicted in Fig. 2b, in which three compact entities (corresponding to Fragments

I–III) are joined flexibly by two regions of the A chains. A hybrid combining some of the features of both of these models is shown in Fig. 2c. A decision between models a and b could be made if the dimensions of the individual fragments were known. One must, of course, assume that papain cleavage does not appreciably alter the conformation of the fragments. This appears to be the case, since Nisonoff, Wissler, and Woernley (27) have demonstrated that the average hapten binding constant of Fragment I is the same as that for one

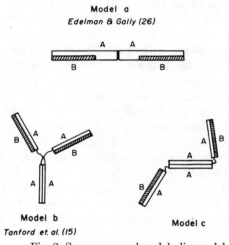

Fig. 2. Some proposed γ-globulin models.

site of the parent molecule, and Kern, Helmreich, and Eisen (28) have shown that Fragment III has about the same effectiveness as the parent molecule in causing the elution of antibody γ-globulin from microsomes. These observations indicate that the sites have the same biological activity, whether on the isolated fragments or on the intact molecule, which is fairly good evidence that the conformations of the three portions remain unaltered upon splitting. We have undertaken a study of γ-globulin and Fragment I in an attempt to resolve the conformation problem.

EXPERIMENTAL

Measurements were performed on a relative scale using a Rigaku-Denki diffractometer and a Philips small-focus copper target tube.

A nickel β filter was used in the incident beam, and 4000 second counts were taken with a Geiger tube at each angle. Rabbit γ-globulin having $M = 156,000$ was used as obtained from Pentex, Inc., Kankakee, Ill. We are indebted to Dr. C. Tanford for providing us with a sample of Fragment I obtained by papain hydrolysis. For this material $M = 50,000$ and $\bar{v} = 0.74$. An aqueous buffer at pH 4.7 was prepared.

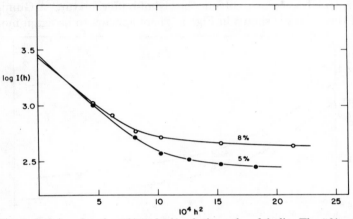

Fig. 3. Guinier plots for 5% and 8% solutions of γ-globulin. The 5%-solution is flowing through a capillary.

This was 0.1 M in both acetic acid and sodium acetate, and contained 0.05 M NaCl as a supporting electrolyte. Tanford, Buckley, De Paretosh and Lively (29) have reported that the conformation of γ-globulin is unchanged in solutions containing up to 80% ethylene glycol. In order to increase the electron density difference, we have used as a solvent for our measurements 60% ethylene glycol and 40% of the aqueous buffer just described.

RESULTS

Our first measurements were performed upon a 5% solution of γ-globulin flowing through a thin capillary. We had hoped to be able to orient the molecules in this manner. For comparison, an 8% solution was studied in our normal scattering cell. From the Guinier plot shown in Fig. 3 one sees that the two curves are similar, and that both have the form expected for a rod-like scattering particle. The radii of gyration obtained from the initial slopes were 73.9 and 72.5 Å for the 5% and 8% solution, respectively. It is evident that our flow gradient was not sufficient to orient the molecules. Since these values stand in rather good agreement with 72 Å as reported by Kratky and coworkers (24), this confirms the conclusion of Tanford *et al.* (29) concerning the absence of a conformation change of γ-globulin in the presence of ethylene glycol.

Measurements were therefore performed upon 3 and 6% solutions of Fragment I using as a solvent the glycol-buffer mixture. A Guinier plot of these data is shown in Fig. 4. There appears to be again more

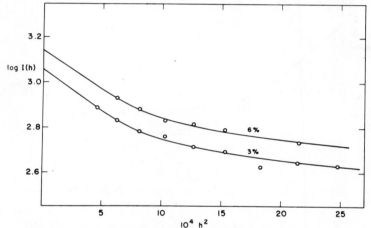

Fig. 4. Guinier plots for 3% and 6% solutions of Fragment I in glycol-buffer mixture.

curvature than one would expect for spherical particles. The radius of gyration deduced for the 3 and 6% solutions was 35.8 and 35.4 Å, respectively.

It would have been desirable to obtain similar data for Fragment III; however, its limited solubility and lack of stability prevented this. In the discussion to follow we will assume its radius of gyration is the same as that measured for Fragment I, which is probably an underestimation.

According to the rigid cylindrical model of Edelman and Gally (26) shown in Fig. 2a, the radius of gyration of intact γ-globulin should be approximately three times that for Fragment I, whereas for the flexible model proposed by Tanford *et al.* (15), and shown in Fig. 2b, this ratio should be nearly two. The ratio of the experimental radii is very close to two, which would appear to eliminate the rigid rod from consideration. On the other hand, the three fragments need not be flexibly connected in the star shape shown in Fig. 2b. The flexibly-jointed chain of three segments shown in Fig. 2c has a radius of gyration ratio 2.2, which would be in reasonable accord with our observations.

We wish to thank Dr. Charles Tanford for providing us with a sample of Fragment I, and for his helpful discussions and encouragement. We also wish to express our appreciation to the National Institutes of Health for support of this work on Grant GM-09767.

BIBLIOGRAPHY

1. Edelman, G. M. and Poulik, M. D., J. Exptl. Med., **113,** 861 (1961).
2. Edelman, G. M. and Gally, J. A., ibid., **116,** 207 (1962).
3. Edelman, G. M. and Benacerraf, B., Proc. Nat. Acad. Sci., **48,** 1035 (1962).
4. Porter, R. R., *Basic Problems in Neoplastic Disease*, Columbia University Press, New York, 1962; p. 177.
5. Fleishman, J. B., Pain, R. H. and Porter, R. R., Arch. Biochem. Biophys., suppl. **1,** 174 (1962).
6. Small, P. A., Jr., Kehn, J. E. and Lamm, M. E., Science, **142,** 393 (1963).
7. Pain, R. H., Biochem. J., **88,** 234 (1963).
8. Edelman, G. M., *Abstracts*, 2A, 143rd Meeting of the American Chemical Society, Cincinnati, Ohio, Jan., 1963.
9. Edelman, G. M., Olms, D. E., Gally, J. A. and Zinder, N. D., Proc. Nat. Acad. Sci. **50,** 753 (1963).
10. Marler, E., Nelson, C. A. and Tanford, C., Biochemistry, **3,** 279 (1964).
11. Roholt, O., Onoue, K. and Pressman, D., Proc. Nat. Acad. Sci., **51,** 173 (1964).
12. Porter, R. R., Biochem. J., **73,** 119 (1959).
13. Palmer, J. L., Mandy, W. J. and Nisonoff, A., Proc. Nat. Acad. Sci., **48,** 49 (1962).
14. Nelson, C. A., J. Biol. Chem., **239,** 3727 (1964).
15. Noelken, M. E., Nelson, C. A., Buckley, C. E. and Tanford, C., ibid., **240,** 218 (1965).
16. Jirgensons, B., Arch. Biochem. Biophys., **74,** 57 (1958).
17. Winkler, M. and Doty, P., Biochim. Biophys. Acta, **54,** 448 (1961).
18. Lafferty, K. J. and Oertelis, S. J., Nature, **192,** 764 (1961).
19. Almeida, J., Cinader, B. and Howatson, A., J. Exptl. Med., **118,** 327 (1963).
20. Chowdhury, F. H. and Johnson, P., Biochim. Biophys. Acta, **53,** 482 (1961).
21. Steiner, R. F. and Edelhoch, H., J. Am. Chem. Soc., **84,** 2139 (1962).
22. Kratky, O., Porod, G., Sekora, A. and Paletta, B., J. Polymer Sci., **16,** 163 (1955).
23. Kratky, O. and Paletta, B., Angew. Chem., **67,** 602 (1955).
24. Kratky, O., Pilz, I., Schmitz, P. J. and Oberdorfer, K., Z. Naturforschg., **18b,** 180 (1963).
25. Holasek, A., Kratky, O., Mittelbach, P. and Wawra, H., J. Mol. Biol., **7,** 321 (1963).
26. Edelman, G. M. and Gally, J. A., Proc. Nat. Acad. Sci., **51,** 846 (1964).

27. Nisonoff, A., Wissler, F. C. and Woernley, D. L., Arch. Biochem. Biophys., **88,** 241 (1960).
28. Kern, M., Helmreich, E. and Eisen, H. N., Proc. Nat. Acad. Sci., **47,** 767 (1961).
29. Tanford, C., Buckley, C. E., De Paretosh, K. and Lively, E. P., J. Biol. Chem., **237,** 1168 (1962).

Structure of Nucleoproteins Studied by Means of Small-Angle Scattering*

ANATOLE NICOLAÏEFF
Centre de Recherches sur les Macromolécules, Strasbourg, France

INTRODUCTION

Desoxyribonucleic acid (DNA) is present in all living cells and is important as the carrier of genetic information (1). In all higher cells, it is associated with basic proteins: in the spermatozoa of some fishes these are protamines, but in most nuclei they are histones, which are believed to play a role in the regulation of gene action (2, 3, 4). It would therefore be of interest to obtain some information about the structure of DNA + histone (DNH) and DNA + protamine (DNP) complexes under conditions similar to those existing in the living cells. The X-ray scattering techniques developed in our laboratory by Luzzati and our group (5, 6, 7) permit the study of structure in the 150 to 20 Å region with control of the amount of water and salt, and of pH. These techniques also make possible some direct observations on nuclei.

In Fig. 1 are represented the types of X-ray diagrams observed at different concentrations of DNH in water. In two ranges of concentration we obtain diagrams which are easy to interpret in terms of DNH organization: the range from 1 to 20% represents a micellar solution and the range from 48 to 68% represents a two-dimensional hexagonal packing.† The structure at intermediate concentrations, such as exist in nuclei, is more complicated than a simple combination of solution and hexagonal packing. X-ray diagrams for concentrations higher than 70% have been obtained, but not studied in detail.

---

* Part of this work was done in collaboration with Dr. Stoeckenius at the Rockefeller Institute and with Dr. V. Luzzati at the Centre de Recherches sur les Macromolécules.

† Concentrations are expressed as the ratio of the weight of DNH to the weight of solvent + DNH.

THE ISOTROPIC PHASE

We usually used DNH extracted in water without dissociation. With salt-extracted samples (i.e., after dissociation of the protein from the DNA in 1 M NaCl) we obtained homogeneous solutions only by using a mixture (prepared in 1 M NaCl) of DNH + DNP or DNH + DNA. However, even with a change either in the ratio of DNA to protein or in the nature of the protein associated with DNA,

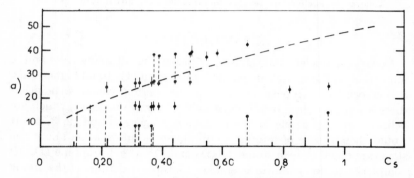

Fig. 1. DNH gels. Chicken erythrocyte in 0.15 M NaCl. The spacing of the lines and bands are plotted as ordinate, concentration as abscissa.

$$C_s = \frac{\text{weight of substance}}{\text{weight of gel}}$$

A dot defines the position of a sharp reflection. A broken line indicates the presence of scattering. A dot and a bar show the position of the center of a band and its half-width. The broken line corresponds to the change of spacing as a function of the concentration in the case of a two-dimensional hexagonal packing of DNA molecules.

the results were similar to those obtained with DNH extracted without dissociation.

The DNH was dissolved in water at concentrations of 1 to 32%. For the lower concentrations the scattered intensity was not high enough to permit precise measurements. For concentrations higher than 6%, particle interactions are too important to allow a simple interpretation of the scattering curves.

We have used small-angle X-ray scattering on an absolute intensity scale. This method, developed in our laboratory (5, 8), involves the measurement of both the scattered intensity and the intensity of the direct beam. Each experiment allows the determination of a normalized function $j_n(s)$ of the scattered intensity,

$$\left(s = \frac{2 \sin \theta}{\lambda} = \frac{h}{2\pi}, \quad \lambda = \text{wavelength}, \quad 2\theta = \text{scattering angle}\right)$$

a function dependent only on the structure within the sample (8). In the case of rod-like particles, Luzzati (5) has proposed a graphical method to determine two parameters: R_c, the radius of gyration, around the axis of the rod, of the electron density distribution, that is the difference in electron densities of the rod and of the solvent; and $A = c_e\mu(1 - \rho_0\psi)^2$ where c_e is the electron concentration defined as the ratio of the number of electrons of the solute to that of the solvent + solute, μ is the electronic mass per unit length in electrons per Å, ρ_0 is the electronic density of the solvent (electrons per Å^3), and ψ is the electronic partial specific volume of the solute in the solvent (Å^3 per electron).

In this method the experimental plot of log j_n against log s is brought into coincidence with the theoretical curve of a rod (5), drawn to the same scale, by a translation of axes. The shifts of abscissa and ordinate give respectively the values of R_c and $A = c_e\mu(1 - \rho_0\psi)^2$. Since c_e, ρ_0 and ψ can be determined independently, the mass per unit length can be calculated.

The main source of error in this type of experiment is not in the determination of j_n, because the scattering is concentrated around the center, and, as we shall see, the particles are large, permitting a precise determination of R_c and A. However, the values of c_e and ψ are difficult to establish with great precision, because in the mixture DNA + histones, the ratio of DNA to protein varies from one preparation to another (13).

In Fig. 2 some results are shown in a double logarithmic plot. The theoretical curve can be superimposed on each experimental curve. The discrepancies observed at large s are due to the fact that the theoretical approximation of a Gaussian distribution of electronic density in the rod, made by Luzzati (5), is valid only for small s, the difference between the experimental and theoretical intensities being a function of the difference between the actual distribution of the electron density and the Gaussian model.

The best agreement between the theoretical and experimental

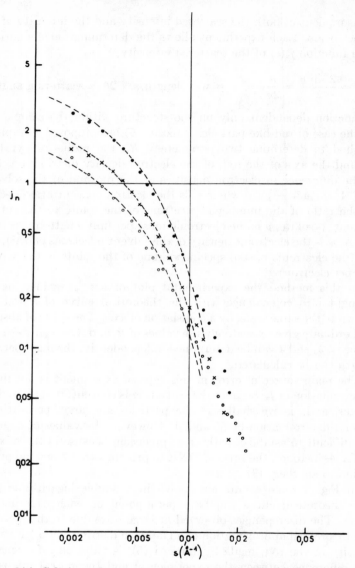

Fig. 2(a). Solutions of nucleoproteins. Logarithmic plot of the experimental points j_n and theoretical function corresponding to each experimental curve. The numerical data relevant to each experiment are given in Table I. Mixture 85% DNH and 15% DNP.

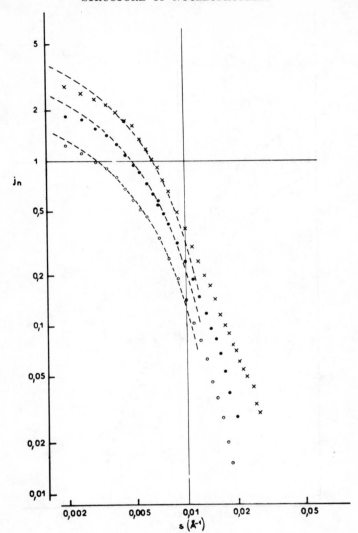

Fig. 2(b). Solutions of nucleoproteins. Logarithmic plot of the experimental points j_n and theoretical function corresponding to each experimental curve. The numerical data relevant to each experiment are given in Table I. Mixture 75% DNH and 25% DNA.

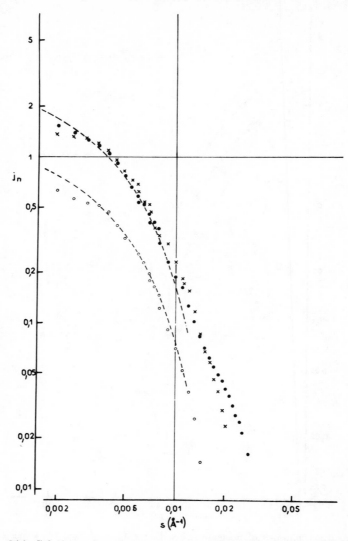

Fig. 2(c). Solutions of nucleoproteins. Logarithmic plot of the experimental points j_n and theoretical function corresponding to each experimental curve. The numerical data relevant to each experiment are given in Table I. Calf thymus DNH.

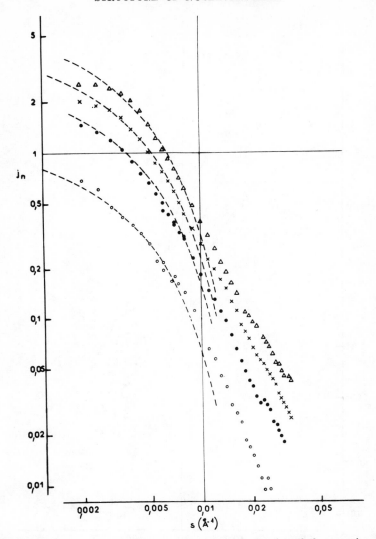

Fig. 2(d). Solutions of nucleoproteins. Logarithmic plot of the experimental points j_n and theoretical function corresponding to each experimental curve. The numerical data relevant to each experiment are given in Table I. Chicken erythrocyte DNH.

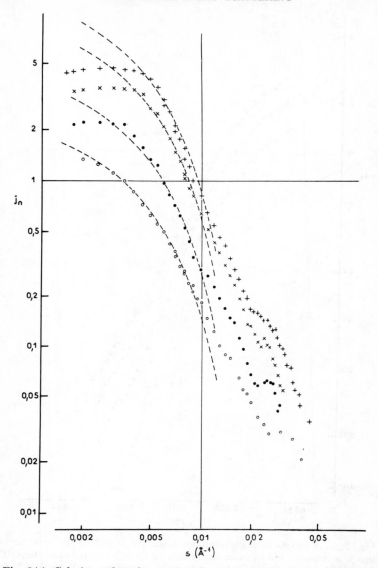

Fig. 2(e). Solutions of nucleoproteins. Logarithmic plot of the experimental points j_n and theoretical function corresponding to each experimental curve. The numerical data relevant to each experiment are given in Table I. Chicken erythrocyte DNH.

STRUCTURE OF NUCLEOPROTEINS

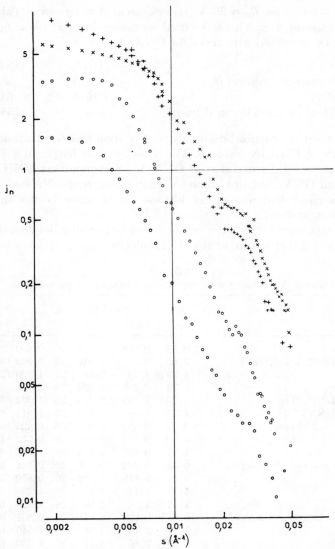

Fig. 2(f). Solutions of nucleoproteins. Logarithmic plot of the experimental points j_n and theoretical function corresponding to each experimental curve. The numerical data relevant to each experiment are given in Table I. Chicken erythrocyte DNH.

curves is found for $R_c = 26$ Å. The values of A are given in Table I; the average of A/c_e, 61, can be used for determining μ. These figures should be compared with those for DNA (6):

	DNA	DNH
Radius of gyration R_c	8.5	26
A/c_e	15.5	61
Electron mass per unit length	105	755

The rod-like organization involves more than just the coating of a molecule of DNA by histones. The mass per unit length of a DNH rod is about eight times that of DNA. If we assume that DNH contains half DNA and half histone a rod containing four DNA molecules and an equivalent amount of histone would agree best with the experimental data.

The experimental data are not sufficient to permit a detailed study of the internal structure of the DNH micelle. Since, for large values

TABLE I
Value of the various parameters relevant to the isotropic solution.

	% c_{DNA}	% $c_{e\,DNA}$	% $c_{e\,DNH}$	Å R_c	A	$\dfrac{A}{c_{e\,DNH}}$
Mixture	0.64	0.608	1.216	26	0.690	57.0
85% DNH + 15% DNP	1.05	0.955	1.910	26	0.944	49.4
Fig. 2a	1.35	1.28	2.56	26	1.490	58.2
Mixture	0.73	0.698	1.396	26	0.736	52.7
75% DNH Fig. 2b	1.10	1.045	2.090	26	1.232	59.1
25% DNA	1.37	1.305	2.610	26	1.818	69.7
DNH − calf thymus	0.50	0.480	0.960	26	0.426	44.5
Fig. 2c	0.90	0.854	1.708	26	0.952	55.7
	1.02	0.952	1.904	26	1.088	57.2
DNH − chicken erythrocytes	0.32	0.304	0.608	26	0.345	56.8
Fig. 2d	0.65	0.618	1.236	26	0.870	70.5
	1.20	1.14	2.28	26	1.280	56.2
	1.41	1.34	2.68	26	1.785	66.7
DNH − chicken erythrocyte	0.64	0.608	1.216	26	0.807	66.5
Fig. 2e	1.19	1.13	2.26	26	1.540	68.2
	2.59	2.46	4.92	26	3.330	67.8
	3.35	3.18	6.36	26	4.76	75.0
Fig. 2f	0.81	0.769	1.538			
	2.59	2.46	4.92			
	9.61	9.12	18.24			
	20.4	19.4	38.8			

of s, the intensity (see Fig. 2) is greater than the theoretical, a concentration of scattering material in the center (9, 10) is suggested.

Little can be said about the length of the rod. Since X-rays give evidence of the structure of the micelle only up to a size of 1000 Å, it is possible that some protein or DNA interconnects the rods to form a larger gel-like network. This may be a reason why dissociated and recombined DNH does not give homogeneous solutions.

We notice from the experimental curves (Fig. 2e,f) that the ordering of the micelles gives a maximum in intensity at the position $s = (110 \text{ Å})^{-1}$ for a concentration $c = 18\%$. If we assume a two-dimensional hexagonal packing of the micelles when they are in close contact, then we can determine from this first-order spacing of $(110 \text{ Å})^{-1}$, and the concentration, the mass per unit length of the rods (9, 10): this calculated value is in good agreement with the one determined from the scattering curves.

THE HEXAGONAL STRUCTURE

In the concentration range between 50 and 70% the X-ray diagrams show a band, whose position is a function of the concentration. Assuming that the DNA is in the B form, we have calculated the change of spacing as a function of the concentration in the case of a two-dimensional hexagonal packing of DNA molecules (9, 10); the curves are shown as broken lines in the figures. The experimental points, calculated under the assumptions that the arrangement of single molecules of DNA is hexagonal and that water and protein fill the space between molecules, lie on the curve shown as a broken line. Confirmation of this structure is obtained in the study of the mixtures of DNH + DNA + water (Fig. 3) and DNH + DNP + water (Fig. 4). Evidence that the initial structure of DNA + water (Fig. 5) and DNP + water (Fig. 6) is hexagonal comes not only from the fact that the experimental points lie on the broken curve. This model also is supported by the observation that X-ray diagrams from the highest concentrations show a set of sharp lines with spacings in the ratio $1 : \sqrt{3} : \sqrt{4}$, which indicates a two-dimensional hexagonal packing. Results from intermediate mixtures containing increasing amounts of DNH (Figs. 3 and 4) show that there is no discontinuity in the hexagonal packing in going from either DNA + water or DNP + water to DNH + water.

We notice that the closest packing of the DNA molecules (DNH + water) is $(22 \text{ Å})^{-1}$, about the same as in DNP + water

232   SMALL-ANGLE X-RAY SCATTERING

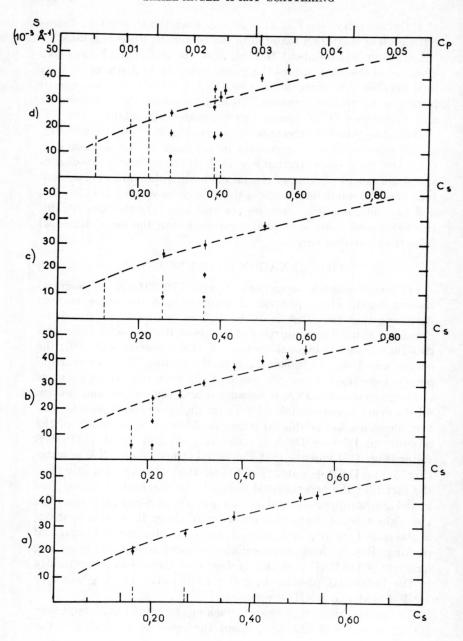

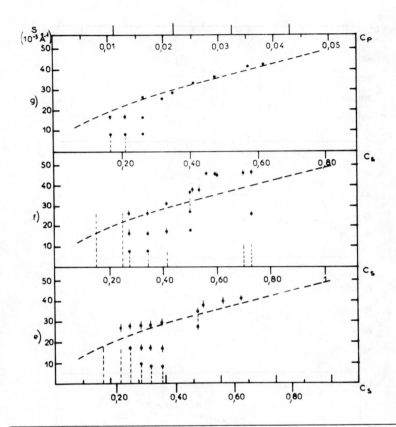

Fig. 3. The change of structure with different mixtures is followed by comparing X-ray diagrams for the same concentration of DNA; this is easily determined by dosing the phosphorus (there is roughly 10% phosphorus in DNA). C_p is the ratio of phosphorus to that of the gel.
(a) 33% DNH and 66% DNA 7.3% P in the mixture
(b) 80% DNH and 20% DNA 6.99% P in the mixture
(c) 66% DNH and 33% DNA 6.15% P in the mixture
(d) 50% DNH and 50% DNA 5.9% P in the mixture
(e) 75% DNH and 25% DNA 4.9% P in the mixture
(f) 80% DNH and 20% DNA 5.4% P in the mixture
(g) Calf thymus DNH with a partial extraction of histones 5.85% P.

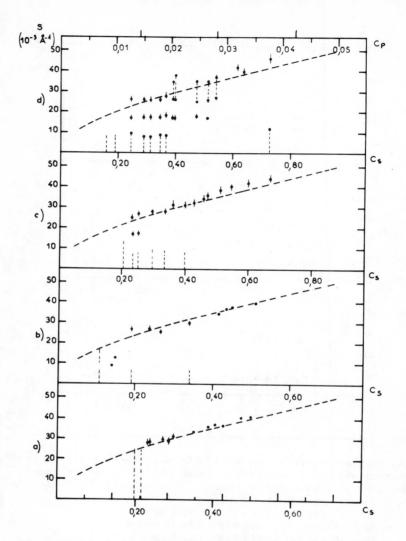

Fig. 4. Mixtures of DNH + DNP + water. (See legends Figs. 1 and 3.)
(a) 15% DNH 85% DNP (7.05% P)
(b) 33% DNH 66% DNP (7% P)
(c) 60% DNH 40% DNP (5.68% P)
(d) 85% DNH 15% DNP (5.2% P).

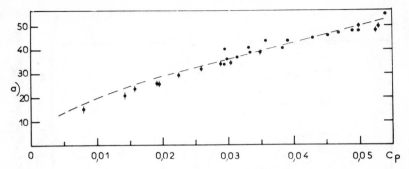

Fig. 5. DNA gels (see legends Figs. 1 and 3). Chicken erythrocyte DNA in 0.15 M NaCl.

(Figs. 1 and 6). That this packing is looser than that of DNA + water, $(20 \text{ Å})^{-1}$ (Fig. 5) can be explained by the presence of protein, which prevents the close-packing of DNA.

Another interesting observation regarding the mixtures DNH + DNA or DNH + DNP is that the hexagonal packing extends to lower concentrations than with DNH alone, apparently up to a limiting spacing of $(37 \text{ Å})^{-1}$ instead of $(27 \text{Å})^{-1}$, but for high amount of DNH there is a discontinuity between the two spacings (see Fig. 4d).

THE INTERMEDIATE STRUCTURE

The spacings observed in the concentration range between 30 and 50% are shown in Fig. 7. It is clear that there is no simple superposi-

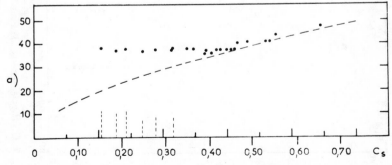

Fig. 6. DNP gels (see legends Figs. 1 and 3). Trout sperm DNP in 0.15 M NaCl. Note the superposition of solution and hexagonal packing up to 48% concentration.

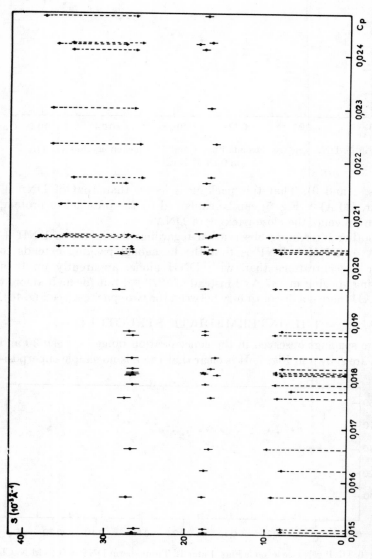

Fig. 7. DNH gels (see legends Figs. 1 and 3). Note the change of X-ray diagrams around C_p 0.020 or 40% DNH gel.

tion of solution and hexagonal packing as in DNP + water (see Fig. 6). The scattering disappears at a concentration of about 40% and at the same concentration there is a broadening of the $(37 \text{ Å})^{-1}$ spacing to a $(27 \text{ Å})^{-1}$ limit. There are two types of diagrams (we shall see later that they are related to the structure in nuclei) in that range of concentrations.

A simple way to fit the experimental data would be to suppose an intermediate structure (9, 10) characterized by the $(54 \text{ Å})^{-1}$ and $(37 \text{ Å})^{-1}$ spacings with a concentration of about 40%. The best model is a two-dimensional tetragonal array with two DNA molecules per unit cell. The spacings of the first two reflections of this lattice are $s_{100}(54 \text{ Å})^{-1}$, $s_{110}(38 \text{ Å})^{-1}$.

A more recent study (see the section on samples for electron microscopy) suggests that these spacings are not different orders from the same structure but correspond to two different structures or, at least, two independent spacings in the same structure.

The gels we use are not oriented. We may find additional information from X-ray diagrams of DNH fibers obtained under controlled relative humidity conditions, published by Wilkins and coworkers (12). Here the DNH molecules are presumably oriented parallel to the long fiber axis. Their X-ray diagrams show two strong equatorial spacings, $(60 \text{ Å})^{-1}$ and $(37–30 \text{ Å})^{-1}$; similar but weaker spacings appear in a meridional direction. For a lower relative humidity the equatorial spacing moves to $(22 \text{ Å})^{-1}$. The equatorial reflections must be due to the lateral arrangement of the DNH molecules. Because these are the strongest reflections in the diagram the highest proportion of the DNH must be ordered in this direction. We conclude that in the unoriented gels the $(54 \text{ Å})^{-1}$ spacing also reflects the lateral organization of the DNH molecules. However we cannot exclude the possibility that a single spacing belongs to different structures. Even these data are not sufficient to give an unequivocal model of structures present in the intermediate range of concentrations of DNH gels. We have tried to obtain more data with the electron microscope.

SAMPLES FOR ELECTRON MICROSCOPY

The first step was to set up a technique for the preparation of DNH gel samples suited for electron microscopic observations. The samples must be fixed and the water replaced by a monomeric mixture which is then polymerized. We followed these steps with X-ray diffraction in order to be sure that the initial structure was preserved.

Most of the usual fixatives such as those based on osmium tetroxide destroy the structure. Formaldehyde in a 10% concentration with pH between 5 and 8, used for several hours in the cold ($+2$–$4°C$), preserves the initial structure best. That is, if after fixation the gels are put in an excess of water, they do not swell and the two spacings, $(54 \text{ Å})^{-1}$ and $(37 \text{ Å})^{-1}$, are preserved. The same fixative used on a viscous solution (5%) of DNH does not destroy the micelles: we have measured the same radius of gyration after fixation, $R_c = 26$ Å.

The replacement of water by analogous compounds indicates (see Table II) that hydroxyl groups are necessary to preserve the struc-

TABLE II
Dehydration of fixed DNH.

	30%	66%	80%	100%
Methanol	+	+	+	−
Ethanol	+	−		
Isopropanol	+	−		
Glycol	+	+	+	+
Glycerol	+	+	+	+
"Durcupan"	−			
m-Cellosolve	+	+	−	
p-Dioxane	−			
Acetone	+	+	−	
Propylene oxide	−			

+ structure preserved.
− structure changed.

ture after it has been fixed. Table II shows the percentage of water which can be replaced without changing the fixed structure. In fact, it is possible to use the pure solvent directly, rather than exchanging the water stepwise. Although we have found that glycerol is best, for practical reasons, i.e., lower viscosity and more rapid exchange of water, we have used 1:1 mixtures of glycerol and methanol, or glycerol and dimethylsulfoxide, which preserve the fixed structure. We cannot say anything about the isotropic solution which is part of the gel and responsible for the scattering of X-rays (we used a concentration of 36%). When the 5% viscous solution, which after fixation does not swell, is put in an excess of a mixture containing $\frac{1}{3}$ glycerol and $\frac{2}{3}$ water the scattering curve is changed: the micelles are destroyed.

The replacement of the solvent by a monomeric mixture such as

"Vestopal" (14), "Araldite" (15), "Epon" (16) or glycol methacrylate (17), changes the spacings from $(54 \text{ Å})^{-1}$ and $(37 \text{ Å})^{-1}$ to $(70 \text{ Å})^{-1}$ and $(35 \text{ Å})^{-1}$, the same as those observed in the dry form of DNH gels.

We observe the same change in the X-ray diagram after drying a fixed sample of DNH. When put in an excess of water, the gel swells and the scattering is more intense, which indicates that part of the structure is destroyed in an irreversible way and that part of the gel gives the two spacings $(54 \text{ Å})^{-1}$ and $(37 \text{ Å})^{-1}$.

It appears that fixation only prevents the swelling of the gel but is not sufficient to prevent a change of structure on drying or, what appears to be the same thing, during dehydration and embedding.

The electron microscopic observations on sections of DNH gel have not shown any structure that can easily be related to the X-ray observations. The X-ray diagrams of polymerized samples show the two bands characterizing the dry DNH gels. Analogous work by Huxley and Zubay (18) gave a final spacing of $(66 \text{ Å})^{-1}$. It is difficult to relate this either to the intermediate form $(54 \text{ Å})^{-1}$ or dry state $(70 \text{ Å})^{-1}$.

Besides having an interest for the electron microscopist, this study shows how to dehydrate DNH gels under non-equilibrium conditions. The sharpness and the relative intensity of the two bands [$(54 \text{ Å})^{-1}$ and $(37 \text{ Å})^{-1}$] on the X-ray diagrams during our study clearly show that these bands are not related or at least that they are not merely two successive orders of the same structure.

THE STRUCTURE IN THE DRY STATE

We have already indicated in the last section that the final product in samples prepared for electron microscopy has a structure characteristic of the dry state. It is difficult to handle samples of gels with a concentration higher than 70% and to reach equilibrium (our samples had a cross-section of a few tenths of a millimeter and a length of 5–6 millimeters).

We were only able to characterize the organization by two bands with $s = (70 \text{ Å})^{-1}$ and $(35 \text{ Å})^{-1}$. We did not try to interpret the diagrams because this structure is not found in nuclei, and because the oriented fibers studied by Wilkins and his group (12) may give more precise information.

However, our study clearly indicates that between the dry state and the intermediate range of concentrations there is a two-dimensional hexagonal packing of DNH; in other words, there is a discontinuity between the two ranges of concentration in the organization of DNH gels.

GENERALITY OF THE OBSERVED STRUCTURES

We have already indicated that small amounts of DNA or DNP do not change the structure of DNH gels. The results we have obtained in our study on the dependence of the structure on the amount of water do not depend on the way DNH is prepared. The same results are obtained with DNH prepared by the Zubay-Doty technique; even mixing DNA and histones in 1 M NaCl and precipitating in 0.15 M NaCl gives a nucleoprotein with the same characteristics (9, 10).

This study was made with DNH extracted either from chicken erythrocytes or calf thymus. The study of samples of nucleoproteins extracted from different sources gives the same X-ray diagram at the same concentration (11).

The structure of the DNH gels is not altered by changing the NaCl concentration of the solvent until it is about 2 M as (NaCl). At low concentrations of DNH, a dissociation into DNA and histones occurs and an equilibrium is established between the components.

We cannot, of course, exclude more complex relations in the nuclei than in extracted DNH. More detailed studies are needed with compounds of better known compositions by using fractions of histones and by using complementary techniques such as electron microscopy to obtain a better knowledge of the structure of DNH gels, and that of the structure *in situ*.

CONCLUDING REMARKS

Our study with X-ray diffraction techniques shows that DNH gels possess a water-dependent organization. It is possible, by using appropriate fixatives and dehydrating agents, to produce changes in the structures present for a given amount of water in the gels. The intermediate range of concentrations (30 to 50%) has two, and possibly three, different structures coexisting. Beside the micelles in the solution and a two-dimensional hexagonal packing of DNH there is an intermediate structure which at present cannot be characterized without ambiguity.

A comparison with nuclei indicates that DNH in chicken erythrocytes gives an X-ray diagram identical to DNH gel of $c = 30\%$.

With thymus nuclei the diagram corresponds to a concentration of about 45%. The nuclei were kept cold during the experiment, and light-microscope observations suggest that they were morphologically intact after the experiment. Nothing can be said at the molecular level. An X-ray diagram of thymus nuclei after only four hours exposure time already shows the typical pattern of extracted DNH. There is no indication of any structure in the nuclei which is different from that of extracted DNH.

To be sure that observed organization of DNH is due to the association of DNA and proteins, we have destroyed each component in the gel by an enzymatic action. We have used respectively desoxyribonuclease and chymotrypsin. In both cases DNH structure disappears (9, 10). This involvement of histones in the organization of DNH is in agreement with their role (2, 3, 4) in the cell. A change in the ratio or nature of histones associated with DNA changes the equilibrium between the coexisting structures. This may be a way to regulate the functioning of DNA in the nuclei.

The author is indebted to Dr. Brumberger for a careful revision of the manuscript.

BIBLIOGRAPHY

1. Hershey, A. D. and Chase, M., J. Gen. Physiol., **36,** 177 (1959).
2. Stedman, E. and Stedman, E., Nature, **166,** 780 (1950).
3. Huang, R. C. and Bonner, J., Proc. Nat. Acad. Sci., **48,** 1216 (1962).
4. Allfrey, V. G., Littau, V. C. and Mirsky, A. E., ibid., **49,** 414 (1963).
5. Luzzati, V., Acta Cryst., **13,** 939 (1960).
6. Luzzati, V., Nicolaïeff, A. and Masson, F., J. Mol. Biol., **3,** 185 (1961).
7. Luzzati, V. and Baro, R., J. Phys. Radium, **22,** 186 A (1961).
8. Luzzati, V., Witz, J. and Baro, R., J. Physique, **24,** 141 A (1963).
9. Luzzati, V. and Nicolaïeff, A., J. Mol. Biol., **7,** 142 (1963).
10. Nicolaïeff, A., Thèse, Strasbourg (1962).
11. Nicolaïeff, A., Mazen-Knobloch, A., Vendrely, R. and Luzzati, V., Compt. Rend., **248,** 2805 (1959).
12. Wilkins, M. H. F., Zubay, G. and Wilson, H. R., J. Mol. Biol., **1,** 179 (1959).
13. Davison, P. F., Conway, B. E. and Butler, J. A. V., Progr. Biophys. and Biophys. Chem., **4,** 148 (1954).
14. Kellenberger, E., Ryter, A. and Sechaud, J., J. Biophys. Biochem. Cytol., **4,** 671 (1958).
15. Glauert, A. M. and Glauert, R. H., ibid., **4,** 191 (1958).
16. Luft, J. H., ibid., **9,** 409 (1961).
17. Leduc, E. H., Marinozzi, V. and Bernhard, W., J. Roy. Microscop. Soc., **81,** 119 (1963).
18. Huxley, H. E. and Zubay, G., J. Biophys. Biochem. Cytol., **11,** 273 (1961).

X-Ray Scattering from Small RNA Viruses in Solution

J. W. ANDEREGG

Physics Department & Biophysics Laboratory, University of Wisconsin, Madison, Wis. 53706

INTRODUCTION

For several years we have studied the small-angle X-ray scattering from a number of small "spherical" viruses in solution. The viruses we have studied are all just under 300 Å in diameter, they range in molecular weight from 3.6 to 7.0 million, and they all are RNA*-containing viruses. Under the electron microscope these viruses look roughly spherical but closer examination of freeze-dried or negatively-stained preparations frequently show that they have polygonal contours and a regular pattern of subunit structure on their surface. All the viruses to be discussed here are plant viruses except for one bacterial virus.

The intent of this paper is to review some already published work and to report some new results on the X-ray scattering from these viruses and to try to draw from these results some general conclusions about the structure of the viruses. The previously published work includes that on WCMV* (1) and BMV* (2). The results on the bacterial virus R17 are to be published soon (3). The work on BBMV* by White and Beeman has not yet been published but is reported in a Ph.D. thesis (4). The results on SMV* are from unpublished data by Anderegg, Kaesberg, and Fischbach.

The knowledge that the small plant viruses consist of a shell of protein surrounding a core of RNA goes back to the work of Markham on TYMV.* Markham and Smith (5) had found that plants infected with TYMV produced in addition to this virus another particle which was antigenically similar to the virus but which contained no nucleic acid and was uninfectious. Since this particle sediments more slowly than the virus it was called the top component and the virus was

---

* Abbreviations used: RNA—ribonucleic acid; WCMV—wild cucumber mosaic virus; BMV—bromegrass mosaic virus; BBMV—broad bean mottle virus; SMV—squash mosaic virus; TYMV—turnip yellow mosaic virus.

referred to as bottom component. Markham (6) found that the top component had the same diameter, the same surface properties, and the same molecular weight of protein as the virus and concluded on this basis that the virus consisted of a protein shell with a nucleic acid core and that the top component consisted of only the protein shell. X-ray scattering work by Schmidt, Kaesberg, and Beeman (7) bore out this view. The scattering from the virus indicated a uniform sphere of diameter 280 Å and the scattering from the top component indicated a hollow shell 280 Å in outside diameter and 35 Å thick.

Yamazaki and Kaesberg (8) later found that infection with WCMV causes the production of two virus-like particles in addition to virus. These are designated top components a and b and they contain no nucleic acid and 0.48×10^6 molecular weight of RNA respectively. The virus has a molecular weight of 7.0×10^6 and contains 2.4×10^6 molecular weight of RNA. X-ray scattering studies were made on the virus and on a mixture of the top components (1) and in this case for the first time radial electron density distributions were calculated for both samples from a Fourier inversion of the scattered amplitudes.

Mazzone, Incardona, and Kaesberg (9) found that plants infected with SMV produce three kinds of macromolecular particles which contain 0, 1.6×10^6, and 2.4×10^6 molecular weight of RNA respectively. Again only the bottom component is infectious. All three components have been separated by density-gradient sedimentation and we have studied the X-ray scattering and calculated the radial density distribution for each component. The results will be discussed below.

R17 is one of the smallest viruses known and it is one of the recently discovered group of RNA-containing bacteriophages. It has a molecular weight of 3.6×10^6 and an RNA content of 1.1×10^6 (10). BMV and BBMV viruses are plant viruses with essentially the same weight of RNA as R17 but with somewhat greater protein contents. BMV has a total molecular weight of 4.6×10^6 (11) and BBMV has a molecular weight of 5.2×10^6 (12). Incardona and Kaesberg (13) have discovered that BMV undergoes a structural transition as the pH is raised above 6.7 and we have therefore made a study of the scattering from this virus at pH 6 and at pH 7.

Our knowledge of the structure of the small "spherical" or isometric viruses has been advanced greatly by the single crystal diffraction work of Caspar and Klug and their coworkers on a number of these viruses, which showed that these viruses have the symmetry properties to be expected from structures made up of a large number of subunits in equivalent or quasi-equivalent positions. This work, as well as the electron microscope observations on the subunit structure of these viruses, has been reviewed by Caspar and Klug (14). We will discuss below the evidence on subunit structure in our X-ray scattering results.

THEORETICAL

We shall review very briefly the X-ray scattering theory pertinent to the work being discussed here. The intensity, $i(h)$, of X-rays scattered by a random collection of N identical particles is just N times the intensity scattered by a single particle. That is

$$i(h) = Ni_e(h)\overline{F^2(h)} \qquad 1.$$

where $i_e(h)$ is the intensity scattered by a single electron; $F(h)$, the structure factor of the particle, is the ratio of the amplitude scattered by the particle to the amplitude scattered by a single electron; $h = 4\pi \sin \theta/\lambda$, 2θ is the scattering angle, and λ is the X-ray wave length. The bar over $F^2(h)$ indicates that this quantity is averaged over all orientations of the particle with respect to the X-ray beam. $\overline{F^2(h)}$ is given by

$$\overline{F^2(h)} = \int_V \int_V \rho_m \rho_n \frac{\sin hr_{mn}}{hr_{mn}} dv_m\, dv_n = \int_0^\infty D(r) \frac{\sin hr}{hr} 4\pi r^2\, dr \qquad 2.$$

where ρ_m is the electron density in volume element dv_m at a distance r_m from the origin and r_{mn} is the separation of volume elements. The quantity $4\pi r^2 D(r)\, dr$ is the number of electron pairs in the particle with a separation between r and $r + dr$. $D(r)$ can be determined by a Fourier inversion of the scattered intensity as follows:

$$D(r) = \frac{1}{2\pi^2 r} \int_0^\infty h\, \overline{F^2(h)}\, \sin hr\, dh \qquad 3.$$

For all particles at small angles

$$\overline{F^2(h)} = n^2 \exp(-h^2 R_g^2/3) \qquad 4.$$

where n is the number of electrons in the particle and R_g is the radius of gyration of the particle. For particles in solution n must be replaced by n_{eff}, the number of electrons in the particle minus the number of electrons in the solvent which the particle displaces. That is

$$n_{\text{eff}} = n - \frac{M\bar{V}\rho_0}{N_0} \qquad 5.$$

where M and $\bar{V}$ are the molecular weight and partial specific volume of the particle, ρ_0 is the electron density of the solvent, and N_0 is Avogadro's number.

For particles with spherical symmetry $F(h)$ is independent of the orientation of the particle and we can write

$$F(h) = \int_0^R \rho(r) \frac{\sin hr}{hr} 4\pi r^2 \, dr \qquad 6.$$

R is the radius of the particle and $\rho(r)$ is the electron density at a distance r from the center of the particle. $\rho(r)$ can be determined from the Fourier transform of $hF(h)$ as follows:

$$\rho(r) = \frac{1}{2\pi^2 r} \int_0^\infty hF(h) \sin hr \, dh \qquad 7.$$

The magnitude of $F(h)$ can be determined from the square root of the scattered intensities (Eq. 1) but the sign must be determined by some additional information.

From Eq. 6 we see that the area under the $\rho(r) \, 4\pi r^2$ curve is equal to the value of the structure factor in the forward direction and is equal to n or, if the particle is in solution, to n_{eff}. Thus if we normalize the curve of $hF(h)$ so that $F(0)$ is equal to n_{eff} then when we calculate the transform of $hF(h)$ we will have the density scale on an absolute basis and the densities will represent electron densities above that of water. We will call electron densities above that of water "effective electron densities," ρ_{eff}.

$$F(0) = \int_0^R \rho_{\text{eff}} \, 4\pi r^2 \, dr = n_{\text{eff}} = n - \frac{M\bar{V}\rho_0}{N_0} \qquad 8.$$

EXPERIMENTAL

The experimental equipment and procedures used in this work are discussed in more detail by W. W. Beeman in another paper in this volume. Most of the scattering data were taken with the 50 cm slit assembly with slit heights of 1 cm and slit widths of 0.015 cm when taking data in the radius of gyration region and 0.045 to 0.06 cm when taking data in the peak region. The data were corrected for smearing effects due to the finite height and width of the slits, and

Fourier inverted with the help of programs written for the CDC 1604 electronic computer. Background scattering was eliminated by making a separate run over the same angular range with only solvent in the sample holder and subtracting this curve from that obtained with the virus solution.

In order to eliminate interparticle interference effects from the data, a series of runs was made at small angles with decreasing virus

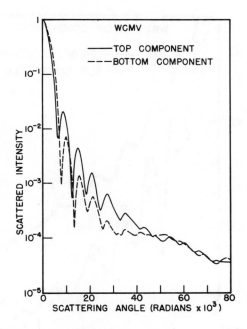

Fig. 1. The slit-corrected scattering curves for top and bottom component of wild cucumber mosaic virus [Anderegg et al., Biophys. J., **1,** 657 (1961)].

concentration down to approximately 5 mg/ml. Very concentrated solutions (up to 200 mg/ml) were used to study the weak scattering at larger angles. These runs were then all combined into a composite curve free of interparticle interference effects. The purity of all virus preparations was checked in the model E analytical ultracentrifuge. Samples were kept near 4°C during the X-ray runs by circulating a cold liquid through a cooling shield surrounding the sample.

RADIAL ELECTRON DENSITY DISTRIBUTIONS
WCMV

Most of the results to be discussed here are derived from the radial electron density distributions for the viruses as calculated by Fourier inversion of the scattering amplitudes. We will use the data on WCMV to illustrate the details and difficulties of the procedure for calculating these density distributions.

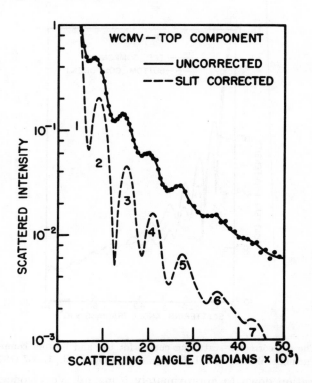

Fig. 2. The scattering curve for the top component of wild cucumber mosaic virus before and after correction for the finite width and height of the collimating slits.

As mentioned earlier the top component of WCMV is actually a mixture of two components, one of which contains no nucleic acid and one of which contains 0.48×10^6 molecular weight of RNA. The complete, slit-corrected scattering curves for the mixture of the two top components and for the bottom components of WCMV are shown in Fig. 1. These curves are composites of data taken with two different slit widths —0.13 mm and 0.51 mm. The narrow slits were necessary to get sufficient resolution for the low-angle part of the curve; the wider slits were necessary to get sufficient intensity out in the region where the intensity has dropped down by several orders of magnitude and the highest resolution is no longer required. The two sets of data were then corrected independently for the smearing effect of the slit widths, combined into a smooth curve, and then corrected for the effects of the slit heights. The width correction was normally very small, the height correction appreciable. Figure 2 shows the wide-slit data on the top component of WCMV both before and after correcting for the effects of slit widths and heights in order to illustrate the magnitude of the correction.

The scattering amplitudes can be obtained from the square root of the intensity curve. One must decide by some independent means, however, whether to take the positive or negative square root at any given point. The scattering curves for uniform spheres, uniform spheres with a hollow center, and other simple density distributions all have a positive central maximum and subsidiary maxima which alternate in sign. Assuming that at low resolution the viruses will have some correspondingly simple density distribution, we have proceeded to use the same sign choice in calculating all the density distributions of which we shall make use. In a number of cases we have tried alternative sign choices and some of these will also be discussed.

Another difficulty in these calculations is that the minima in the intensity curves do not go to zero. If the amplitude curve is to have successive sections which alternate in sign then the amplitude curve and the intensity curve must go through zero between these sections. As can be seen in Fig. 1 the minima in the experimental intensity curve do not go to zero. This is very likely due to the lack of true spherical symmetry in the virus structure. Some calculations on models indicate that the deviations from spherical symmetry would have the primary effect of filling in the minima without seriously affecting the position or magnitude of the maxima. For this reason we ignored when necessary the data near the intensity minima and drew an amplitude curve that fits the data near the maxima and passes through zero in between. This is illustrated in Fig. 3. The dots are the slit-corrected amplitudes plotted both positively and negatively and the solid line is the smooth curve that was used in calculating the Fourier transform.

250   SMALL-ANGLE X-RAY SCATTERING

A final difficulty involves the cut-off of the experimental data. Although the curves were carried out further, we could only make use of six to eight maxima in calculating the transforms because the maxima become too weak and ill defined after that. An abrupt cut-off in the amplitude curve results in spurious diffraction effects in the transform calculated from that curve. To minimize these effects we

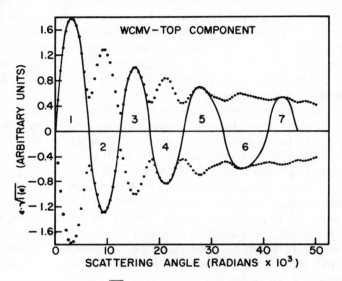

Fig. 3. The curve of $\epsilon\sqrt{i(\epsilon)}$ for the top component of wild cucumber mosaic virus. The dots represent the slit-corrected data plotted as both positive and negative numbers. The solid line is the smooth curve that was used to calculate the transform in Fig. 4. ϵ = scattering angle.

multiplied the amplitude curve in each case by an artificial temperature factor consisting of a Gaussian with the constant chosen so as to multiply the data at the cut-off point by 0.1.

Using these procedures we calculated the Fourier transforms of the amplitude curves for top and bottom component of WCMV using six maxima (the central one plus five subsidiary maxima) in each case. The resultant radial density distributions are shown in Fig. 4a. The electron densities are on an arbitrary scale in this figure. The resolution in the transforms is of the order of 40 Å. For comparison purposes the theoretical scattering curve for a uniform sphere of 140 Å radius was cut off after six maxima, multiplied by the same temperature factor, and the Fourier transform calculated. The result is shown in Fig. 4b. The dashed line represents the uniform density distribution which would have resulted if the entire amplitude curve had been used in the inversion. This gives an indication of the magnitude of the spurious diffraction effects and the smearing due to poor resolution which are present in the density distributions.

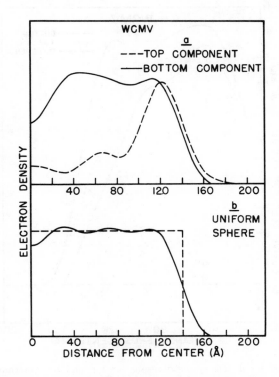

Fig. 4(a). The radial density distributions for top and bottom components of wild cucumber mosaic virus as calculated from the Fourier transforms of the scattered amplitudes. Six maxima with alternating signs were used in the transform. This makes use of data out to a Bragg spacing of approximately 40 Å. An artificial temperature factor was used which multiplied the data at the cut-off point by 0.1. (b). The Fourier transform (solid curve) of the theoretically calculated scattering amplitudes from a uniform sphere of 140 Å radius. For comparison with the transform in (a) the amplitude curve was cut off at the same angle and multiplied by the same temperature factor. The dashed line represents the uniform density distribution which would have resulted if the entire amplitude curve had been used in the inversion [Anderegg et al., Biophys. J., **1**, 657 (1961)].

The transform for bottom component looks very much like a uniform sphere of 140 Å radius. Top component on the other hand has a transform very similar to that for a hollow sphere of outer radius 140 Å and inner radius 100 Å. There is some electron density extending in nearly to the center in top component but it must be remembered

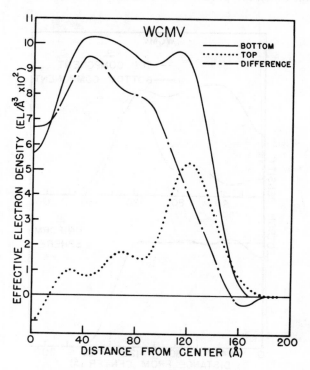

Fig. 5. The radial density distributions for top and bottom components of wild cucumber mosaic virus normalized so that the ordinate represents "effective electron density" or electron density above that of the solvent. The dash-dot curve represents the difference between the curve for bottom component and the curve for top component and hence represents the distribution of RNA.

that one of the top components does have a small amount of RNA. One of the first things to be noticed is that the dry volume of the virus (molecular weight times partial specific volume) is much less than the volume of a 140 Å sphere. (These quantities are tabulated for WCMV and the other viruses in Table I at the end of the article.)

The additional volume must be occupied by water. If we assume that the protein occupies the region from 100 to 140 Å and that the RNA is located inside 100 Å then we can calculate the amount of water in each region and the average electron density of that region. This calculation indicates that average electron density in the protein region would be 0.067 electrons/Å^3 above that of water and in the RNA region would be 0.127 electrons/Å^3 above that of water. The

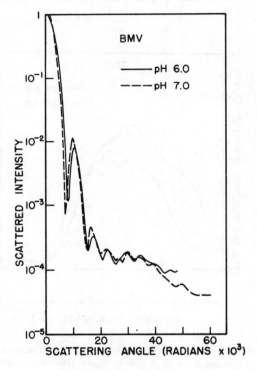

Fig. 6. The slit-corrected scattering curves for bromegrass mosaic virus at pH 6 and at pH 7.

transform for bottom component shows no such large density difference but indicates instead an almost uniform electron density throughout the virus. We would have to extend the region available to the RNA out to 113 Å in order to make the density of the RNA region match that of the protein region. Thus we are forced to conclude that the protein and RNA must overlap.

The situation is perhaps made clearer by putting the transforms on an absolute scale, that is by normalizing them so that the area under the $4\pi r^2 \rho(r)$ curve corresponds to the correct number of effective electrons. This has been done in Fig. 5. If we now calculate how far out we must go on the curve for bottom component before we have enough effective electrons to account for those contributed by the RNA, we find that the answer is again 113 Å. So if there is no overlapping of RNA and protein, RNA must extend out to 113 Å.

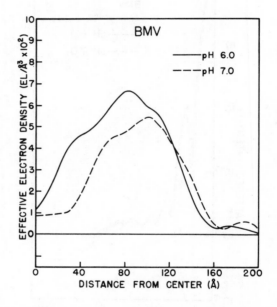

Fig. 7. The radial density distributions for bromegrass mosaic virus at pH 6 and at pH 7 normalized so that the ordinate represents "effective electron density."

Yet the top component curve clearly indicates that protein extends in to at least 100 Å. Hence RNA must extend out beyond 113 Å and there must be at least 20 Å of overlap of protein and RNA. The curve in Fig. 5 which represents the density difference between top and bottom component should represent the distribution of RNA and it shows RNA extending out essentially to the outer edge of the virus. Thus the effective electron density for bottom component is everywhere higher by a considerable amount than the effective electron density of top component. Because of the poor resolution we cannot say exactly how much overlap of RNA and protein there is; some of the increased density in the outer region of bottom component may just be due to a smearing out of the RNA density. It would appear, however, that at least 20 Å overlap is necessary to account for the data.

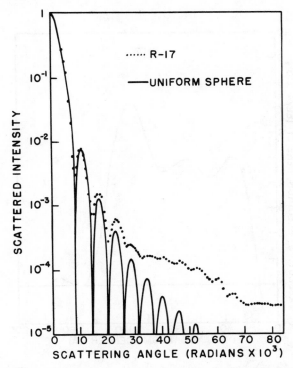

Fig. 8. The slit-corrected scattering curve for the bacteriophage R17. The dots are representative points from the slit-corrected curve; the solid line is the theoretical scattering curve for a uniform sphere of 133 Å radius [Fischbach et al., J. Mol. Biol., in press].

BMV

Bromegrass mosaic virus is interesting because of the already mentioned structural transition that takes place as the pH is raised above pH 6.7. Incardona and Kaesberg (13) found an abrupt increase in sedimentation contact of about 10% taking place at this pH. We found the radius of gyration to increase from 106 Å to 128 Å on going from pH 6 to pH 7. The scattering curves for this virus at pH 6 and pH 7 are shown in Fig. 6. The differences are small but significant. The electron density distributions on an absolute scale are shown in Fig. 7. These were calculated using eight maxima in the amplitude curve and assuming alternating signs. The pH 6.0 form of the virus

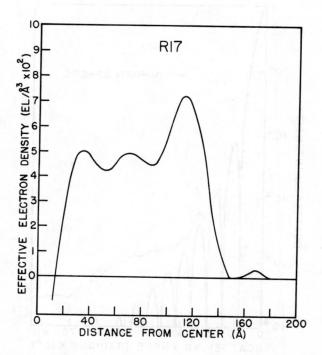

Fig. 9. The radial density distribution for the bacteriophage R17 normalized so that the ordinate represents "effective electron density." Six maxima with alternating signs were used in calculating the transform. This included data out to a scattering angle of 0.037 radians corresponding to a Bragg spacing of 40 Å.

has an outer radius of approximately 130 Å and shows considerable hollowness out to 40 Å or more. The pH 7.0 form shows an increase in outer radius to about 138 Å and an increase in the hollowness—a movement of material from inside 100 Å into the region outside that radius. Both curves show a considerable tail extending out to approximately 200 Å. This tail is more pronounced for the pH 7.0 form than for the lower pH and much of the increase in radius of gyration of the virus is associated with this increased tail. It seems possible that this tail represents material that is extending out beyond the outer edge

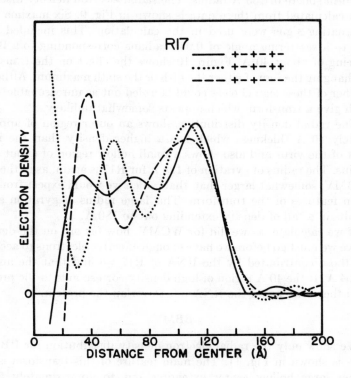

Fig. 10. Fourier transforms of the scattering amplitudes from the bacteriophage R17. Six maxima were included in the calculation of each transform and an artificial temperature factor was used which multiplied the data at the cut-off point by 0.1. The solid curve represents the alternating sign choice which is assumed to give the correct radial density distribution. The other two curves show the effect of changing the sign of either the fifth or sixth maximum.

of the major part of the virus and its increase at pH 7.0 could mean that the structural transition involves RNA moving out of the center of the virus until it is actually extending out beyond the protein coat. The pH 7.0 form has been found by Incardona and Kaesberg (13) to be more susceptible to degradation by ribonuclease.

R17

The slit-corrected scattering curve for the bacterial virus R17 is shown in Fig. 8 where it is compared to the scattering curve for a uniform sphere of 133 Å radius. The radial electron density distribution calculated from these data is shown in Fig. 9. Six maxima with alternating signs were used in the calculation. This included data out to a scattering angle of 0.037 radians corresponding to a Bragg spacing of about 40 Å. Figure 10 shows the effect on the transform of changing the sign of either the fifth or the sixth maximum. Although neither of these sign choices could be ruled out as unreasonable, they both give a transform which appears somewhat unlikely.

The radial density distribution shows an outer region of approximately 40 Å thickness which has a higher density than the inner part of the virus and also shows a small hollow region of about 15 Å radius. The radius of gyration of 128 Å for R17 is again, as in the case of BMV, somewhat larger than that which we would expect from the main features of the transform. This large radius of gyration again results in a tail of density extending out to 180 Å.

If we calculate, as we did for WCMV, how far out on the density curve we must go before we have enough effective electrons to account for those contributed by the RNA of R17 we find that the answer is 104 Å. If the 40 Å region of high density corresponds to the protein shell then once again the RNA must overlap the protein.

BBMV

We show only the radial electron density distribution for BBMV. This is shown in Fig. 11. The main feature of this transform is the rather large hollow center extending out to approximately 50 Å radius. As for the other viruses, some of the properties of BBMV are listed in Table I.

SMV

Squash mosaic virus is interesting because of its three components—top, middle, and bottom—which contain 0, 1.6×10^6, and 2.4×10^6 molecular weight of RNA respectively. The scattering curves for all three components are shown in Fig. 12 and the corresponding radial density curves in Fig. 13. Six maxima were used in calculating the transform giving a resolution again of about 40 Å. The increasing hollowness with decreasing RNA content is obvious.

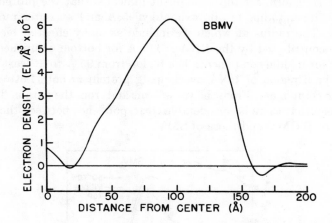

Fig. 11. The radial density distribution for broad bean mottle virus normalized so that the ordinate represents "effective electron density." Nine maxima with alternating signs were included in the calculation of the transform. This includes data out to an angle of about 0.05 radians corresponding to a Bragg spacing of 30 Å. An artificial temperature factor was used which multiplied the data at the cut-off angle by 0.1 [White, R. A., Ph.D. Thesis, Univ. of Wisconsin (1962)].

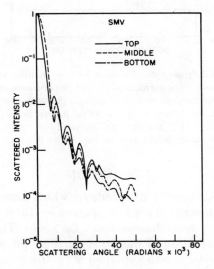

Fig. 12. The slit-corrected scattering curves for top, middle, and bottom components of squash mosaic virus.

The transform for top component indicates that the protein coat has an inner radius of approximately 95 Å and an outer radius of 140 Å. The radius at which we include as many effective electrons as are contributed by the RNA is 111 Å for bottom component and 110 Å for middle component. The region from 95 Å to at least 110 Å must be a region of RNA overlapping protein in both bottom and middle component. From the transforms and from the data in Table I it is seen that there is considerable correspondence between the properties of WCMV and those of SMV.

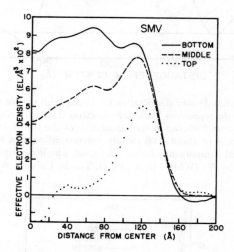

Fig. 13. The radial density distributions for top, middle, and bottom components of squash mosaic virus normalized so that the ordinate represents "effective electron density." Six maxima with alternating signs were used in calculating the transform, giving a resolution of about 40 Å. An artificial temperature factor was used which consisted of multiplying the amplitude curve by a Gaussian so that the data at the cut-off angle were multiplied by 0.1.

DISCUSSION

From the radial density distributions which have been discussed and from the summary of virus properties in Table I at the end of this article a few generalizations can be made. The first of these concerns the comparison of the dry volume of the virus (molecular weight times partial specific volume) with the volume of a sphere having the radius indicated by the transform. This comparison is made in the last two columns of Table I. In every case the volume of the sphere is considerably greater than the dry volume. The additional volume must be occupied by water and the water lowers the average electron density of the virus. For BBMV, BMV, and R17 the ratio of sphere volume to dry volume is roughly 2:1, for WCMV and SMV the ratio is closer to 1.5:1. Thus the density distributions for WCMV and SMV show higher effective electron densities than the other viruses. It is interesting that the two viruses with the most tightly packed RNA are also the two which occur in nature in the presence of an RNA-free component.

Secondly, we notice that in those cases where an estimate can be made the protein coat has a thickness of approximately 40 Å. The problem of the resolution in the density distributions is involved in this determination. Since the data used in calculating the transforms only go out to a Bragg spacing of about 40 Å, we obviously cannot measure accurately a shell which is only 40 Å thick. In fact the total width of the high density region in the top component transform for both SMV and WCMV is nearly 80 Å. However if we assume that the protein shell has approximately uniform density and that this distribution has been smeared out in our curves by the low resolution, then we can estimate that the actual shell width is 40 Å. If the shell density is not uniform the thickness may be greater than 40 Å.

Also in those cases where a top component is available (WCMV and SMV) there is evidence that the RNA overlaps at least half of the protein shell. The evidence for this has already been discussed in the sections on WCMV and SMV. Once again the extent of overlap is uncertain because of lack of resolution. The curves indicate nearly 100% overlap but allowing for the smearing effect of the poor resolution one would estimate that the nucleic acid overlaps at least half of the protein shell or approximately 20 Å. For the other viruses where no top component is available there is little evidence for the thickness of the shell and the extent of RNA overlap into the shell, except in the case of R17. In this case there is a high density region approximately 40 Å thick around the outside of the virus that could well correspond to the protein shell. If this is so then the RNA must again overlap since we calculate from the transforms that the RNA must extend out to at least 104 Å.

All the viruses except SMV show at least a small hollow center. BBMV and BMV have quite large hollow centers. One is tempted to associate these hollow centers with the excess volume of the virus over its "dry volume." In all cases, however, the hollow center is

TABLE I

Virus	Molecular weight × 10⁻⁶			Dimensions (Å)				Volume (10⁶ Å³)	
	Virus	Protein	RNA	Inner radius of protein coat	Radius of RNA*	Outer radius	Radius of gyration	Dry volume	Volume of sphere
WCMV	7.0	4.6	2.4	100	113	140	112	7.8	11.5
SMV (Bottom)	6.9	4.5	2.4	95	111	140	112	7.7	11.5
SMV (Middle)	6.1	4.5	1.6	95	110	140	116	7.1	11.5
BBMV	5.2	4.1	1.1	—	100	147	117	6.3	13.3
BMV (pH 6)	4.6	3.6	1.0	—	92	130	106	5.4	9.2
BMV (pH 7)	4.6	3.6	1.0	—	101	138	128	5.4	11.0
R17	3.6	2.5	1.1	96	104	133	128	4.0	9.8

* "Radius of RNA" refers to that radius inside of which the area under the curve of $4\pi r^2 \rho_{eff}$ is just equal to the total number of "effective electrons" contributed by the RNA of the virus.

much too small to account for more than a small fraction of the excess volume. In R17, for example, the 30 Å hollow center occupies only 0.1% of the volume of the virus and the 80 Å hole in BMV has a volume equal to only 3% of the total volume of the virus. There

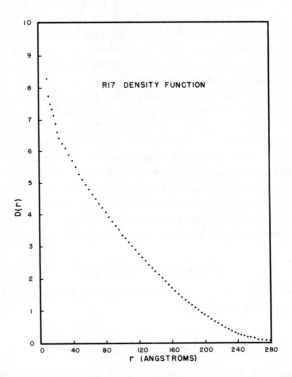

Fig. 14. The density function, $D(r)$, for the bacteriophage R17 as calculated by a Fourier transformation of the scattered intensity curve. The calculation used data out to a scattering angle of 0.1 radians corresponding to a Bragg spacing of 15 Å. An artificial temperature factor was used which multiplied the data at the cut-off angle by 0.1.

appears to be no relation between the size of the hole and the amount of RNA. R17 has the same amount of RNA as BMV and it has more space available for the RNA since its outer radius is slightly larger and it has a smaller amount of protein. Yet the hollow center is larger in BMV than it is in R17. The hollow centers do indicate that the RNA is not just expanding to fill the entire space available to it but rather interacts with itself or the protein in such a way as to be forced away from the center of the virus.

Finally, in connection with the subunit structure of the protein coat we can say that the scattering curves for the top components of WCMV and SMV give no indication of subunits. If there were spherical subunits approximately 40 Å in diameter the scattering curve for top component should be modulated by the transform of a 40 Å sphere, which means that there should be a minimum in the scattering curve at approximately 55 milliradians and a maximum at 71 milliradians. The scattering curve for top component of WCMV does not show these features; the curves for SMV were not carried out far enough. One might also expect subunits in the protein coat to produce maxima in the "density function," $D(r)$, at a spacing corresponding to the spacing between subunits. Using Eq. 3 we have calculated $D(r)$ for R17 from scattering data extending out to an angle of 100 milliradians corresponding to a Bragg spacing of 15 Å. The result is shown in Fig. 14. The curve is seen to be very featureless. We conclude from both of these results that the subunits in the protein coats of these viruses must pack together more closely than spheres would pack so that the subunits are not well distinguished.

ACKNOWLEDGEMENT

I wish to thank Dr. F. A. Fischbach for his valuable assistance in the preparation of this paper.

BIBLIOGRAPHY

1. Anderegg, J. W., Geil, P. H., Beeman, W. W. and Kaesberg, P., Biophys. J., **1,** 657 (1961).
2. Anderegg, J. W., Wright, M. and Kaesberg, P., ibid., **3,** 175 (1963).
3. Fischbach, F. A., Harrison, P. M. and Anderegg, J. W., J. Mol. Biol. (1965); in press.
4. White, R. A., Ph.D. Thesis, University of Wisconsin (1962).
5. Markham, R. and Smith, K. M., Parasitology, **39,** 330 (1949).
6. Markham, R., Discussions Faraday Soc., **11,** 221 (1951).
7. Schmidt, P., Kaesberg, P. and Beeman, W. W., Biochim. Biophys. Acta, **14,** 1 (1954).
8. Yamazaki, H. and Kaesberg, P., ibid., **51,** 9 (1961).
9. Mazzone, H. M., Incardona, N. L. and Kaesberg, P., ibid., **55,** 164 (1962).
10. Gestland, R. F. and Boedtker, H., J. Mol. Biol., **8,** 496 (1964).

11. Bockstahler, L. E. and Kaesberg, P., Biophys. J., **2,** 1 (1962).
12. Yamazaki, H., Bancroft, J. and Kaesberg, P., Proc. Nat. Acad. Sci., **47,** 979 (1961).
13. Incardona, N. L. and Kaesberg, P., Biophys. J., **4,** 11 (1964).
14. Caspar, D. L. D. and Klug, A., Cold Spring Harbor Symp. Quant. Biol., **27,** (1962).

Myoglobin: Low-Angle X-Ray Scattering Properties As Calculated from the Known Structure

H. C. WATSON

Medical Research Council Laboratory of Molecular Biology, Hills Road, Cambridge, England

INTRODUCTION

As the structure of myoglobin began to emerge from the high resolution X-ray studies (1) the author was asked by several groups if it would be possible to calculate the low-angle scattering properties of the protein prior to the publication of the complete structure. The reasons for these requests seemed twofold:

(1) To use the myoglobin molecule as a model system to test the low-angle scattering technique as applied to proteins.

(2) To establish by direct experimental methods the relationship between the structure of the protein in the crystal and that in solution.

The purpose of this paper is therefore to describe the gross structural parameters of the myoglobin molecule and to detail calculations which have been made with both grid point electron density data and atomic coordinates. These calculations permit the structure, as found in the crystal, to be compared with the structure indicated by low-angle X-ray scattering measurements made from dilute myoglobin solutions. The results of such a direct physical comparison will complement the conclusions which can be drawn from chemical experiments carried out both in solution and in the crystal (2).

THE MYOGLOBIN MOLECULE

Sperm whale myoglobin is composed of a single polypeptide chain containing 153 amino acid residues and a single prosthetic group. The molecular weight as calculated from the known amino acid sequence is 17,816 (3). The complete X-ray structure determination (4) has revealed that 118 residues are arranged in eight helical sections varying in length from 7 to 24 residues. The helical segments are themselves arranged to form a compact structure (see Fig. 1) in which the interior of the molecule is filled with closely packed hydrophobic groups.

The overall shape of the molecule is best described as a flattened triangular prism which closely approaches an oblate ellipsoid with dimensions:

$$a = 22 \text{ Å} \qquad b = 12.5 \text{ Å}$$

Fig. 1. A space-filling model of myoglobin showing the overall compact shape of the molecule.

Assuming uniform density throughout the molecule, values for the surface area S, volume V, and radius of gyration R_g can be calculated from Eqs. 1, 2 and 3, respectively:

$$S = 2\pi a^2 + \frac{\pi b^2}{\epsilon} \log_e \frac{(1 + \epsilon)}{(1 - \epsilon)} \qquad 1.$$

$$V = \tfrac{4}{3}\pi a^2 b \qquad 2.$$

$$R_g^2 = \frac{2a^2 + b^2}{5} \qquad 3.$$

where the eccentricity $\epsilon = (1 - b^2/a^2)^{\frac{1}{2}}$. The calculated volume of 25,300 Å^3 is in good agreement with the value of 26,000 Å^3 derived by direct measurement on electron density sections, but the value of 4,400 Å^2 for the surface area must be lower than the actual value since the surface of the molecule is considerably less regular than an ellipsoid would imply. Direct measurement indicates a maximum value of 5,500 Å^2. For this reason the ratio of S/V of 0.18 calculated from Eqs. 1 and 2 must be considered slightly lower than the actual value.

The radius of gyration from these dimensions is 15 Å and must be slightly lower than the actual value since the axial parameters given do not necessarily include the long charged side chains of arginine and lysine, which often protrude outwards from the surface of the molecule in an ill-defined manner.

CALCULATION OF THE RADIUS OF GYRATION

The electronic radius of gyration R_g is normally defined by the equation:

$$R_g^2 = \frac{\int_v \rho(\mathbf{r}) r^2 \, dV}{\int_v \rho(\mathbf{r}) \, dV} \qquad 4.$$

where $\rho(\mathbf{r})$ is the electron density a distance $\mathbf{r}$ from the center of charge and V is the volume of the molecule.

When working from a set of atomic coordinates or a set of electron density grid points it is more convenient to use the expression for R_g defined by Eq. 5:

$$R_g^2 = \sum_{n=1}^{N} \frac{f_n\{(x_n - x_0)^2 + (y_n - y_0)^2 + (z_n - z_0)^2\}}{\sum_{n=1}^{N} f_n} \qquad 5.$$

where N is either the number of atoms in the molecule or grid points in the electron density distribution, f_n is the electron density associated with the nth atom or grid point, whose coordinates are x_n, y_n, z_n. The center of electron density is at the point x_0, y_0, z_0 defined by Eqs. 6, 7 and 8:

$$x_0 = \frac{\sum_{n=1}^{N} f_n x_n}{\sum_{n=1}^{N} f_n} \qquad 6.$$

$$y_0 = \frac{\sum_{n=1}^{N} f_n y_n}{\sum_{n=1}^{N} f_n} \qquad 7.$$

$$z_0 = \frac{\sum_{n=1}^{N} f_n z_n}{\sum_{n=1}^{N} f_n} \qquad 8.$$

Calculation of R_g from the Grid Point Density Data

The 6 Å resolution electron density distribution of sperm whale myoglobin shown in Fig. 2 (5) was used as data for this calculation. The molecular boundary for each section of the Fourier synthesis was determined by a curve, following the zero contour wherever possible, drawn so as to enclose the known molecular features. The positions of each of the 3,264 grid points and their associated electron densities were then used in Eqs. 5, 6, 7 and 8 to derive a value of 16.0 Å for the radius of gyration of myoglobin.

The value of R_g derived from these data should be an overestimate of the true value since sulfate ions bound to the surface of the protein could fall within the defined molecular boundary.

Calculation of R_g from the Atomic Coordinates

The 1.4 Å resolution refinement of the structure of myoglobin (4) has led to the determination of the coordinates of some 1,200 out of a total of 1,260 non-hydrogen atoms in the molecule. Using these coordinates and assuming point atoms with the appropriate number of electrons, the radius of gyration derived from Eqs. 5, 6, 7 and 8 is 14.9 Å.

Since the hydrogen atoms appear to be evenly distributed throughout the molecule their omission from the calculation will have little if any effect on the result. The omission of some 60 surface atoms will, however, result in a slightly low value for R_g.

RADIAL DISTRIBUTION CURVES

From the atomic coordinates of a known structure it is possible to calculate the radial intensity function $I(h)$ (6):

$$I(h) = \sum_{j=1}^{N} \sum_{k=1}^{N} \frac{f_j f_k \sin h r_{jk}}{h r_{jk}} \qquad 9.$$

Fig. 2. The 6 Å resolution electron density Fourier sections. Reproduced by kind permission of Dr. J. C. Kendrew.

f_j, r_{jk} and N are as previously defined and h is equal to $4\pi \sin \theta / \lambda$ where 2θ is the scattering angle and λ the X-ray wavelength. For myoglobin, the number of atoms to be considered is close to 1,200 and an exact evaluation of Eq. 9 at small intervals of h would take even the largest computers a considerable time. Two approximations have therefore been made in the calculation of $I(h)$ from the atomic coordinate data given in Table I. r_{jk}, the distance between the j and kth atoms, was found by using the value of r_{jk}^2 truncated to the nearest integer as the address of r_{jk} in a precomputed square root table. r_{jk} multiplied by h and an arbitrary scale factor (100) was then truncated to the nearest integer and used to find $I(h)_{jk}$ from a precomputed table of $(\sin x)/(x)$.

For the purpose of comparison the calculated $I(h)$ curve is shown in Fig. 3 together with similar curves prepared by forming the mean

TABLE I

The normalized radial intensity distribution $I(h)$ of sperm whale myoglobin calculated from the atomic coordinate data.

h	0.00	0.02	0.04	0.06	0.08
0.0	1.000	0.970	0.890	0.767	0.622
0.1	0.476	0.342	0.229	0.148	0.0862
0.2	0.0504	0.0308	0.0209	0.0162	0.0135
0.3	0.0116	0.0102	0.00926	0.00865	0.00824
0.4	0.00780	0.00730	0.00673	0.00625	0.00591
0.5	0.00558	0.00527	0.00507	0.00490	0.00473
0.6	0.00456	0.00436	0.00415	0.00392	0.00368
0.7	0.00351	0.00326	0.00306	0.00285	0.00265
0.8	0.00248	0.00232	0.00218	0.00206	0.00197
0.9	0.00189	0.00183	0.00177	0.00174	0.00171
1.0	0.00169	0.00169	0.00169	0.00169	0.00169
1.1	0.00169	0.00169	0.00169	0.00170	0.00171

over ranges of h of the square of the observed and calculated structure amplitudes derived from single crystal studies of myoglobin. The peak at 10 Å, which has been interpreted "as a manifestation of the tendency of coiled peptide chains to fold at a fixed distance apart" is pronounced in the curve derived from the experimental results but much less so in the other two curves.

The radial Patterson functions calculated from the myoglobin grid point and coordinate data are shown in Fig. 4. Neither curve shows an abnormally high vector distribution at 10 Å from the origin. Furthermore the large differences between observed and calculated structure amplitudes at low scattering angles arises because the waves diffracted from the solution between molecules in the crystal destructively interfere with the waves diffracted by the myoglobin molecules. These results therefore seem to indicate that the nature of the solution should be carefully considered when drawing conclusions about the internal structure of protein molecules from the radial intensity distribution curves.

CONCLUSIONS

Single crystal X-ray studies have shown that the conformations of sperm whale myoglobin and seal myoglobin, and the subunits of human and horse hemoglobin are all similar, although the crystal lattices, and hence the forces acting on them, are quite different in each case. Experiments carried out in solution indicate that a standard

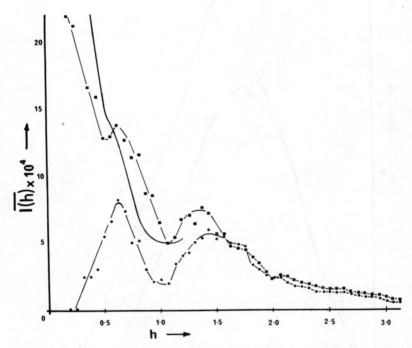

Fig. 3. Radial intensity curves. Squares represent the results obtained from the calculated structure factors. Circles represent the results obtained from single crystal intensity measurements. Full line is a plot of part of the calculated radial intensity function $I(h)$ given in Table I.

conformation exists under a variety of conditions (7) and that this conformation is the one which is thermodynamically most stable. Gurd and his co-workers (2) have examined the relationship between the surface of the molecule in the crystal and in solution and have come to the conclusion that the molecule is very similar, if not identical, in the two states.

The results presented in this paper suggest that the electronic radius of gyration of the myoglobin molecule as found in the crystal is 15.5 ± 0.5 Å. This value is significantly lower than the value derived from low-angle X-ray scattering measurements by Luzzati

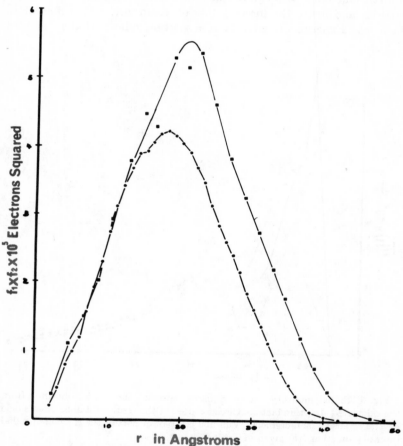

Fig. 4. The radial Patterson function calculated from the grid point electron density data (squares) and the atomic coordinate data (circles).

and his co-workers (personal communication) and also 16.4 ± 0.4 Å reported by Krigbaum and Brierre (9). Sztankay, Anderegg and Beeman (10) have shown however that solutions of sperm whale myoglobin contain stable aggregates and when positive steps are taken to remove these aggregates they obtain a value of 15.8 ± 0.8 Å for the radius of gyration from their low-angle X-ray scattering measurements.

The agreement thus obtained between the calculated and observed values of the myoglobin radius of gyration is in accordance with the work already mentioned which strongly suggests that the potential minimum representing the "native" form of the protein is sufficiently wide and deep to maintain its conformation under widely differing conditions.

It is hoped that the results presented in this paper will enable further use to be made of the myoglobin molecule as a model system for low-angle X-ray studies.

BIBLIOGRAPHY

1. Kendrew, J. C., Watson, H. C., Strandberg, B. E., Dickerson, R. E., Phillips, D. C. and Shore, V. C., Nature, **190,** 666 (1961).
2. Banaszak, L. J., Andrews, P. A., Burgner, J. W., Eylar, E. and Gurd, F., J. Biol. Chem., **238,** 3307 (1963).
3. Edmundson, A. B., Nature, **205,** 883 (1965).
4. Kendrew, J. C. and Watson, H. C., to be published.
5. Bodo, G., Dintzis, H. M., Kendrew, J. C. and Wyckoff, H. W., Proc. Roy. Soc., **A253,** 70 (1959).
6. Debye, P., Ann. Physik, **46,** 809 (1915).
7. Urnes, P. J., Ph.D. Thesis, Harvard University (1963).
8. Harrison, S. C. and Blout, E. R., J. Biol. Chem., **240,** 299 (1965).
9. Krigbaum, W. R. and Brierre, R. T., Nature, **206,** 396 (1965).
10. Sztankay, Z. G., Anderegg, J. W. and Beeman, W. W., American Biophysical Society Abst., **145** (1965).

Application of Small-Angle X-Ray Scattering to Problems in Physical Metallurgy and Metal Physics

VOLKMAR GEROLD

Max-Planck-Institute for Metal Research, Stuttgart, Germany

TABLE I
List of symbols

A = distance between sample and detector slit
$\mathbf{b}$ = Burgers vector of a dislocation
b = modulus of $\mathbf{b}$
c = volume fraction of the particles
D = interparticle distance
E = measured energy flux in the detector
E_0 = total energy flux of the primary beam
$\mathbf{h} = 2\pi\mathbf{s}$ = reciprocal lattice vector
h = modulus of $\mathbf{h}$
i_p = scattering function from a single particle (point collimation)
I = scattering function (point collimation), electrons/atom
$I_e = 7.9 \cdot 10^{-26}$ cm^2
j_p = scattering function from a single particle (line collimation)
J = scattering function (line collimation), electrons/atom
$\mathbf{l}$ = directional vector of the dislocation lines
$2l$ = line collimation length
L_p = characteristic length
m_i = atomic fraction of the solute atom in the phase i.
$n = V_a^{-1}$ = number of atoms per unit volume
n_L = number of dislocation loops per unit volume
n_p = number of particles (zones) per unit volume
q = area of the detector slit
Q_i = Fourier transform of the intensity
$\mathbf{r}$ = distance vector in real space
R_g = radius of gyration
$s = \dfrac{2 \sin \theta}{\lambda}$
t = sample thickness
v = axial ratio
V = irradiated volume
V_a = atomic volume

TABLE I (*Continued*)

V_p = particle volume
w_i = shape factor
x = vector in real space
z = atomic number
2θ = scattering angle
λ = X-ray wavelength
μ = linear absorption coefficient
ν = Poisson ratio
ρ = local electron density
$\bar{\rho}$ = locally averaged electron density
$\langle \rho \rangle$ = average electron density
$\boldsymbol{\tau}$ = reciprocal vector perpendicular to both the primary beam and **s**
τ = modulus of $\boldsymbol{\tau}$
$\tau_0 = \dfrac{l}{A\lambda}$
φ = inter-particle interference function (point collimation)
Φ = inter-particle interference function (line collimation)

INTRODUCTION

Small-angle X-ray scattering (SAS) has a broad application in the field of macromolecules, high polymers, colloidal solutions etc., as is shown by most of the papers given at this conference.

In contrast, the application of SAS to problems of physical metallurgy is more restricted. There are three reasons for this: (i) in most cases the inhomogeneities in metallic alloys have linear dimensions exceeding 1000 Å and therewith exceeding the resolution of normal X-ray SAS cameras (ii) inhomogeneities can only be detected in the small-angle region if the differences in electron densities are large enough (iii) because of the high absorption of metals, thin foils with thicknesses of only 20 to 100 μ can be traversed by X-rays. Therefore, the effective volume of diffraction is very small. Nevertheless, there remains an interesting field of research with this special diffraction method.

SAS techniques were introduced by Guinier to solve physical metallurgical problems. He was the first to study in this way heterogeneities of unstable metallic solid solutions (1). The detected heterogeneities are well-known under the name Guinier-Preston zones (G.-P. zones). They have the same crystallographic structure as the matrix solution but differ in the alloy composition. Their sizes are of the order of 15 to 200 Å. Sometimes they produce elastic distortion fields around themselves due to different sizes of the segregated atoms and the matrix atoms inside and outside the zones.

The analysis follows the scheme of particle scattering analysis. The advantage of the application of small-angle scattering is the fact that lattice strains cause only minor changes of the diffracted intensity. This enormously simplifies the analysis of the diffraction diagrams. But as will be shown, the influence of the lattice strains cannot be neglected.

Recently, SAS has been applied to the analysis of stable solid solutions (2). It is possible to detect clusters of solute atoms which normally have sizes below 10 Å where the particle model is no longer valuable. Details are given by Guinier (3) in this volume.

Several years ago, SAS from deformed metals was widely discussed. Unfortunately it came out that the observed scattering was mainly due to double Bragg scattering (4). The Bragg reflections from lattice planes and therewith the double Bragg scattering can only be avoided by using slow neutrons with a sufficiently long wavelength (5). In that case the remaining scattering from deformed metals in the small-angle region is given by lattice dilatations due to the strain fields of dislocations (6). Recently, SAS of X-rays has been used to study the agglomeration of vacancies in pure metals after quenching.

The next chapter will give the theoretical background and the general formulae for SAS. The chapter following that deals with undesired diffraction effects. Subsequently the main chapter gives the experimental results for unstable solid solutions. After that, the experiments on clustering of vacancies are described and finally the results of the very few neutron diffraction experiments on deformed metals are summarized.

THEORETICAL BACKGROUND

General Formulae for Small-Angle Scattering

In small-angle diffraction experiments a thin flat specimen of thickness t is traversed perpendicularly by a primary X-ray beam of integral flux E_0. If the beam and the window of the detector have point collimation the resulting diffracted energy flux E measured by the detector behind the window of area q is given by

$$E(\mathbf{s}) = \frac{I_e}{A^2} nqte^{-\mu t}E_0 I(\mathbf{s}) \qquad 1.$$

The integral primary flux E_0 is measured by the same detector by direct or indirect means. The definition of symbols used in Eq. 1 and in the following equations is given in Table I.

In most cases line collimation is used for the primary beam as well as for the detector window instead of point collimation. Then Eq. 1 has to be altered by inserting the quantity $J(\mathbf{s})$ instead of $I(\mathbf{s})$:

$$E(\mathbf{s}) = \frac{I_e}{A^2} nqte^{-\mu t} E_0 J(\mathbf{s}) \qquad \text{1a.}$$

Under proper experimental conditions (8), $J(\mathbf{s})$ is given by

$$J(\mathbf{s}) = \frac{1}{\tau_0} \int_0^{\tau_0} I(\mathbf{s} + \boldsymbol{\tau}) \, d\tau \qquad 2.$$

The factor τ_0 depends either on the length of the detector slit parallel to it or on the length of the primary beam collimation, depending on which one is the larger. This length is given by $2l$. Then

$$\tau_0 = \frac{l}{A\lambda} \qquad 3.$$

The most important case is that of centrosymmetry of the diffracted intensity I depending only on the modulus of $\mathbf{s}$. In this case Eq. 2 changes to

$$J(s) = \frac{1}{\tau_0} \int_0^{\tau_0} I(\sqrt{s^2 + \tau^2}) \, d\tau \qquad 4.$$

There exists a mathematical relationship which transforms the function $J(s)$ back into $I(s)$ (9), but many structural parameters can be analyzed directly from $J(s)$. Both I and J are measured in units of electrons/atom (e.u.).

In small-angle scattering techniques the very small angles are cut off from experimental observations as well as the large scattering angles. Therefore, for the present purpose, reciprocal space may be divided into three different regions:

(i) The region of the very small angles.
(ii) The small-angle region.
(iii) The wide-angle region.

These three diffraction regions correspond to three different "inhomogeneity dimensions" in the sample. The first inhomogeneity dimension is the dimension of the sample itself. Its intensity contribution is called volume diffraction I_v. It includes also the diffraction of very large particles (1000 Å and above). The next inhomogeneity may be given by some other electron density fluctuations in the sample with dimensions from 10 to 1000 Å causing particle scattering I_p. The presence of such fluctuations depends on the kind of metallic sample and its heat treatment. The third inhomogeneity range has atomic dimensions and is always present in the sample. It gives rise to a diffraction I_a.

The volume diffraction I_v is concentrated in region (i) which is cut off by the primary beam stop. Occasionally, its tail can be observed in region (ii).

The main intensity contribution in the small-angle region is given by electron density fluctuations of the order of 10 to 1000 Å. The atomic diffraction I_a occurs mainly in the wide-angle region (iii). Occasionally, it may also be observed as additional intensity background in the small-angle region. In the absence of particle scattering I_p it may be the only intensity contribution in that angular range.

Now we define $\rho(\mathbf{x})$ as the exact electron density distribution function, $\bar{\rho}(\mathbf{x})$ as the locally averaged distribution function and $\langle \rho(\mathbf{x}) \rangle$ as the average value of the electron density function which is constant in the sample and zero outside of it.* Then the following relationship exists:

$$I_i(\mathbf{h}) = \iiint Q_i(\mathbf{r}) \exp(-i\mathbf{r}\mathbf{h}) \, dv_{\mathbf{r}} \qquad 5.$$

with the abbreviation $\mathbf{h} = 2\pi \mathbf{s}$. The index i stands for v, p or a, with

$$Q_v(\mathbf{r}) = \frac{V_a}{V} \iiint_V \langle \rho(\mathbf{x}) \rangle \langle \rho(\mathbf{x} + \mathbf{r}) \rangle \, dv_{\mathbf{x}} \qquad 6a.$$

$$Q_p(\mathbf{r}) = \frac{V_a}{V} \iiint_V [\bar{\rho}(\mathbf{x})\bar{\rho}(\mathbf{x} + \mathbf{r}) - \langle \rho \rangle^2] \, dv_{\mathbf{x}} \qquad 6b.$$

$$Q_a(\mathbf{r}) = \frac{V_a}{V} \iiint_V [\rho(\mathbf{x})\rho(\mathbf{x} + \mathbf{r}) - \bar{\rho}(\mathbf{x})\bar{\rho}(\mathbf{x} + \mathbf{r})] \, dv_{\mathbf{x}} \qquad 6c.$$

The integrals are taken over the irradiated volume V. The quantities are normalized to the average atomic volume V_a which is a convenient unit in crystal lattices.

The definition of the locally averaged electron density is not exact. The average has to be taken over the volume of one to several atoms.

---

* More exactly, the average $\langle \ \rangle$ has to be taken over volume sizes of the order of the resolution volume of the small-angle apparatus.

Figure 1a gives an one-dimensional example of an idealized heterogeneity in a solid solution. The solution has segregated into regions of different solute concentrations, where one region has a concentration of 50% and shows long-range order. Figure 1b demonstrates the influence of strain fields from dislocations on $\bar{\rho}$ and ρ. In both cases differences occur between $\bar{\rho}$ and $\langle\rho\rangle$ giving rise to SAS.

There are two useful relationships between $Q_p(\mathbf{r})$ and $J_p(\mathbf{s})$ which are given below:

$$Q_p(0) = \iiint J_p(\mathbf{s})\, dv_\mathbf{s} \qquad 7.$$

$$J_p(0) = \iiint Q_p(\mathbf{r})\, dv_\mathbf{r} \qquad 8.$$

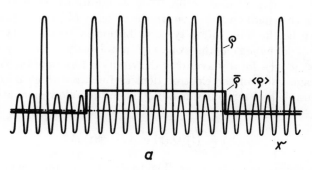

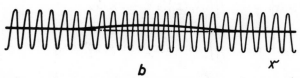

Fig. 1. The electron densities ρ, $\bar{\rho}$ and $\langle\rho\rangle$ for (a) a particle in a solid solution (b) an internal strain field.

In all following paragraphs the index p for quantities such as I_p and Q_p will be omitted for simplification. The symbols I and Q represent the scattered intensity in the small-angle region and its Fourier transform.

The aim of each diffraction analysis is to find from the intensity distribution some information about the structure of the sample. As the knowledge of the X-ray diffraction is restricted to the small-angle region, information can be found only about the average density function $\bar{p}(\mathbf{x})$. This is a severe restriction which will give information only on:

(i) Sizes and shapes of heterogeneities.
(ii) The quantity of heterogeneities and the average electron densities inside and outside of these heterogeneities.
(iii) Strain fields in special cases.

The next paragraphs will deal with three different diffraction sources:

(i) Particles.
(ii) Solid solutions.
(iii) Dislocations.

The Analysis of Small-Angle Scattering from Particles

1. *General Formulae*

In many cases the shape of a particle can be described by an ellipsoid of revolution with axial ratios 1, 1, v. If $v = 1$, the particle is a sphere. For $v < 1$ the ellipsoid is oblate, otherwise it is prolate. Then the size and the shape of a particle can be described by two parameters:

(a) The radius R perpendicular to the axis of revolution.
(b) The ratio v of the unequal axes.

In the case of many particles there are the following additional factors influencing the intensity in a qualitative manner:

(c) Size distribution.
(d) Shape distribution.
(e) Orientation relationships.
(f) Inter-particle distance relationships.

Orientation relationships of particles in crystals depend in most cases on crystallographic directions. These relationships can be found only in studying the diffraction from single crystals, e.g., Guinier-Preston zones in Al-Cu.

In the following paragraphs the analysis of specimens shall be discussed where the orientation relationship does not play a dominant role. In this case the intensity distribution $I(\mathbf{s})$ depends only on the modulus of $\mathbf{s}$. Ignoring at first a possible size distribution the formula for $I(s)$ is given by

$$I(s) = n_p V_a V_p^2 (\Delta\rho)^2 i_p(s)[1 - \varphi(s)] \qquad 9.$$

where

$\Delta\rho = \bar{\rho}_1 - \bar{\rho}_2 =$ the difference of the local averages of the electron density inside and outside the particle,

$i_p(s) =$ average intensity function from a single particle normalized to $i_p(0) = 1$. The average is taken over all directions of the vector **s**.

The inter-particle interference function $\varphi(s)$ has its maximum value at $s = 0$ and drops to zero for large angles. It can be neglected in dilute systems with random distribution of the particles. In that case, i_p is the only function showing an angular dependence.

As the product $n_p V_p$ is equal to the volume fraction of the particles, c, Eq. 9 can be written as

$$I(s) = cV_a V_p (\Delta\rho)^2 i_p(s)[1 - \varphi(s)]$$
$$\equiv I^*(s)[1 - \varphi(s)] \qquad 10.$$

The corresponding formula for the case of line collimation is

$$J(s) = cV_a V_p (\Delta\rho)^2 j_p(s)[1 - \Phi(s)]$$
$$\equiv J^*(s)[1 - \Phi(s)] \qquad 10a.$$

where j_p resp. $j_p\Phi$ follows from i_p resp. $i_p\varphi$ via the transformation, Eq. 4. In all following equations, I, i and φ refer to the point-collimated beam whereas J, j and Φ are connected with the proper line collimation. The quantities I^* and J^* are the (theoretical) intensity functions neglecting the inter-particle function.

2. *Approximations and Parameters of the Scattering from Particles*

In this section all needed approximations and parameters will be listed:

(a) Guinier approximation (8).

$$i_p(h) = \exp\left[-\frac{R_g^2 h^2}{3}\right] \quad \text{for} \quad R_g h < 1.2 \qquad 11.$$

$$j_p(h) = \frac{\sqrt{3\pi}}{4\pi R_g r_0} \exp\left[-\frac{R_g^2 h^2}{3}\right] \qquad 11a.$$

If the Guinier approximation is a good one over a large angular range the line-collimated function j_p will also be a good approximation giving the same slope as i_p. In unfavorable cases the slope will be different. A more exact method of finding the approximation i_p from j_p has been given by Luzzati (10).

For spherical particles with radius R_s

$$R_g{}^2 = \tfrac{3}{5}R_s{}^2 \qquad 12.$$

For rotational ellipsoids with axes R, R and vR

$$R_g{}^2 = \left[\frac{2+v^2}{5}\right]R^2 \qquad 12a.$$

Figure 2 shows the functions i_p and j_p for a spherical particle with their approximations.

(b) The asymptotic approximation (11).

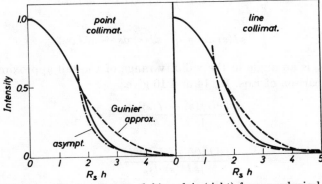

Fig. 2. The intensity functions i_p (left) and j_p (right) from a spherical particle. The oscillating tail has been approximated by a smooth curve.

If the particle has a sharp boundary with surface area S_p the approximation for large angles becomes (see Fig. 2)

$$i_p(h) = \frac{2\pi S_p}{V_p{}^2 h^4} \quad \text{for} \quad hR_g > 2.5 \qquad 13.$$

resp.

$$j_p(h) = \frac{\pi S_p}{4V_p{}^2 h^3 \tau_0} \qquad 13a.$$

(c) The integrated intensity (11).
From Eqs. 6b and 7 it follows that

$$Q(0) = 4\pi \int s^2 I(s)\, ds \qquad 14.$$
$$= 4\pi\tau_0 \int s J(s)\, ds \qquad 14a.$$

and

$$Q(0) = V_a[\langle \bar{\rho}^2 \rangle - \langle \rho \rangle^2] \qquad 15.$$

which is
$$Q(0) = c(1-c)(\Delta\rho)^2 V_a \qquad 16.$$
or
$$Q(0) = (\bar{\rho}_1 - \langle\rho\rangle)(\langle\rho\rangle - \bar{\rho}_2) V_a$$
in the particle model.

The exact evaluation of the integral in the denominator is very important. The contribution of the tail of the intensity curve gives the main trouble. This difficulty can easily be overcome in using the tail approximation, Eq. 13. It follows simply that

$$\int_0^\infty s^2 I(s)\, ds = \int_0^{s_0} s^2 I(s)\, ds + s_0{}^3 I(s_0) \qquad 17.$$

resp.

$$\int_0^\infty s J(s)\, ds = \int_0^{s_0} s J(s)\, ds + s_0{}^2 J(s_0) \qquad 17a.$$

where s_0 is an angle in the validity range of the tail approximation.

Comparison of Eqs. 10, 14 and 16 gives

$$\frac{\int s^2[I^*(s) - I(s)]\, ds}{\int s^2 I^*(s)\, ds} = c \qquad 18.$$

resp.

$$\frac{\int s[J^*(s) - J(s)]\, ds}{\int s J^*(s)\, ds} = c \qquad 18a.$$

(d) *The characteristic length* (11).

The characteristic length L_p in a particle is given by

$$L_p = \frac{\int s I_p{}^*(s)\, ds}{2 \int s^2 I_p{}^*(s)\, ds} \qquad 19.$$

resp.

$$L_p = \frac{\int J^*(s)\, ds}{\pi \int s J^*(s)\, ds} \qquad 19a.$$

For a sphere, $L_p = 1.5\, R_s$.

For an oblate ellipsoid L_p is given by (12)

$$L_p = 1.5\, R\, \frac{\sinh^{-1}\sqrt{1/v^2 - 1}}{\sqrt{1/v^2 - 1}} \qquad 20.$$

(e) *The particle volume* V_p.

This volume is given by

$$V_p = \frac{I^*(0)}{4\pi \int s^2 I^*(s)\, ds} \qquad 21.$$

resp.

$$V_p = \frac{R_g J^*(0)}{\sqrt{3\pi} \int s J^*(s)\, ds} \qquad 21\text{a}.$$

(f) The particle shape factors.

If one compares different parameters, it is possible to get information on the shape of the particles. This can be expressed by shape factors w_i which are normalized to unity for spherical particles. Two of them will be given here

$$w_1 = V_p/V_s \qquad 22.$$

with

$$V_s = \frac{4\pi}{3}\left(\frac{5R_g^2}{3}\right)^{\frac{3}{2}} = \frac{4\pi}{3} R_s^3$$

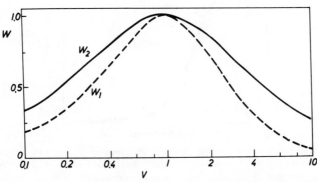

Fig. 3. The shape factors w_1 and w_2 as function of the axial ratio v.

where V_p is taken from Eq. 21 and R_g from the Guinier approximation. If the particle is an ellipsoid of revolution with axial ratio v, it is found with Eq. 12a that

$$w_1 = v\left(\frac{3}{2 + v^2}\right)^{\frac{3}{2}} \qquad 23.$$

The other factor is defined by

$$w_2 = \frac{L_p}{1.5\, R_s} \qquad 24.$$

where R_s is taken from the Guinier approximation and Eq. 12. The relation between w_1 resp. w_2 and the axial ratio v is given in Fig. 3.

(g) The inter-particle interference function.

With the exception of the integrated intensity all other parameters are influenced by the inter-particle interference function φ resp. Φ. The main problem is the evaluation of I^* out of I (resp. J^* out of J). Figure 4 shows two examples of the influence of φ on I for a volume fraction $c = 0.1$ and for identical spherical particles. Curve (a) represents a nearly random distribution of particles whereas curve (b) demonstrates the case of ordering between the particles. In both cases the tail of I^* is not influenced. Ordering between particles restricts the influence of φ on I^* to small angles where it is very pronounced.

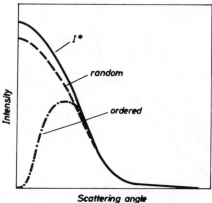

Fig. 4. The influence of the inter-particle interference function on the intensity curve $I(h)$ from spherical particles with the volume fraction $c = 0.1$. The function I is simply the sum of all individual intensity curves $i_p(h)$.

3. *The Influence of the Size Distribution of Particles*

In the preceding paragraphs the scattering from a set of identical particles was discussed. In practice the particles will be distributed over a range of different sizes. Therefore, the measured parameters R_s, V_p, etc. will be some average value depending on the size distribution and the method of parameter evaluation.

Without going into detail, the influence of the size distribution can be characterized as follows:

At small diffraction angles the intensity contribution of the larger particles dominates. Therefore, parameters evaluated in this angular range will overestimate the size of the particles. At large diffraction angles the intensity contribution of the smaller particles is more important. Parameters evaluated from the tail of the curve underestimate the size of the particles.

The size distribution has a similar effect on the scattering curve as the deviation of the particles from spherical shapes. Therefore, the shape factors w_i are diminished in their values compared to the shape factors from identical particles. It is now impossible to determine the shape of the particles in an exact way.

4. *The Determination of the Atomic Concentrations in the Particles*

In the preceding paragraphs the qualitative analysis of the diffraction from particles was outlined. This paragraph deals with the information from quantitative investigations. The most important formulae for this type of analysis are Eqs. 14 to 16 giving the integrated intensity

$$Q(0) = c(1-c)(\Delta\rho)^2 V_a$$
$$= (\bar{\rho}_1 - \langle\rho\rangle)(\langle\rho\rangle - \bar{\rho}_2) V_a \qquad 16.$$

In this equation, only quantitative parameters are involved:

(a) The volume fraction c of the particles.

(b) The change $\Delta\rho = \bar{\rho}_1 - \bar{\rho}_2$ of the electron density inside and outside the particles.

In most cases both quantities are unknown. Their determination seems to be difficult, because they cannot be separated. Fortunately, it can be assumed by thermodynamic arguments that $\bar{\rho}_1$ and $\bar{\rho}_2$ depend only on the aging temperature but that they are independent of the solute concentration of the alloy. Then it is possible to find $\bar{\rho}_1$ and $\bar{\rho}_2$ from the measurement of the integrated intensities $Q(0)$ from two alloys of the same system with different $\langle\rho\rangle$. This gives two equations with the two unknown variables $\bar{\rho}_1$ and $\bar{\rho}_2$ which can be computed.

If the average atomic volume is the same inside and outside the particles Eq. 16 can be expressed as

$$Q(0) = (m_1 - m_A)(m_A - m_2) \frac{(\Delta z)^2}{V_a} \qquad 25.$$

where m_A is the average solute concentration of the alloy, m_1 and m_2 are the solute concentrations inside and outside the particles. The difference of the atomic number between solute and solvent atoms is given by Δz. In this case the concentrations m_1 and m_2 are the unknown parameters which can be computed from two measurements.

If the average atomic volume inside and outside the particles is different, Eq. 25 must be changed. With the assumption that the average atomic volume of the solid solution with the solute concentration m_i is given by

$$V_i = V_0(1 + \kappa m_i)$$

the following result is obtained:

$$Q(0) = (m_1 - m_A)(m_A - m_2) \frac{(\Delta z)^2}{V_a} \left[1 - \kappa \left(m_1 + m_2 + 2 \frac{z_B}{\Delta z} \right) \right] \quad 26.$$

with z_B = atomic number of the solvent.

This equation is an approximation in the first order of κ which is sufficient for practical cases. The volume change has a big influence on $Q(0)$ because of the last additive term in the bracket of Eq. 26. This follows from the fact that the change in atomic volume changes the total electron density in the particular case whereas $Q(0)$ is proportional to the square of differences in electron densities.

SAS from Dislocations

The diffraction from dislocation lines is outlined in detail in the paper by Seeger and Brand (13). Therefore, only two simple formulae will be given. From the general linear elasticity theory of dislocations [Kröner (14)] it follows that the scattering amplitude (15) is given by

$$F(h) = \langle \rho \rangle \left(\frac{V_a}{V} \right)^{\frac{1}{2}} \frac{1 - 2\nu}{1 - \nu} \frac{2\pi \mathbf{b} \times \mathbf{h}}{h^2} \int_{\text{dislocation line}} \exp(-i\mathbf{h}\mathbf{x}) \, d\mathbf{l} \quad 27.$$

For dislocation loops (radius R) of pure edge dislocations with random orientations the following approximate intensity formula (valid for small angles only) is found (7, 15, 16):

$$I(h) = n_L \frac{z^2}{V_a} \left(\frac{1 - 2\nu}{1 - \nu} \right)^2 \frac{8\pi^2 b^2 R^4}{15} \exp \left[-\frac{3}{14} h^2 R^2 \right] \quad 28.$$

The theory neglects

(a) surface terms (the internal cut-off radius has been put at zero which gives infinite values for the strain)

(b) non-linear elastic effects near the core of the dislocation.

In their paper, Seeger and Brand (13) improve the theory of diffraction from dislocation loops. The new results differ especially for small loop sizes. For example, the intensity is increased up to 30% for a loop radius $R = 2.5 \, b$.

UNDESIRED DIFFRACTION EFFECTS

Before discussing the experimental results two undesired diffraction effects should be mentioned: Diffraction from polished surfaces and double Bragg scattering.

Diffraction from Polished Surfaces

During their investigation of small-angle scattering from aluminum alloys, Freise, Fine and Kelly (17) found an influence of the surface treatment on the diffraction pattern. In addition to the usual diffraction they found an intensive scattering at angles below 1°. The intensity increases rapidly with decreasing angle (CuKα radiation). This effect was also found from pure aluminum foils polished electrolytically and quenched from high temperatures.

According to the authors, the polished surface no longer acts as sink for the quenched-in vacancies. The vacancies agglomerate in the sample (pure Al or Al-Ag) leading to additional dislocation loops and affecting the diffraction curve by additional double Bragg scattering.

This explanation could not be confirmed by Zürn (18). His experiments show that the additional diffraction effects are independent of quenching (pure Al and Al-Zn). The effect is always observed after electrolytic polishing and disappears with etching of the surface. The only explanation is that electropolishing itself leads to the diffraction effect.

Double Bragg Scattering

The normal Bragg diffraction from lattice planes leads to intensive secondary beams which may be diffracted again at lattice planes. This double scattering can occur in different crystallites forming a more or less homogeneous background intensity which increases in the small-angle range (19, 20). In most experiments this intensity is small and can be neglected. If double Bragg diffraction occurs within the same grain which is slightly distorted, more intense diffraction is observed in the small-angle region. Therefore, cold rolled metal sheets show relatively high SAS.

Fortunately, this high diffracted intensity is still small compared to the diffraction by inhomogeneities in solid solutions. In more sensitive experiments, however, double Bragg scattering must be taken into account. The best way to do this is to avoid the primary

Bragg reflection. This can be accomplished with single crystal foils and monochromatic radiation. The other possibility is to use a wavelength $\lambda > 2d_0$ where d_0 is the largest interplanar distance in the crystal structure. Then there is no Bragg diffraction at all. This leads to X-rays with wavelengths of the order of 6 Å, which are highly absorbed in metal foils. Very thin specimens are needed and it appears nearly impossible to carry out quantitative experiments. This difficulty can be avoided by the use of slow neutrons (21).

EXPERIMENTS ON SOLID SOLUTIONS

Introduction

Since the first famous experiments of Guinier in 1939 (22) on the small-angle diffraction phenomena of aluminum alloys many investi-

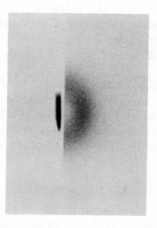

Fig. 5. Small-angle scattering from an Al-Ag alloy containing 5.9 at.% Ag. Aging for 4 h at 150°C. Monochromatic Cu-Kα radiation. The intensity maximum occurs at a diffraction angle of 1.2° (23).

gators have used this tool to study the changes in the structure of solid solutions with aging treatments.

Three different types of diffraction effect have been observed. In the case of Al-Ag and Al-Zn and some other ternary alloys, the diffraction is a concentric ring around the primary beam which is independent of the crystal orientation (Fig. 5) indicating voluminous particles of solute atoms. Similar diffraction halos occur around every lattice point of the reciprocal lattice.

The second type of diffraction pattern consists of intensity rods in the reciprocal lattice. They originate from very thin platelets of solute atoms (probably one atomic layer thick) and have been observed in Al-Cu and Cu-Be alloys. The third type shows intensity planes in the reciprocal lattice originating from thin needles (probably chains of solute atoms). For details see Guinier (24).

In all cases the diffraction occurs after quenching the sample from the homogenization temperature to room temperature where a two-phase field exists in the phase diagram (Fig. 6). The explanation of the

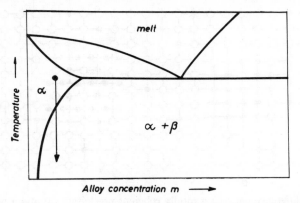

Fig. 6. Schematic phase diagram for a binary alloy. The samples are quenched from the single-phase region into the two-phase region.

observed diffraction phenomena is given by segregation effects within the solid solution leading to the so-called pre-precipitation state. The aggregates enriched with solute atoms are known as Guinier-Preston zones. After annealing at higher temperatures this state changes and finally leads to the precipitation state where real precipitates occur having a different crystalline lattice structure than the solid solution. Small-angle scattering then has nearly vanished. Therefore, the application of small-angle scattering is restricted mainly to the study of these pre-precipitation states, i.e., Guinier-Preston zones. The modern view describes the zones as metastable precipitates having the same lattice structure as the matrix. There exist only coherent phase boundaries which ease the nucleation of the zones (Fig. 7).

In many cases the study of the diffraction at large angles is more convenient. For instance, the quantitative determination of the amount of copper atoms precipitated into monatomic layers in Al-Cu alloys has been performed by studying the cross section of the intensity rod (11l) at the position (110) (26). Other alloys, such as Au-Pt or Cu-Ni-Fe, give nearly no contrast in electron densities and therefore nearly no small-angle scattering during segregation processes. As the segregated areas have slightly different lattice constants, they give rise to so-called side-bands accompanying the Scherrer lines on powder patterns (27, 28).

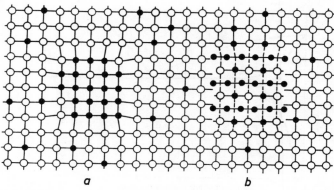

Fig. 7. The difference between a totally coherent precipitate (a) and a partially coherent precipitate (b) (25).

For these reasons the following review gives the results primarily for two alloys where intensive SAS has been found.

The Binary Systems Al-Ag and Al-Zn

1. *Introduction*

Both systems show small-angle diffraction patterns which are very similar. They consist of a diffuse diffraction ring around the primary spot (Fig. 5). It can be observed immediately or shortly after quenching the alloy into the two-phase field. Because the scattering is centro-symmetric, polycrystalline specimens are convenient for experimental investigation.

The diffraction was explained as being due to spherical zones enriched by the solute atoms. Some difficulties arose because of the profile of the diffraction pattern. Instead of the normal intensity distribution (Fig. 4) a diffraction ring occurs. Extrapolation to $s = 0$

leads to the intensity $I(0) \approx 0$. The first interpretation was due to Guinier (29) who explained the decrease with the inter-particle interference function $\varphi(s)$. The angular width s_w of the whole diffraction was used as a reciprocal measure of the size $2R$ of the particle ($2R \approx s_w^{-1}$) whereas the angle s_m of maximum intensity was interpreted as a reciprocal measure of the average distance between the particles, D:

$$D \approx s_m^{-1}$$

Further investigation of the change of the zone state with annealing treatments showed an increase of the size $2R$ as well as of the distance D. The ratio between both quantities remained nearly constant during isothermal treatments. This led to another explanation of the diffraction, given by Walker and Guinier (30). Each spherical zone enriched by solute atoms is surrounded by a depleted shell. The distance D of the first model is now interpreted as the outer diameter of the depleted spherical shell. In this case the line profile is given by the variation of the electron density within a single particle.

Because the angle s_m defining D is an easy measure, the quantity D is sometimes used as the particle size (31). It is larger by roughly a factor of 2 than the diameter $2R$ of the sphere enriched by solute atoms. As will be shown, the latter model is questionable. Therefore $2R$ is the more important quantity.

Soon interest arose in the study of physical properties in connection with the structural changes in solid solutions. Age-hardening is the most important property. It occurs in two steps. Zone-hardening (cold-hardening) can be distinguished from precipitation hardening (heat-hardening) (32). The first occurs at lower temperatures and can be reverted by short heating periods at about 200°C. It was believed that at this temperature the alloy reverts to the supersaturated homogeneous solid solution. SAS showed this is true for Al-Zn (31) but not for Al-Ag (33, 34). In the first case small-angle scattering disappears, whereas in the second case it is only lowered in intensity. This was very difficult to understand and the reason was not quite clear. Two different interpretations were given:

(i) During reversion small zones disappear whereas large zones still remain. The small zones are responsible for age-hardening and make only a minor contribution to SAS (33).

(ii) Silver atoms from the enriched spherical zone go into the depleted shell diminishing the concentration variation (34).

The next paragraph will show how further quantitative study of SAS on an absolute scale leads to a plausible description of the zone state.

Besides the zone state the beginning of the γ'-precipitation in Al-Ag alloys also gives a diffraction effect in the small-angle range. It consists of short streaks parallel to $\langle 111 \rangle$ directions (35).

At high temperatures, the system Al-Zn has a stable miscibility gap within the solid solution. The critical point occurs at 39.5 at.% Zn and 351.5°C (36). Münster and Sagel (37) investigated the phenomenon of critical opalescence which gives a pronounced SAS effect at the critical concentration just above the critical temperature.

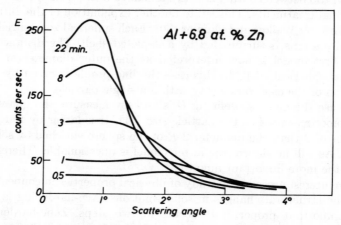

Fig. 8. Typical sequence of SAS curves for isothermal heat treatments of Al-Zn alloys. The intensity was measured at fixed angles as a function of time for repeated aging treatments at room temperature after quenching from 350°C (39).

2. *Miscibility Gaps in Al-Ag and Al-Zn Alloys*

In a preliminary investigation, it was found by Gerold (34) that the integrated intensity remained constant during isothermal treatment but changed reversibly with the aging temperature. Similar results were found by Jan (38). More precise measurements on different Al-Ag and Al-Zn alloys on an absolute scale (39, 40, 41) led to the conclusion that the integrated intensity, $Q(0)$, is a function of alloy concentration and of temperature only. As an example, Fig. 8 shows a sequence of SAS curves taken during isothermal annealing of an Al-Zn alloy. With increasing annealing times the

SAS curves shift to smaller angles and increase their intensity values. The shift indicates the increase in zone size. The integrated intensities of all these curves are the same. Figure 9 shows the radius of the zones and the integrated intensity as a function of annealing time for an Al-Ag alloy.

The experimental determination of the integrated intensity is not very accurate. The main trouble comes from the tail of the SAS curve. The accuracy of the measurement is not very high in this region. Secondly, some other diffraction effects may give an additional background which becomes noticeable in the tail. This point will be

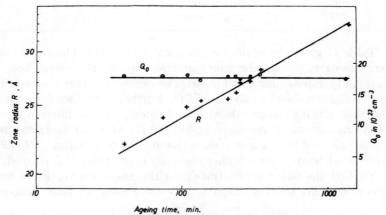

Fig. 9. The variation of the spherical radius R_s and the integrated intensity $Q(0)$ as a function of aging time. Al + 6 at.% Ag, aged at 140°C (41).

discussed in the next paragraph. To evaluate the integrated intensity the following procedure is adopted (see Eq. 17a):

The quantity

$$N(s_0) = \int_0^{s_0} sJ(s)\,ds + s_0^2 J(s_0) \qquad 29.$$

is calculated as a function of s_0, with s_0 being an angle in the tail region of the SAS curve. Normally this function has a minimum or a flat plateau. This special value of $N(s_0)$ is chosen as the right value and converted to $Q(0)$ according to Eqs. 1 and 14 by standard methods. The structure model (previously outlined) which assumes the total segregation of the alloy into two phases with compositions m_1 and m_2 was adopted. Then Eq. 25 holds.

TABLE II
Average silver concentrations m_1 and m_2 of the limits of the metastable miscibility gap in Al-Ag (41).

Aging temp., (°C)	Silver concentrations	
	m_1 (at.%)	m_2 (at.%)
23	59.5 ± 2.0	0.20 ± 0.35
100	56.8 ± 1.8	0.68 ± 0.33
140	53.9 ± 0.3	0.71 ± 0.04
175	41.9 ± 2.0	1.07 ± 0.60
226	34.7 ± 0.3	0.83 ± 0.15
300	28.5 ± 3.3	1.33 ± 1.30

Table II gives the result for the alloy Al-Ag (41). The silver concentrations m_1 inside the zones and m_2 outside the zones form a miscibility gap as shown in Fig. 10. The characteristic feature of this gap is its contraction above 170°C. Further investigation of the diffraction at high angles showed the existence of two different kinds of zones existing above resp. below 170°C. At lower temperatures the atoms inside the zones show some sort of ordering (29, 42), whereas at temperatures higher than 170°C the order is destroyed.

To find the silver concentration in the zones Freise, Kelly and Nicholson (43) tried a different way by combining SAS methods

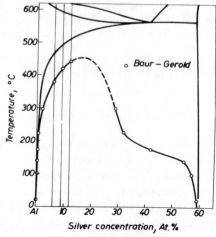

Fig. 10. The metastable miscibility gap in the system Al-Ag (41, 52).

with electron transmission microscopy. Instead of the integrated intensity they used Eq. 9 where the number n_p and the square of the particle volume V_p play the dominant role. Both quantities were taken from electron transmission patterns. The size of the particles found from electron microscopy and SAS differed by a factor 2. Because of this discrepancy and the high dependence of their results on the particle size used their findings* seem to be very doubtful.

In the case of Al-Zn alloys measurements were made only at room temperature (39, 40). The following Zn concentrations inside and outside the zones were found by application of Eq. 25:

$$m_1 = 0.78** \qquad m_2 = 0.018 \qquad\qquad 30.$$

These values are in good agreement with measurements of Herman, Cohen and Fine (44) on an alloy with 5.3 at.% Zn. From the integrated intensity they find experimentally $(m_1 - m_A)(m_A - m_2) = 0.025$ which fits exactly with the m-values given above.

Both values, m_1 and m_2, can be interpreted as the limit of a metastable miscibility gap which is the extension of the stable gap found above 275°C. However, there is a complication. Because of the differences in atomic diameters the lattice constant in the zinc-enriched zones is smaller than the lattice constant of the depleted matrix. Recent measurements of the diffraction halos around the Bragg diffraction spots (45, 45a, 45b) showed a difference in lattice constants of 1.17%. Therefore, Eq. 26 has to be applied instead of Eq. 25, with κ equal to -4%. As the correction is small the bracket of Eq. 26 can be calculated using the m_1 and m_2-values evaluated from Eq. 25. Dividing Eq. 26 by this factor leads again to an Eq. 25 with a modified $Q(0)$-value which is now smaller by 10%. This influences mainly the m_1-value, reducing it to 0.72. Thus the final result is

$$m_1 = 71.9 \text{ at.}\% \qquad m_2 = 1.75 \text{ at.}\% \qquad\qquad 31.$$

which is plotted in the phase diagram of Fig. 11. Another value for the limit of the miscibility gap has been given by Bonfiglioli and Levelut (46) for the aging temperature $-45°C$. They give their result in electron densities: $\rho_1 = 1.69$ el/Å^3 and $\rho_2 = 0.792$ el/Å^3 which leads to $m_1 = 78$ at.% and $m_2 = 1$ at.%.

* After aging an alloy with 5.6 at.% Ag at 125°C, only 10% of the alloy has segregated into zones with 94 at.% Ag and a depleted matrix of pure Al.

** In the original paper (40) the influence of the anomalous dispersion of X-ray diffraction on Δz has been neglected. The m-values given above are corrected.

As can be seen from Fig. 11, the miscibility gap of the zone state (dashed line) is different from the stable miscibility gap at high temperatures. The reason is due to J. W. Cahn (46a) who found from theoretical arguments that there exists a temperature difference between the gaps of the coherent phase (G.-P. zones) and the incoherent phase (α'). For the critical point this difference is proportional to the square of the change of the lattice constant with the concentration. Cahn gave a temperature difference of $40°$ for the Al-Zn system. Experimentally, Lašek (46b) found a difference of $25°$ which has been plotted in Fig. 11.

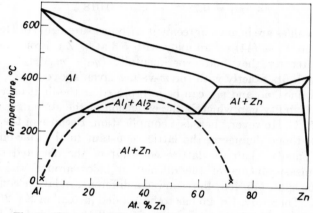

Fig. 11. The metastable miscibility gap in the system Al-Zn.

It should be pointed out that the amount of lattice distortion cannot be found from SAS alone. The differences of the lattice parameters inside and outside the zones follow only by comparing quantitatively SAS with the diffraction halo around a matrix spot (hkl) at large scattering angles (45, 45b).

3. *The Profile of the SAS Curves*

The result of the preceding paragraph suggests a two-phase model for the zones in both Al-Ag and Al-Zn alloys. Then the intensity formulae for the diffraction from particles should be applicable. Figure 12 shows three typical diffraction curves from Al-Ag and Al-Zn alloys, which were obtained by counter techniques with slit collimation. Instead of an intensity ring (Fig. 5) the plotted intensity profiles were measured. Besides the experimental curves the best extrapolated Guinier approximations and the h^{-3}-asymptotes are also shown on the figure.

All these curves are to be compared with the theoretical curve for the diffraction from a single spherical particle shown in the right part of Fig. 2. The following differences may be noted:

(i) The experimental curves show a totally different behavior at small angles.

(ii) The Guinier curve (plotted as dashed curve) is valid to much larger angles compared to the theoretical curve.

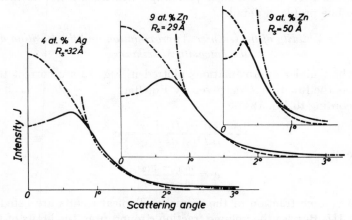

Fig. 12. The profile of SAS curves from Al-Ag and Al-Zn alloys containing spherical Guinier-Preston zones.

(iii) The h^{-3}-asymptote is valid to much smaller angles compared to the theoretical curve.

(iv) The intensity of the h^{-3}-asymptote is higher in the experimental curves compared with the theoretical curve. It is still higher for Al-Zn compared to Al-Ag. The theoretical curve, Fig. 2, has a large overlap between asymptote and Guinier curve. In the case of Al-Ag there is still a small overlap, whereas there is no overlap in the case of Al-Zn.

(v) For Al-Ag there exists a deviation of the experimental curve from the asymptote at large angles.

While the deviation (i) is due to the inter-particle interference function, deviations (ii) to (iv) may have different reasons:

(a) As the inter-particle interference function is not known the method of constructing the Guinier curve may be doubtful.

(b) There exists a particle size distribution. Because of this distribution the tail of the diffraction curve is raised. In this case the extrapolated Guinier curve should deviate at smaller angles to higher intensity values.

(c) There exists a deviation from the spherical particle shape.

The deviation (v) for the tail of the intensity curve for Al-Ag in Fig. 12 can have its reason in inhomogeneities of the electron density inside the particles. The next paragraph will discuss the deviations of the line profile in detail.

4. *The Validity of the Guinier Approximation and the Position of the Intensity Maximum*

If the Guinier approximations plotted in Fig. 12 are correct then one has the function J^* as given by Eq. 10a.

According to Eq. 18a

$$\frac{\int s[J^*(s) - J(s)] \, ds}{\int s J^*(s) \, ds} = c \qquad 18a.$$

where

$$c = \frac{m_A - m_2}{m_1 - m_2} \qquad 32.$$

is the volume fraction of the particles. Typical results are listed in Table III. Besides the volume fraction c found from the limits of the miscibility gap, Eq. 32, the experimental results from the left side of Eq. 18a are given. They are denoted by c^* for abbreviation. The spherical zone radius R_s evaluated from the Guinier approximation and the shape factor w_2 are also included in the table. As can be seen from the table the results for Al-Ag and Al-Zn are quite different. For example, Al-Ag alloys containing 9 or 12 at.% Ag and showing small zone radii have c^*-values which are smaller than c by a factor 2. It can be concluded that in these cases the Guinier approximation leads to intensity values J^* which are far too low for small angles. Another big discrepancy between c and c^* exists for the Al-Zn alloy at large zone sizes ($R_s > 30$ Å). In this case, c^* is smaller by a factor of nearly 3. Again, the Guinier approximation leads to much too low J^*-values at small angles.

The reason for the discrepancy between c and c^* seems to be very different for both alloys. In the case of Al-Ag at the beginning of the

TABLE III
Values of c, c^*, and w_2 from different alloys.

Alloy (at.%)	Aging temp., (°C)	Aging time (hours)	R_s (Å)	c (%)	c^* (%)	w_2
4 Ag	175	0.5	26	7	9	0.81
		2	32		9	0.82
		10	44		7	0.80
6 Ag	150	0.2	20	11	11	0.81
		4	29		14	0.87
		39	39		14	0.90
9 Ag	190	0.02	23	20	13	0.81
		0.13	34		21	0.87
		1	55		24	0.91
12 Ag	140	0.13	21	22	10	0.87
		1	28		17	0.95
		4.5	36		25	0.92
		24	48		26	0.93
9 Zn	25	0.6	22	11	11	0.80
		1.2	29		8	0.80
		3.2	39		4	0.76
		7	51		4	0.73
		22	60		4	0.68

aging treatment a wide size distribution probably causes the small c^*-values. They are observed only for alloys with a high silver concentration and for the as-quenched condition (resp. short annealing times after quenching). As the limit of the miscibility gap is higher than 400°C for these alloys, zone formation already occurs during the quench and continues for about 5 minutes after quenching to room temperature. Therefore, it may be reasonable that the distribution of zone sizes increases in width with increasing silver content. During the aging at elevated temperatures the average zone size increases, leading at the same time to a standard size distribution which gives c^*-values about 20% larger than c. For large zone sizes the Guinier approximation leads to intensity values J^* which are a bit too high. Thus, the zone radius evaluated from this approximation is a bit too high, too.

In the case of Al-Zn alloys the discrepancy between c and c^* has its cause in the change of the particle shape. Details will be given in the following section.

Another possibility for discussing the influence of the inter-particle interference function is the interpretation of the position of the maximum of the line profile which has been given by Syneček (47, 48). He adopted a close-packed spatial distribution of the spherical zones with the distance D between next-nearest zones. From the two-phase model the ratio between the distance D and the zone diameter $2 R_s$ is given by

$$\left(\frac{2R_s}{D}\right)^3 = \frac{3\sqrt{2}}{\pi} c \qquad 33.$$

If allowance is made for some deviation ΔD from the exact distance D, the long-range order of the zone distribution changes to short-range order leading to an intensity ring on SAS diagrams.

For point-collimated beams the intensity maximum is calculated to be at

$$s_m = \sqrt{\frac{3}{2}} \frac{1}{D} \qquad 34.$$

Syneček and Sebo (48) measured the quantities s_m and R_s for the alloys Al + 6 at.% Ag and Al + 11.8 at.% Zn for various aging temperatures. Then they compared the measured s_m values with the calculated ones using Eqs. 32 and 33. For the alloy Al-Ag they found good agreement within 3% which is an additional proof for the validity of the miscibility gap of Fig. 10.

The Al-Zn alloy gave controversial results. The experimental values s_m were found to be about 10% smaller than the calculated ones. This would indicate that the volume fraction c of the zones should be smaller. For the calculation the authors used the limits of the miscibility gap, at room temperature Eq. 30. If the change in the atomic volume is taken into account the gap limits are given by Eq. 31 and Fig. 11. Then the discrepancy in s_m increases to 17%.

At the moment it is difficult to explain these large differences. The zone radius was given as approximately 30 Å which still produces a nearly correct c^*-value as shown in Table III. This would suggest a nearly correct position of s_m. For larger zone sizes the maximum should appear at too small angles in accordance with the much too small c^*-values observed in our investigation.

Two investigators compared the zone sizes from both SAS and electron transmission microscopy for Al-Ag alloys. The results disagreed. Baur and Gerold (49) found satisfactory agreement between both methods. For example, zone diameters were measured microscopically in a particular case in the range from 35 to 65 Å. The main contribution to the volume fraction occurred from zones with a diameter of 55 Å. SAS gave a diameter of 58 Å. In contrast, Freise et al. (43) evaluated zone diameters from 19 to 47 Å by the microscope with the main contribution at 28 Å. The SAS value was 56 Å which is outside the microscopic range.

5. The Tail of the SAS Curves

The change of the tail intensity with respect to the rest of the curve is expressed by the different shape factors w. The best factor for the present purpose seems to be the factor w_2 because it does not involve

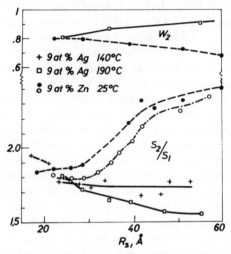

Fig. 13. The shape factor w_2 and the angle ratio s_2/s_1 as a function of zone radius R_s for Al-Ag and Al-Zn.

the unknown quantity $J^*(0)$ which is needed for w_1. Instead, the ratio of two integrals over J^* and sJ^* are involved which reduces to some extent the influence of the unknown part of J^*.

A more practical measure of the height of the tail is the ratio at two scattering angles s_1 and s_2 where the intensities have a distinct ratio J_1/J_2. If J_m is the maximum intensity, then $J_1 = 0.75\,J_m$ and $J_2 = 0.125\,J_m$ have been chosen. The ratio s_2/s_1 is shown in Fig. 13 for an alloy Al-Ag and Al-Zn, together with the shape factor w_2. These values are also listed in Table III.

In all cases, the shape factor is smaller than unity. This indicates either polydispersity or deviation from spherical shapes of the particles or both. The shape factor is higher for Al-Ag alloys where it ranges from 0.8 to 0.9. SAS experiments with single crystals and electron transmission microscopy indicate a spherical shape of the zones in Al-Ag alloys. Therefore, the w_2-values given above are related to polydispersity. For Al-Zn alloys there exists a variation of w_2 from 0.8 down to 0.68. This can no longer be explained by polydispersity alone. A change of the particle shape with increasing zone size seems to be probable. If one assumes w_2 to be the product of two factors related to polydispersity and to particle shape, the first factor can be estimated to be 0.8 to 0.9 by comparison with w_2 from Al-Ag alloys. Then the second factor has to change from unity ($R_s \lesssim 30$ Å) down to approximately 0.75 ($R_s = 60$ Å). Because of the strain field, an oblate ellipsoid is more likely than a prolate one. From Fig. 3 the change of the axial ratio v follows from unity down to approximately 0.35. This suggestion has been proved by other experimental facts.

A more prominent change occurs with the ratio s_2/s_1 which starts increasing at a zone radius of 30 Å. This is paralleled by the decrease of c^* below the correct c-value (Table III).

In single crystals of Al-Zn, SAS is centro-symmetric up to the zone size given above. At larger zone sizes the diffraction becomes anisotropic. All these effects together with the investigation of the diffraction around the Bragg spots and electron transmission microscopy lead to the following result: below the critical radius of $R_s = 30$ Å the zones are spheres. Larger zones become oblate ellipsoids which reach an axial ratio of about 3:1 (45, 45a).

The spherical zones in Al-Zn have a lattice constant which is $1.17 \pm 0.1\%$ smaller than that of the depleted matrix (45, 45b). Coherency strains give a spherical strain field. The ellipsoids have a preferred {111} orientation and show a rhombohedral distorted structure with a rhombohedral strain field (50). More details will be given elsewhere (45a).

Another interesting difference in SAS profiles from Al-Ag and Al-Zn is the better fit of the Al-Zn curves with the theoretical asymptote. As is shown for Al-Ag in Fig. 12, at diffraction angles higher than 2.5° the experimental curve is higher than the asymptote. This has been reported elsewhere (41). In a recent paper Bonfiglioli (51) discussed a similar topic, but obviously his discussion is concerned with a different part of the diffraction curve which is not the tail. Compared to Fig. 12, it is the region from 1 to 1.7°, where the experimental curve (Al-Ag) is above the asymptote. This line profile he compares with the corresponding part of the Al-Zn curve where the asymptote already fits in this region. His finding that for Al-Ag the quantity $s^3 J(s)$ decreases in the angular range given above is in agreement with the present results. The tail of Fig. 12 which starts at an angle of 2.5° has the opposite behavior, namely an increase of $s^3 J(s)$. This increase depends on the aging temperature (42) and is correlated to the distribution of the atomic scattering, J_a, which is due to electron density changes inside the zones on a more atomic scale. It is more pronounced for aging temperatures from 200 to 300° and after quenching from the homogenization temperature to room temperature. It is correlated with the appearance of additional diffraction effects around the matrix spots on single crystal patterns which have not been investigated in detail.

6. *Physical Properties and the Zone State*

To study the physical properties SAS is a useful tool for finding the structural parameters of the zone state. It is now possible to compare the change of these properties with the change of the structural parameters. The kinetics of competing zone growth have been studied for both Al-Ag and Al-Zn alloys (39, 52). Especially in Al-Zn alloys the quenching treatment can be varied easily because of the large temperature range of the homogeneous solution. The density of quenched-in vacancies and the density of vacancy sinks (dislocations) play an important role. The electrical resistivity was found to be affected by the total boundary surface (53) in both alloys. Resistivity changes were found to be proportional to the change of the boundary area. In addition, the resistivity is influenced by the atomic order inside the zones (Al-Ag) (53) and by the change of the strain field from spherical to rhombohedral symmetry (Al-Zn) (50).

Comparison of isothermal calorimetric measurements on an Al-Zn alloy at room temperature with the change of the zone state led to the following conclusion: With the exception of the first three minutes the evolution of heat stems from the change of the integral boundary energy of the zones which is proportional to the boundary area (53, 54). This follows from the fact that the volume fraction c of the zones remains constant. The calculated heat agrees very well with the experimental data. The volume fraction c changes only in the first minutes where most of the heat is produced.

The elastic modulus diminishes during zone growth in Al-Ag and

Al-Zn as long as the zones remain spherical. When the Al-Zn zones reach their critical radius where they change their shape to ellipsoids, the modulus starts to increase with further zone growth (50). The critical resolved shear stress of Al-Zn single crystals depends on the zone size up to radii of 30 Å. For larger zones the shear stress remains nearly constant (55).

The influence of plastic deformation on the shape of the zones has been investigated by Jan (38), Ohta, (56), and Sato and Kelly (57). The first two authors used cold-rolled polycrystals and found less elongation of the zones compared to the elongation of the sheets.

Sato and Kelly deformed Al-Ag single crystals by tensile tests and concluded from their experiments that the elongation of the zones in the slip direction was essentially the same as the elongation of the crystal given by the glide strain. This does not contradict the results of the other experiments because cold-rolling is a complicated process. The results of Sato and Kelly clearly show that dislocations cut through the zones.

Graf (31) observed the creation of zones in an alloy Al + 4.4 at.% Zn after reversion and cold rolling. The reversion at 175°C dissolves all zones which do not reappear after days of aging at room temperature because of the low concentration of vacancies. Cold-rolling produces point defects in the lattice which accelerate the atomic diffusion processes to such an extent that zones reappear within minutes.

SAS from Other Alloys

For all other alloys only qualitative evidence has been given for the occurrence of SAS indicating the presence of small particles (zones or precipitates). The following results have been reported:

Room temperature aging of Al-Mg-Zn alloys containing approximately 3 at.% Zn and 3 at.% Mg shows SAS from voluminous zones, the diameter not exceeding 26 Å (58). The zones contain Zn and Mg atoms in an ordered arrangement which can be concluded from diffraction at large angles. Similar results have been reported from other ternary Al-Mg-Zn alloys (59, 60).

An alloy Mg + 4 at.% Pb shows SAS from voluminous zones with diameters of 100 Å. These zones have an ordered Mg_3Pb structure of the Mg_3Cd type (61). They appear after annealing the specimen for approximately 500 hours at 80 to 100°C.

Mg-Zn alloys with 3 at.% Zn show streaks in the small-angle

range after annealing the alloy for 140 hours at 150°C. They are due to the metastable β' precipitate $MgZn_2$ which forms either needles or small platelets parallel to the [0001] direction of the hexagonal matrix (62).

An alloy of Ni + 12 at.% Al shows SAS from voluminous zones which consist of Ni_3Al. The zones appear after annealing the quenched specimen at 400 to 600°C (63).

THE CLUSTERING OF POINT DEFECTS IN PURE METALS

Under special quenching conditions the quenched-in vacancies anneal out of the matrix by forming small clusters which are nearly

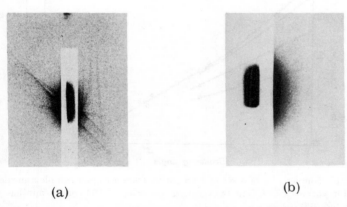

(a)   (b)

Fig. 14. Comparison of double Bragg scattering (a) with the SAS from vacancy clusters (b). The primary beam (monochromatic CuKα) is shielded by a copper foil (65).

invisible in the electron microscope. SAS experiments are a useful tool for studying the quenching and annealing conditions for such vacancy clusters. As the number of vacancies is small compared to the number of solute atoms in an alloy, the diffraction effects are very weak. Special techniques are necessary to get quantitative information. Special care has to be taken to avoid double Bragg diffraction in the small-angle region. For this purpose single-crystal foils and monochromatized X-rays are necessary. Figure 14 gives an example of the observed diffraction effects from a single-crystal foil of copper. Figure 14a represents the typical pattern of double Bragg scattering from a slightly distorted single crystal. Some Kossel lines are also visible. They probably have their origin from diffuse scattering maxima near the Bragg spots (temperature scattering) which are diffracted again at lattice planes. These lines are a useful tool for determining the exact orientation of the foil with respect to the primary X-ray beam. The best incident direction for avoiding all these secondary diffraction effects is [100]. Figure 14b represents the SAS from clustered vacancies which is quite different from secondary diffraction effects.

Because of the very weak diffracted intensity, the background scattering has to be carefully accounted for. The copper foils gave

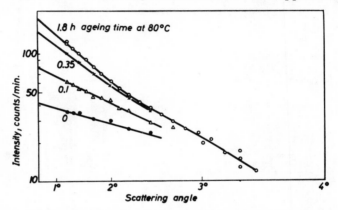

Fig. 15. Guinier plot of a set of SAS curves from a copper sample annealed at 80°C for various times. The background scattering of 183 counts/min has been subtracted (65).

an angle-independent intensity which was measured at $-150°C$ to be 10.2 e.u. on an absolute scale. It consists of incoherent, fluorescent, and temperature scattering. All measurements were performed at the same low temperature. As an example Fig. 15 shows the Guinier plots from a quenched foil which was annealed at 80°C for various times. The background scattering (183 counts/min which equals 10.2 e.u.) has been subtracted. It is higher than the SAS which demonstrates the difficulty of these experiments. The observed scattering was interpreted to originate from spherical or nearly spherical tiny voids formed by clustered vacancies. From Eq. 16, the total concentration c of the clustered vacancies could be calculated. The quantity $(\Delta\rho)^2 V_a$ was put equal to z^2/V_a, where z is the number of electrons of the copper atoms corrected for the dispersion effect of CuKα radiation. The concentration c was found to be 10^{-5} after quenching the foil from 1050°C to room temperature. Further annealing at elevated temperatures (40 to 100°C) increased this concentration to 6.10^{-5} (7). These concentrations have a reasonable order of magnitude.

The alternative explanation for the diffraction is the occurrence of small dislocation loops formed by condensed vacancies. This possibility can be ruled out by the following reasoning. If one compares the intensity, Eq. 10, for voids (I_1) with the intensity, Eq. 28, for loops (I_2), they can be compared for the same vacancy concentration c and for the same radius of gyration, R_g. For loops this concentration is given by

$$c = n_L \pi R^2 b$$

with $R^2 = \tfrac{14}{9} R_g^2$. The ratio of both intensities finally becomes

$$I_1/I_2 = \frac{45}{28}\left(\frac{5}{3}\right)^{\frac{3}{2}}\left(\frac{1-\nu}{1-2\nu}\right)^2 \frac{R_g}{b} \qquad 35.$$

which gives with $\nu = 0.35$

$$I_1/I_2 = 12 R_g/b$$

The ratio R_g/b is of the order of 2.5 to 5. As a result the scattering from voids is 30 to 60 times more intense than the scattering from loops. Therefore, a condensed vacancy concentration of 3.10^{-3} to 3.10^{-4} is needed to explain the experiments by dislocation loops. This concentration of vacancies is far too high for quenched samples.

Table IV shows some typical results. Because the concentration c is very low the functions φ resp. Φ are equal to zero. From the extrapolated value, $J(0)$, and from R_g the quantity $I(0)$ has been calculated, Eqs. 10 and 11. Then the void volume V_p is given by the ratio $I(0)/Q(0)$. The shape factor w_1 follows from Eq. 22.

As can be seen from the table, the experimental results depend highly on the quenching speed. At speeds lower than 10^4 °C/s the quenched-in vacancy concentration is too low to be detected. At speeds higher than 5.10^4 °C/s the quenched-in vacancies anneal out as larger dislocation loops which give no SAS.

The shape factor w_1 given in Table IV is very inaccurate because of the inaccuracy of $Q(0)$. Therefore, no conclusion can be drawn from w_1. The error for $Q(0)$ is estimated to be between 15 and 40%

TABLE IV
Voids in quenched and annealed copper foils.

Exp. No.	Quenching speed (10^4 °C/sec)	Aging temp. (°C)	Aging time (min)	R_g (Å)	$I(0)$ (e.u.)	$Q(0)$ (e.u./Å³)	n_p (10^{15}/cm³)	c (10^{-5})	w_1
1	2	23	10	6.6	3.4	1150	6.7	1.8	1.14
2	2	80	120	11.9	35.4	3500	6.0	5.9	0.66
3	3.5	23	11	5		not measurable			
4	3.5	80	120	14	8.4	640	0.8	1.2	0.54
5	10	variable				no diffraction effect			

depending on the profile of the curve. If the concentration c of the clustered vacancies is calculated from $I(0)$ and R_s, assuming spherical voids, it differs by the factor w_1 from the values given in the table.

As the experiments are near the limit of observation the dependence of this limit on c, w_1 and R_s shall be given. The intensity equation in the Guinier approximation is

$$J(h) = \frac{\sqrt{3\pi}}{3}\left(\frac{5}{3}\right)^{\frac{3}{2}} \frac{z^2}{\tau_0 V_a} cw_1 R_g^2 \exp\left(-\frac{R_g^2 h^2}{3}\right) \qquad 36.$$

The limiting condition is that the intensity J must exceed the value 1 electron/atom at the diffraction angle $1°$ ($h = 0.071$ Å^{-1}).

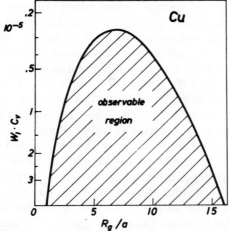

Fig. 16. The observable region of SAS curves from vacancy clusters in copper foils.
w_1 = shape factor
c_v = clustered vacancy concentration
R_g = radius of gyration
a = lattice constant.

This value is 10% of the background scattering. The result is shown in Fig. 16 where R_g/a is plotted against cw_1 (a = lattice constant). The minimum observable concentration of vacancies clustered into spherical voids is $6 \cdot 10^{-6}$ if the radius of gyration is 20 Å. The experiments No. 1 and 4 of Table IV are close to the observation limit. Therefore, the application of SAS techniques is restricted to special quenching conditions. Figure 16 shows qualitatively that for the detection of small dislocation loops ($w \ll 1$) a very high concentration c is needed.

The investigation led to a better understanding of the annealing process in quenched copper foils. In another investigation Galligan and Washburn (64) studied the change of the yield stress and the strain hardening of quenched single crystal copper foils. SAS was used to identify the vacancy clusters and to measure their size.

SAS OF SLOW NEUTRONS FROM DEFORMED AND IRRADIATED COPPER

There are only a very few experiments reported where slow neutrons have been used to study defects in metals. Because the wavelength used is of the order of 8 to 10 Å no double Bragg scattering can occur. This is the main advantage of neutron SAS compared to X-ray diffraction experiments, and it is very important for the investigation of plastically deformed metals. Double Bragg scattering cannot be avoided even from single crystals because of the distortions introduced by deformation. The main disadvantage of the neutron experiments are the unfavorable geometrical conditions. The primary beam has a nearly circular cross-section with a diameter of 20 mm which cannot be diminished because the whole intensity is needed for the diffraction experiment. Because of this collimation it is difficult to compare the experiments with theory. The theoretical curves have to be distorted numerically in order to take into account the geometrical conditions of the experiment. More details are given in the review paper by Christ et al. (66). The earlier experiment by Atkinson (21) gave only qualitative evidence for the scattering from deformed copper. The total neutron flux of the scattered neutrons was measured only in a relatively wide-angle range. Recent experiments by Christ (67) on deformed polycrystalline copper and on deformed single crystals give more quantitative results. There the diffraction has been measured as a function of the scattering angle. For larger angles a s^{-3} law has been found with polycrystalline copper in accordance with the theory by Atkinson and Hirsch (68). A total dislocation length of 8.10^{11} cm/cm^3 has been found from a polycrystalline sample strained 17% by tensile test. A similarly deformed single crystal showed a marked anisotropy in the diffraction experiment. The maximum intensity was obtained when the reciprocal diffraction vector **s** pointed into the direction parallel to the normal of the slip plane. This follows from Eq. 28 if it is assumed that most of the dislocations are located in the slip plane.

Neutron diffraction from neutron-irradiated copper samples has been reported by Schilling and Schmatz (69). The diffracted intensity was so low that the angular dependence could not be measured. The diffracted intensity from the irradiated samples was only about 10% higher than the background intensity from the reference specimen. The authors interpreted their results by two models. The first model is given by voids whereas the second model consists of dislocation loops. The density of these defects was assumed to be one defect per primary impact. The measured intensity could be explained by either 20 vacancies per void or by 40 vacancies per loop (loop diameter approximately 15 Å). As the angle dependence of the diffraction was not measured these results may not be final. The samples were irradiated by $1.1 \cdot 10^{19}$ n/cm² (energy > 1 MeV).

ACKNOWLEDGEMENTS

The valuable assistance of Mr. W. Merz in performing the calculation of Table III is gratefully acknowledged. Most of the experimental work was supported by the Deutsche Forschungsgemeinschaft.

BIBLIOGRAPHY

1. Guinier, A., Nature, **142**, 569 (1938).
2. Levelut, A.-M., Lambert, M. and Guinier, A., Compt. Rend., **255**, 319 (1962).
3. Guinier, A., Conference on Small-Angle X-Ray Scattering, Syracuse University, 1965.
4. Neynaber, R. H., Brammer, W. G. and Beeman, W. W., Phys. Rev., **99**, 615 (1955).
5. Christ, J., Schilling, W., Schmatz, W. and Springer, T., Z. angew. Phys., **18**, 295 (1965).
6. Seeger, A., J. Appl. Phys., **30**, 629 (1959).
7. Seeger, A., Gerold, V. and Rühle, M., Z. Metallk., **54**, 493 (1963).
8. Guinier, A., et al., *Small-Angle Scattering of X-Rays*, J. Wiley and Sons, New York, 1955.
9. Du Mond, J. W. M., Phys. Rev., **72**, 83 (1947).
10. Luzzati, V., Acta Cryst., **13**, 939 (1960).
11. Porod, G., Kolloid-Z., **124**, 83 (1951).
12. Holasek, A., Kratky, O., Mittelbach, P. and Wawra, H., Biochim. Biophys. Acta, **79**, 76 (1964).
13. Seeger, A. and Brand, P., Conference on Small-Angle X-Ray Scattering, Syracuse University, 1965.
14. Kröner, E., *Kontinuumstheorie der Versetzungen*, Springer-Verlag, Berlin, 1958.
15. Seeger, A. and Kröner, E., Z. Naturforschg., **14a**, 74 (1959).
16. Seeger, A. and Rühle, M., Ann. Physik, **7**, 216 (1963).
17. Freise, E. J., Fine, M. E. and Kelly, A., Phil. Mag., **8**, 101 (1960).

18. Zürn, W., Diplomarbeit, Techn. Hochschule, Stuttgart (1962).
19. Fricke, H. and Gerold, V., J. Appl. Phys., **30,** 661 (1959).
20. Warren, B. E., Acta Cryst., **12,** 837 (1959).
21. Atkinson, H. H., J. Appl. Phys., **30,** 637 (1959).
22. Guinier, A., Ann. Phys., **12,** 161 (1939).
23. Gerold, V., Ergeb. exakt. Naturw., **33,** 105 (1961).
24. Guinier, A., Solid State Phys., **9,** 293 (1959).
25. Gerold, V., Altenpohl, D. and Bichsel, H., *Aluminium und Aluminiumlegierungen*, Springer-Verlag, Berlin, 1965; p. 456.
26. Baur, R. and Gerold, V., Z. Metallk. **57,** 181 (1966).
27. Bradley, A. J., Proc. Phys. Soc., **52,** 80 (1940).
28. Tiedema, T. J., Bouman, J. and Burgers, W. G., Acta Met., **5,** 310 (1957).
29. Guinier, A., J. Phys. Radium, **8,** 124 (1942).
30. Walker, C. B. and Guinier, A., Acta Met., **1,** 568 (1953).
31. Graf, R., Compt. Rend., **249,** 1110 (1959).
32. Köster, W., Z. Metallk., **41,** 71 (1950).
33. Belbeoch, B. and Guinier, A., Acta Met., **3,** 370 (1955).
34. Gerold, V., Z. Metallk., **46,** 623 (1955).
35. Guinier, A., ibid., **43,** 217 (1952).
36. Münster, A. and Sagel, K., Z. physik. Chem., **7,** 296 (1956).
37. Münster, A. and Sagel, K., Mol. Phys., **1,** 23 (1958).
38. Jan, J. P., J. Appl. Phys., **26,** 1291 (1955).
39. Gerold, V. and Schweizer, W., Z. Metallk., **52,** 76 (1961).
40. Gerold, V., Phys. Stat. Sol., **1,** 37 (1961).
41. Baur, R. and Gerold, V., Acta Met., **10,** 637 (1962).
42. Auer, H. and Gerold, V., Z. Metallk., **56,** 240 (1965).
43. Freise, E. J., Kelly, A. and Nicholson, R. B., Acta Met., **9,** 250 (1961).
44. Herman, H., Cohen, J. B. and Fine, M. E., ibid., **11,** 43 (1963).
45. Merz, W., Thesis, Technische Hochschule, Stuttgart (1965).
45a. Merz, W. and Gerold, V., Z. Metallk. **57,** 607 (1966).
45b. Merz, W. and Gerold, V., ibid., in press.
46. Bonfiglioli, F. and Levelut, A., Communication à la VIe Assemblée Générale de l'Union Internationale de Cristallographie, Rome (Italie), 9 (1963).
46a. Cahn, J. W., Acta Met., **9,** 795 (1961).
46b. Lašek, J., Phys. Stat. Sol., **5,** K117 (1964).
47. Syneček, V., J. Phys. Radium, **23,** 828 (1962).
48. Syneček, V. and Sebo, P., Czech. J. Phys., **14,** 622 (1964).
49. Baur, R. and Gerold, V., Acta Met., **12,** 1449 (1964).
50. Merz, W., Anantharaman, T. and Gerold, V., Phys. Stat. Sol., **8,** K5 (1965).
51. Bonfiglioli, A. F., J. Phys. Radium, **23,** 817 (1962).
52. Baur, R. and Gerold, V., Z. Metallk., **52,** 671 (1961).
53. Gerold, V., J. Phys. Radium, **23,** 812 (1962).
54. Gerold, V., Z. Metallk., **54,** 257 (1962).
55. Auer, H., Gerold, V., Haberkorn, H. and Zürn, W., Phys. Stat. Sol., **6,** K91 (1964).
56. Ohta, M., Nippon Kinzoku Gakhai-Si, **23,** 177 (1959).
57. Sato, S. and Kelly, A., Acta Met., **9,** 59 (1961).
58. Schmalzried, H. and Gerold, V., Z. Metallk., **49,** 291 (1958).

59. Graf, R., Compt. Rend., **242,** 2834 (1956).
60. Graf, R., ibid., **244,** 337 (1957).
61. Henes, S. and Gerold, V., Z. Metallk., **53,** 703 (1962).
62. Gallot, J., Lal, K., Graf, R. and Guinier, A., Compt. Rend., **258,** 2818 (1964).
63. Starke, E., Gerold, V. and Guy, A., Acta Met., **13,** 957 (1965).
64. Galligan, J. and Washburn, J., Phil. Mag., **8,** II, 1455 (1963).
65. Rühle, M., Diplom-Arbeit, Technische Hochschule, Stuttgart (1962).
66. Christ, J., Schilling, W., Schmatz, W. and Springer, T., Z. angew. Phys., **18,** 295 (1965).
67. Christ, J., Phys. Stat. Sol., **7,** 557 (1964).
68. Atkinson, H. H. and Hirsch, P. B., Phil. Mag., **3,** 213 (1958).
69. Schilling, W. and Schmatz, W., Phys. Stat. Sol., **4,** 95 (1964).

On the Determination of the Metastable Miscibility Gap from Integrated Small-Angle X-Ray Scattering Data*

R. W. HENDRICKS AND B. S. BORIE

Metals and Ceramics Division, Oak Ridge National Laboratory,
Oak Ridge, Tennessee 37830

INTRODUCTION

The earliest stages of the decomposition of quenched Al-Ag, Al-Zn, and NaCl-AgCl solid solutions are characterized by the formation of small, spherical, solute-rich regions known as Guinier-Preston zones. The extensive research on G.-P. zone formation has been reviewed by Hardy and Heal (1), Guinier (2), and Kelly and Nicholson (3), and will not be repeated here. Rather, it is the object of this paper to consider one recent and particularly significant contribution to our understanding of G.-P. zone formation—that of determining the phase boundaries of the metastable miscibility gap by the measurement of the integrated small-angle X-ray intensity, as first suggested by Gerold (4). We intend to clarify a common misunderstanding of this theory by comparing the total scattered X-ray intensity computed from the two-phase model of Guinier-Preston zones with the more general result of Cowley (5).

THE PROBLEM OF RELATING INTEGRATED SAXS INTENSITY TO THE METASTABLE MISCIBILITY GAP

Based on electrical conductivity measurements in quenched Al-Ag alloys, Borelius and Larsson (6) concluded that the formation and reversion of Guinier-Preston zones was controlled by a metastable miscibility gap. More recently Gerold (4) has shown that the phase boundaries of such a metastable miscibility gap may be calculated

* Research sponsored by the U.S. Atomic Energy Commission under contract with the Union Carbide Corporation.

from integrated small-angle X-ray scattering (SAXS) data. To date, this theory has been applied to three solid solutions known to form spherical Guinier-Preston zones on quenching: Al-Zn (4, 7), Al-Ag (8–10), and NaCl-AgCl (11).

In order to relate the integrated small-angle X-ray intensity to the metastable phase diagram, it is necessary to assume some model for the structure of the solid solution. Two different models have been suggested, both of which predict the observed shape of the diffuse intensity distribution. The first model, initially proposed by Guinier (12), considers the decomposed solid solution to consist of solute-rich zones of composition m_1 distributed, with a preferred nearest neighbor separation, in a uniform solute-impoverished matrix of composition m_2. In this model, the coherent zones may be considered to be a second phase in metastable thermodynamic equilibrium with the surrounding matrix. Walker and Guinier (13) proposed a second model in which the zones, considered to be spherical regions enriched in solute atoms surrounded by a concentric region impoverished in solute atoms, were distributed at random throughout a uniform unprecipitated matrix. Gerold (4, 10) has argued convincingly in favor of the first model, and that interpretation will be used in the following discussion.

Using SAXS theory, it may be shown (4, 14) that the integrated intensity Q_0 may be expressed by

$$Q_0 \equiv 4\pi \int_0^\infty h^2 I'(h)\, dh = \frac{V_a}{V} \int_V (\rho^2 - \bar{\rho}^2)\, dV \qquad 1.$$

where

$I'(h)$ = scattered small-angle intensity (in electron units per atom),

$h = \dfrac{2 \sin \theta}{\lambda}$,

λ = wavelength of incident X-irradiation,

V_a = atomic volume,

V = volume irradiated,

ρ = local electronic density, and

$\bar{\rho}$ = mean electronic density.

Using the first model to describe the electronic density of the decomposed solid solution, Gerold (4) has shown that the right-hand side of Eq. 1 may be integrated to give

$$Q_0 = (m_1 - m_A)(m_A - m_2)(f_A - f_B)^2/V_a \qquad 2.$$

where f_A and f_B are the atomic scattering factors of A and B atoms, respectively, and m_1, m_2 and m_A are the compositions (in mole fraction of A atoms) of the zones, the matrix, and the unprecipitated solid solution, respectively.

Equation 2 suggests that the integrated intensity depends on the way in which the atoms are distributed in the segregated solid solution. If the zones are really a thermodynamically metastable phase, m_1 and m_2 are independent of m_A, and therefore by measuring Q_0 for two different compositions m_A it is possible to solve the resulting pair of equations simultaneously to recover the composition of the metastable phase boundary.

On the other hand, Cowley (5) has shown that the total intensity (I_T) scattered by an alloy in any state of clustering or short-range order may be expressed in terms of the intensity due to the fundamental Bragg reflections (I_F), which is not dependent on the way in which the atoms are arranged in solid solution, plus a term known as the order intensity (I_0), which is the Laue monotonic diffuse intensity modulated by the nonrandom arrangement of the atoms. Mathematically,

$$I_T = I_F + I_0$$
$$= (m_A f_A + m_B f_B)^2 \sum_j \sum_k e^{i\mathbf{K} \cdot \mathbf{R}_{jk}}$$
$$+ N m_A m_B (f_A - f_B)^2 \sum_{\substack{lmn \\ -\infty}}^{\infty} \alpha_{lmn} e^{2\pi i (h_1 l + h_2 m + h_3 n)} \qquad 3.$$

where

I = intensity at $\mathbf{K}$ (in electron units),
$\mathbf{R}_{jk}$ = vector from atom at site j to atom at site k,
$\mathbf{K} = 2\pi(h_1 \mathbf{b}_1 + h_2 \mathbf{b}_2 + h_3 \mathbf{b}_3)$,

and

$$\alpha_{lmn} = 1 - \frac{P_{lmn}{}^{AB}}{m_A}$$

In Eq. 3, $P_{lmn}{}^{AB}$ is the probability of finding an A atom at the lattice site

$$l \frac{\mathbf{a}_1}{2} + m \frac{\mathbf{a}_2}{2} + n \frac{\mathbf{a}_3}{2}$$

after first having found a B atom at the origin. The vectors $\mathbf{b}_1$, $\mathbf{b}_2$ and $\mathbf{b}_3$ are reciprocal to half the crystallographic axes $\mathbf{a}_1$, $\mathbf{a}_2$ and $\mathbf{a}_3$, and are twice the usual reciprocal lattice vectors. The procedure for the conversion from the double to the single sum in the last term of Eq. 3 is adequately discussed in the literature (5, 15, 16) and will not be repeated here. In order to compare the above description of the small-angle X-ray scattering with Gerold's model, we integrate the order intensity over one repeat volume in reciprocal space to find

$$\frac{1}{N}\int_{-\frac{1}{2}}^{\frac{1}{2}}\int_{-\frac{1}{2}}^{\frac{1}{2}}\int_{-\frac{1}{2}}^{\frac{1}{2}} I_0(h_1 h_2 h_3)\, dh_1\, dh_2\, dh_3$$

$$= m_A m_B (f_A - f_B)^2 \int_{-\frac{1}{2}}^{\frac{1}{2}}\int_{-\frac{1}{2}}^{\frac{1}{2}}\int_{-\frac{1}{2}}^{\frac{1}{2}} \sum_{lmn} \alpha_{lmn} e^{2\pi i (h_1 l + h_2 m + h_3 n)}\, dh_1\, dh_2\, dh_3 \quad 4.$$

Integrating the right-hand side term by term, Eq. 4 reduces to

$$\frac{1}{N}\int_{-\frac{1}{2}}^{\frac{1}{2}}\int_{-\frac{1}{2}}^{\frac{1}{2}}\int_{-\frac{1}{2}}^{\frac{1}{2}} I_0(h_1 h_2 h_3)\, dh_1\, dh_2\, dh_3 = m_A m_B (f_A - f_B)^2 \quad 5.$$

since α_{000} is unity and all other terms in the sum of Eq. 4 vanish on integration. The left-hand side of Eq. 5 is simply the value of the order intensity averaged over the volume of one unit cell in reciprocal space. Comparing Eqs. 1 and 3, we note that the variables h in Gerold's notation and $h_1 h_2 h_3$ in Cowley's notation differ significantly. Remembering the definition of the diffraction vector, we may find the following relationship

$$|\mathbf{K}| = 2\pi |(h_1 \mathbf{b}_1 + h_2 \mathbf{b}_2 + h_3 \mathbf{b}_3)| = \frac{2\pi}{\lambda} |\mathbf{S} - \mathbf{S}_0| = 2\pi h \quad 6.$$

where $\mathbf{S}_0$ and $\mathbf{S}$ are unit vectors in the direction of the incident and diffracted X-rays. Following Cowley's notation, for a cubic lattice

$$|h_1 \mathbf{b}_1 + h_2 \mathbf{b}_2 + h_3 \mathbf{b}_3| = h' \frac{2}{a_0} = h \quad 7.$$

where $h' = (h_1^2 + h_2^2 + h_3^2)^{\frac{1}{2}}$. If we then compute the average of Eq. 5 in Gerold's notation, assuming that I_0 has spherical symmetry, we obtain

$$\frac{4\pi \int_0^H h^2 I'(h)\, dh}{4\pi \int_0^H h^2\, dh} = m_A m_B (f_A - f_B)^2 \qquad 8.$$

where $I' = I_0/N$ and H is the radius of a sphere in reciprocal space whose volume is equal to the volume per reciprocal lattice point. Examining the limits of integration in Eq. 5, and using the relationship between the Cowley and Gerold notations found in Eq. 7, the upper limit of integration H in Eq. 8 is

$$\frac{4}{3}\pi H^3 = \frac{1}{2} \cdot \left(\frac{2}{a_0}\right)^3 = \frac{1}{V_a} \qquad 9.$$

The factor $\frac{1}{2}$ arises because for the f.c.c. lattice there are two reciprocal lattice points per unit cell. In most small-angle scattering experiments, $I'(h)$ has decreased below the noise level for $h \ll H$, and it is then convenient to extend the limit of integration to infinity. Hence there results

$$Q_0 \equiv 4\pi \int_0^\infty h^2 I'(h)\, dh = m_A m_B \frac{(f_A - f_B)^2}{V_a} \qquad 10.$$

This analysis, based on Cowley's equation, indicates that the total integrated order intensity Q_0 is a function only of the average composition, and is independent of the way in which the atoms are arranged in the alloy. This appears to be in direct disagreement with Gerold's result.

In the preceding discussion, we have outlined two different approaches to the interpretation of integrated small-angle X-ray scattering data. One result suggests that the phase boundaries of the metastable miscibility gap are related to the integrated small-angle X-ray intensity, while the other indicates they are not. The question of the existence of a metastable miscibility gap, and the ability to determine its location experimentally, is important to our understanding of the theory of precipitation in supersaturated solid solutions. Thus, it is worthwhile examining both of the above calculations in detail in an attempt to resolve the present disagreement of interpretation. Although this problem has been considered previously (10), we believe that the following analysis, which approaches the problem from a different point of view, is more easily understood and gives a deeper insight into the origin of the several different modulating effects on the total Laue monotonic intensity.

CALCULATION OF X-RAY SCATTERING FROM THE TWO-PHASE MODEL OF G.-P. ZONES

In this section, we shall calculate the total intensity scattered from a solid solution which is assumed to be segregated into two phases, as described earlier. We shall consider the precipitated solid solution to consist of zones of composition m_1 distributed in a uniform matrix of composition m_2. It will be assumed that the atom sites within the zones can be located by lattice vectors of the matrix, and that the interface between the zones and the matrix is sharp. No assumptions about the size, shape, or spatial distribution of the zones are made. Based on this model, the amplitude of the scattered X-rays may be written as

$$\left.\begin{aligned}
A &= \sum_m f_m e^{i\mathbf{K}\cdot\mathbf{R}_m} \\
&= \sum_p f_p e^{i\mathbf{K}\cdot\mathbf{R}_p} + \sum_r f_r e^{i\mathbf{K}\cdot\mathbf{R}_r} \\
&= \sum_p (f_p - F_1) e^{i\mathbf{K}\cdot\mathbf{R}_p} + \sum_r (f_r - F_2) e^{i\mathbf{K}\cdot\mathbf{R}_r} \\
&\quad + \sum_p (F_1 - F_A) e^{i\mathbf{K}\cdot\mathbf{R}_p} + \sum_r (F_2 - F_A) e^{i\mathbf{K}\cdot\mathbf{R}_r} \\
&\quad + \sum_m F_A e^{i\mathbf{K}\cdot\mathbf{R}_m}
\end{aligned}\right\} \quad 11.$$

where

$$F_1 = m_1 f_A + (1 - m_1) f_B$$
$$F_2 = m_2 f_A + (1 - m_2) f_B$$

and

$$F_A = m_A f_A + (1 - m_A) f_B$$

The subscripts (m,n) denote any lattice site (either zone or matrix), (p,q) denote zone sites only, and (r,s) denote matrix sites only, while F_1, F_2 and F_A are the average atomic scattering factors for the zones, the matrix, and the unprecipitated solid solution, respectively. The total intensity, which is the product of the amplitude with its complex conjugate, is given by

$$\begin{aligned}
I = & F_A{}^2 \sum_m \sum_n e^{i\mathbf{K}\cdot\mathbf{R}_{mn}} \\
& + \sum_p \sum_q (f_p - F_1)(f_q - F_1) e^{i\mathbf{K}\cdot\mathbf{R}_{pq}} \\
& \qquad\qquad + \sum_r \sum_s (f_r - F_2)(f_s - F_2) e^{i\mathbf{K}\cdot\mathbf{R}_{rs}} \\
& + (F_1 - F_A)^2 \sum_p \sum_q e^{i\mathbf{K}\cdot\mathbf{R}_{pq}} \\
& \qquad\qquad + 2(F_1 - F_A)(F_2 - F_A) \sum_p \sum_r e^{i\mathbf{K}\cdot\mathbf{R}_{pr}} \\
& \qquad\qquad + (F_2 - F_A)^2 \sum_r \sum_s e^{i\mathbf{K}\cdot\mathbf{R}_{rs}} \\
& + 2F_A(F_1 - F_A) \sum_p \sum_m e^{i\mathbf{K}\cdot\mathbf{R}_{pm}} + 2F_A(F_2 - F_A) \sum_r \sum_m e^{i\mathbf{K}\cdot\mathbf{R}_{rm}} \\
& + 2 \sum_p \sum_r (f_p - F_1)(f_r - F_2) e^{i\mathbf{K}\cdot\mathbf{R}_{pr}} \\
& \qquad\qquad + 2(F_1 - F_A) \sum_p \sum_q (f_p - F_1) e^{i\mathbf{K}\cdot\mathbf{R}_{pq}} \\
& + 2(F_2 - F_A) \sum_p \sum_r (f_p - F_1) e^{i\mathbf{K}\cdot\mathbf{R}_{pr}} \\
& \qquad\qquad + 2F_A \sum_p \sum_m (f_p - F_1) e^{i\mathbf{K}\cdot\mathbf{R}_{pm}} \\
& + 2(F_1 - F_A) \sum_r \sum_p (f_r - F_2) e^{i\mathbf{K}\cdot\mathbf{R}_{rp}} \\
& \qquad\qquad + 2(F_2 - F_A) \sum_r \sum_s (f_r - F_2) e^{i\mathbf{K}\cdot\mathbf{R}_{rs}} \\
& + 2F_A \sum_r \sum_m (f_r - F_2) e^{i\mathbf{K}\cdot\mathbf{R}_{rm}}
\end{aligned} \qquad 12.$$

Equation 12 describes the total X-ray intensity scattered by the assumed model. The first term gives the fundamental Bragg reflections, but the remaining terms, which are modulations of the total Laue monotonic intensity, must be evaluated and rearranged to put them in a more recognizable form.

The terms involving the atomic scattering factors f_p and f_r may be evaluated directly. Consider first the term

$$2(F_1 - F_A) \sum_p \sum_q (f_p - F_1) e^{i\mathbf{K}\cdot\mathbf{R}_{pq}}$$

Regardless of the way in which the atoms are arranged in the precipitate particles, $(f_p - F_1)$ may be withdrawn from under the double sum and f_p replaced by its spatial average $\langle f_p \rangle$. But by the previous definitions, $\langle f_p \rangle = F_1$, and

$$2(F_1 - F_A) \sum_p \sum_q (f_p - F_1)e^{i\mathbf{K}\cdot\mathbf{R}_{pq}}$$
$$= 2(F_1 - F_A)(\langle f_p \rangle - F_1) \sum_p \sum_q e^{i\mathbf{K}\cdot\mathbf{R}_{pq}} = 0 \quad 13.$$

Similar reasoning shows that the last five terms of Eq. 12 are also zero. Further, if there is no correlation between A atoms in the zones and A atoms in the matrix,* by similar reasoning

$$2 \sum_p \sum_r (f_p - F_1)(f_r - F_2)e^{i\mathbf{K}\cdot\mathbf{R}_{pr}}$$
$$= 2(\langle f_p \rangle - F_1)(\langle f_r \rangle - F_2) \sum_p \sum_r e^{i\mathbf{K}\cdot\mathbf{R}_{pr}} = 0 \quad 14.$$

This result is also independent of the arrangement of the atoms in the zones or the matrix. In Eq. 14, it is to be remembered that the terms $p = r$ do not exist, as it is impossible for the same site to be simultaneously in both the matrix and a precipitate particle.

On the other hand, in the second and third double sums of Eq. 12, not only must the terms $p = q$ and $r = s$ be considered, but we must also account for the possible nonrandom distribution of atoms within both the zones and the matrix. Examining the double sum over zone sites only, we find

$$\sum_p \sum_q (f_p - F_1)(f_q - F_1)e^{i\mathbf{K}\cdot\mathbf{R}_{pq}} = P \sum_p (f_p - F_1)^2$$
$$+ \sum_{p \neq q} \sum (f_p - F_1)(f_q - F_1)e^{i\mathbf{K}\cdot\mathbf{R}_{pq}} \quad 15.$$

which, by following identically the analysis of Cowley (5), reduces directly to

$$= P\{\langle f_p^2 \rangle - 2\langle f_p \rangle F_1 + F_1^2\}$$
$$+ \sum_{p \neq q} \sum m_1(1 - m_1)(f_A - f_B)^2 \left\{1 - \frac{P_{pq}^{AB}}{m_1}\right\} e^{i\mathbf{K}\cdot\mathbf{R}_{pq}}$$
$$= Pm_1(1 - m_1)(f_A - f_B)^2 \sum_p \sum_q \alpha'_{pq} e^{i\mathbf{K}\cdot\mathbf{R}_{pq}} \quad 16.$$

* This assumption is valid if there is a sharp interface between the zones and the matrix, and if there is only a small degree of nonrandomness in either one of the phases.

where P is the total number of lattice sites in the zones, and $\alpha'_{pq} = 1 - P_{pq}^{AB}/m_1$ is the order parameter for zones only. The identity $\langle f_p^2 \rangle = m_1 f_A^2 + (1 - m_1) f_B^2$ has been substituted to give the final expression. A similar expression may be derived for the matrix-matrix double sum if we define a different order parameter $\alpha''_{rs} = 1 - P_{rs}^{AB}/m_2$ for the matrix sites. Combining all of these results, the total diffracted intensity becomes

$$\left. \begin{aligned}
I = {} & F_A{}^2 \sum_m \sum_n e^{i\mathbf{K}\cdot\mathbf{R}_{mn}} \\
& + Pm_1(1 - m_1)(f_A - f_B)^2 \sum_p \sum_q \alpha'_{pq} e^{i\mathbf{K}\cdot\mathbf{R}_{pq}} \\
& + (N - P)m_2(1 - m_2)(f_A - f_B)^2 \sum_r \sum_s \alpha''_{rs} e^{i\mathbf{K}\cdot\mathbf{R}_{rs}} \\
& + (F_1 - F_A)^2 \sum_p \sum_q e^{i\mathbf{K}\cdot\mathbf{R}_{pq}} \\
& + 2(F_1 - F_A)(F_2 - F_A) \sum_p \sum_r e^{i\mathbf{K}\cdot\mathbf{R}_{pr}} + (F_2 - F_A)^2 \sum_r \sum_s e^{i\mathbf{K}\cdot\mathbf{R}_{rs}} \\
& + 2F_A(F_1 - F_A) \sum_p \sum_m e^{i\mathbf{K}\cdot\mathbf{R}_{pm}} + 2F_A(F_2 - F_A) \sum_r \sum_m e^{i\mathbf{K}\cdot\mathbf{R}_{rm}}
\end{aligned} \right\} \quad 17.$$

The last five terms of Eq. 17 may be combined by noting that

$$\sum_p \sum_m e^{i\mathbf{K}\cdot\mathbf{R}_{pm}} = \sum_p \sum_q e^{i\mathbf{K}\cdot\mathbf{R}_{pq}} + \sum_p \sum_r e^{i\mathbf{K}\cdot\mathbf{R}_{pr}} \qquad 18.$$

and

$$\sum_r \sum_m e^{i\mathbf{K}\cdot\mathbf{R}_{rm}} = \sum_r \sum_s e^{i\mathbf{K}\cdot\mathbf{R}_{rs}} + \sum_p \sum_r e^{i\mathbf{K}\cdot\mathbf{R}_{pr}} \qquad 19.$$

Thus, Eq. 17 reduces to

$$\left. \begin{aligned}
I = {} & F_A{}^2 \sum_m \sum_n e^{i\mathbf{K}\cdot\mathbf{R}_{mn}} \\
& + Pm_1(1 - m_1)(f_A - f_B)^2 \sum_p \sum_q \alpha'_{pq} e^{i\mathbf{K}\cdot\mathbf{R}_{pq}} \\
& + (N - P)m_2(1 - m_2)(f_A - f_B)^2 \sum_r \sum_s \alpha''_{rs} e^{i\mathbf{K}\cdot\mathbf{R}_{rs}} \\
& + (F_1{}^2 - F_A{}^2) \sum_p \sum_q e^{i\mathbf{K}\cdot\mathbf{R}_{pq}} + (F_2{}^2 - F_A{}^2) \sum_r \sum_s e^{i\mathbf{K}\cdot\mathbf{R}_{rs}} \\
& + 2(F_1 F_2 - F_A{}^2) \sum_p \sum_r e^{i\mathbf{K}\cdot\mathbf{R}_{pr}}
\end{aligned} \right\} \quad 20.$$

The first term of Eq. 20 describes the sharp fundamental reflections, while the second and third terms are the Laue monotonic scattering from the zones and the matrix, respectively, each modulated by a Cowley order series which is dependent on the local atomic arrangements. The fourth term is scattering due to the size and interparticle interference effects of the zones, and the fifth term is the analogous description of the matrix interactions. The final term arises from the interaction between the matrix sites and the zone sites.

The symmetry of this result with respect to the matrix and the zones should be noted. As the average composition m_A varies from zero to unity, the roles of the matrix sites (r,s) and the zone sites (p,q) continuously interchange with each other. Near equimolar concentrations, where it makes little sense to speak of zones embedded in a matrix, Eq. 20, still describes the scattered X-ray intensity so long as there is a sharp interface between the two phases.

Equation 20 may be simplified further by noting that

$$\sum_r \sum_s e^{i\mathbf{K}\cdot\mathbf{R}_{rs}} = \left(\sum_m e^{i\mathbf{K}\cdot\mathbf{R}_m} - \sum_p e^{i\mathbf{K}\cdot\mathbf{R}_p}\right)\left(\sum_n e^{-i\mathbf{K}\cdot\mathbf{R}_n} - \sum_q e^{-i\mathbf{K}\cdot\mathbf{R}_q}\right)$$
$$= \sum_m \sum_n e^{i\mathbf{K}\cdot\mathbf{R}_{mn}} - 2\sum_p \sum_m e^{i\mathbf{K}\cdot\mathbf{R}_{pm}} + \sum_p \sum_q e^{i\mathbf{K}\cdot\mathbf{R}_{pq}} \qquad 21.$$

and

$$\sum_p \sum_r e^{i\mathbf{K}\cdot\mathbf{R}_{pr}} = \sum_p \sum_m e^{i\mathbf{K}\cdot\mathbf{R}_{pm}} - \sum_p \sum_q e^{i\mathbf{K}\cdot\mathbf{R}_{pq}} \qquad 22.$$

Substituting these identities into Eq. 20, and using the result that, if the crystal is very large,

$$\sum_p \sum_m e^{i\mathbf{K}\cdot\mathbf{R}_{pm}} = \frac{P}{N}\sum_m \sum_n e^{i\mathbf{K}\cdot\mathbf{R}_{mn}} \qquad 23.$$

we find

$$\left.\begin{aligned}I = F_A{}^2 \sum_m \sum_n e^{i\mathbf{K}\cdot\mathbf{R}_{mn}} \\ + Pm_1(1-m_1)(f_A - f_B)^2 \sum_p \sum_q \alpha'_{pq} e^{i\mathbf{K}\cdot\mathbf{R}_{pq}} \\ + (N-P)m_2(1-m_2)(f_A - f_B)^2 \sum_r \sum_s \alpha''_{rs} e^{i\mathbf{K}\cdot\mathbf{R}_{rs}} \\ + (F_1 - F_2)^2 \sum_p \sum_q e^{i\mathbf{K}\cdot\mathbf{R}_{pq}} \\ - \left(\frac{P}{N}\right)^2 (F_1 - F_2)^2 \sum_m \sum_n e^{i\mathbf{K}\cdot\mathbf{R}_{mn}}\end{aligned}\right\} \quad 24.$$

This expression shows that the total diffracted intensity scattered from the two-phase model of coherent zones embedded in a uniform matrix consists of sharp Bragg reflections, two modulated Laue monotonic terms arising from the distribution of the atoms both in the zones and in the matrix, a term due to the size, shape, and interparticle interference effects of the zones, and a (negative) sharp intensity at the reciprocal lattice points.

The reduction of the double summation over all pairs of atoms within the zones to functions involving the scattering from individual zones and interparticle interference effects is very difficult to perform in the most general case where the zones vary in size and shape, especially when there is a preferred spatial orientation. This problem has been considered in detail by Fournet (17), is reviewed by Guinier and Fournet (14), and will not be treated here. For the discussion which follows, it is sufficient to note* that for diffraction vectors which are not equal to reciprocal lattice vectors the intensity due to the double sum over the atom pairs within the zones is given by the number of zones (N_Z) times the scattering from a single zone. The limit, in the case of widely separated particles, as **K** tends toward the reciprocal lattice point is $N_Z(P/N_Z)^2$, where (P/N_Z) is the number of lattice sites per zone. However, if the diffraction vector **K** is a reciprocal lattice vector, the order intensity is P^2. Hence, the graph of order intensity as a function of diffraction vector has a sharp peak of magnitude P^2 at each reciprocal lattice point, while the diffuse intensity approaches the point to a limiting magnitude (P^2/N_Z). Since N_Z is often very large ($N_Z \approx 10^{15}$ zones/cc in quenched Al-Ag alloys), the ratio of these two limits is very small. Since we can observe data experimentally only outside this peak, we may consider that the negative delta function described by the last term of Eq. 24 exactly cancels the peak in the zone-zone double sum. (This is accurate to within one part in N_Z.) Thus, the observed diffuse scattering is described by the sum of the last two terms in Eq. 24.

COMPARISON WITH PREVIOUS WORK

In the discussion of the Cowley representation of the diffuse scattering, the order intensity (I_0) was defined as that portion of the total intensity whose structure is dependent on the way in which the A and B atoms are arranged on the lattice sites, and was shown to be the difference between the total intensity and the intensity in the fundamental reflections. It was also shown that the total integrated order intensity was independent of the atomic arrangements.

* See for example, p. 71 of reference (14).

In comparing the results of the present derivation with Cowley's analysis, it is necessary to include in the order intensity all of the terms of Eq. 24 except the first, which represents the fundamental reflections. The integrated order intensity for the assumed model is then

$$\left.\begin{aligned}
V_a Q_0 &= \frac{1}{N} \int_{-\frac{1}{2}}^{\frac{1}{2}} \int_{-\frac{1}{2}}^{\frac{1}{2}} \int_{-\frac{1}{2}}^{\frac{1}{2}} \Big\{ Pm_1(1-m_1)(f_A - f_B)^2 \\
&\qquad\qquad\qquad\qquad\qquad \times \sum_p \sum_q \alpha'_{pq} e^{i\mathbf{K}\cdot\mathbf{R}_{pq}} \\
&\quad + (N-P)m_2(1-m_2)(f_A - f_B)^2 \sum_r \sum_s \alpha''_{rs} e^{i\mathbf{K}\cdot\mathbf{R}_{rs}} \\
&\quad + (F_1 - F_2)^2 \sum_p \sum_q e^{i\mathbf{K}\cdot\mathbf{R}_{pq}} \\
&\quad - \left(\frac{P}{N}\right)^2 (F_1 - F_2)^2 \sum_m \sum_n e^{i\mathbf{K}\cdot\mathbf{R}_{mn}} \Big\} dh_1\, dh_2\, dh_3 \\
&= \frac{P}{N} m_1(1-m_1)(f_A - f_B)^2 \\
&\qquad + \frac{N-P}{N} m_2(1-m_2)(f_A - f_B)^2 \\
&\quad + \frac{P}{N}(F_1 - F_2)^2 - \frac{N}{N}\left(\frac{P}{N}\right)^2 (F_1 - F_2)^2
\end{aligned}\right\} \quad 25.$$

In this integration, as in the integration of Eq. 4, the order parameters α'_{pq} and α''_{rs} give a contribution only for the terms $p = q$ and $r = s$. Using the equation for the conservation of A atoms,

$$Nm_A = Pm_1 + (N-P)m_2 \qquad 26.$$

Eq. 25 becomes

$$V_a Q_0 = m_A(1 - m_A)(f_A - f_B)^2 \qquad 27.$$

which is identical to the integrated order intensity given by the Cowley theory.

When the order intensity computed for the model given here (Eqs. 20 or 24) is compared directly with the Fourier series representation given by Cowley (Eq. 3), we notice that, by assuming the two-phase model of zone formation, the order intensity is separated into four recognizable terms. The integration performed in Eq. 25

shows that, although each of these terms depends on the local atomic arrangement, their sum is dependent only on the average composition m_A and the difference in the atomic scattering factors. From an experimental point of view, if any one of these four terms is unobservable, then the observed value of Q_0 will no longer be independent of the atomic arrangements. We therefore wish to examine the two modulated Laue monotonic terms of Eq. 24 to determine the conditions under which either one of them might be negligible in a small-angle X-ray scattering experiment, and then to compute the observable value of Q_0 from the remaining terms.

The fraction of the total order intensity which is associated with both Laue monotonic terms depends on the values of m_1, m_2 and m_A, and cannot be evaluated except in specific cases. As an example, using Gerold's data (4) for G.-P. zones formed in quenched Al-9.4 at.% Zn alloys, we find that 28% of the total order intensity arises from the Laue monotonic scattering from the zones, an additional 19% is due to the equivalent Laue term for the matrix, and the remaining 53% is contributed by the size, shape, and distribution of the particles. Since a significant fraction (47%) of the total order intensity scattered by this alloy is due to the two Laue monotonic terms, whether or not they contribute appreciably to the observed small-angle X-ray intensity is dependent on the modulations due to local order.*

If the atoms in the zones and the matrix are randomly distributed, all of the order parameters α'_{pq} and α''_{rs} are zero except those for $p = q$ and $r = s$, which are unity. Therefore, the Laue monotonic intensity is spread uniformly over the entire repeat volume in reciprocal space. On the other hand, for particle sizes of interest in small-angle scattering experiments (10 Å to 1000 Å radius), it is found that the particle-size summation describes a broad peak at the reciprocal lattice points, including the origin. This intensity has decreased below the noise level of the experiment on going a distance h approximately one-tenth of the distance to the edge of the cell. Thus, roughly one-half of the total observable diffuse intensity is piled up in about one-thousandth of the repeat volume, while the remaining half is spread uniformly throughout it. Order of magnitude calculations show that in most small-angle scattering experiments, this flat Laue

* The manner in which the order parameters modulate the Laue monotonic intensity has been discussed in detail by Cowley (5), Warren and Averbach (15), and Sparks and Borie (16).

monotonic intensity is so weak at any point in reciprocal space that it cannot be resolved from the background noise of the experiment.* Thus, these terms give no observable contribution to the diffuse intensity measured in the small-angle region, and may be neglected in computing the integrated small-angle X-ray intensity.

If either phase has a tendency toward short-range order, the associated Laue monotonic intensity is no longer uniform, but is distributed away from the fundamental reciprocal lattice points (including the origin). Therefore, in the small-angle region such Laue monotonic intensity is still unobservable and does not enter the integration performed in Eq. 25, although it may often be detected at larger angles. Such is the case for the formation of Guinier-Preston zones having a high degree of internal short-range order as has been found in both Al-Ag (5, 18) and Ni-Al (19) alloys.

Finally, if the atoms in either the zones or the matrix (especially the latter) show a tendency toward clustering, the diffuse intensity from the appropriate Laue term will be distributed near the reciprocal lattice points (which include the origin). In this case, one needs evidence other than small-angle X-ray scattering data in order to determine whether or not this contribution to Q_0 will be negligible. We believe that the most useful additional evidence for interpreting experiments of this kind can be obtained from high-angle diffuse intensity measurements, as reviewed by Warren and Averbach (15).

In those situations where, on the basis of information (such as high-angle diffuse scattering measurements) in addition to the small-angle scattering data, it is permissible to neglect both of the modulated Laue monotonic terms in Eq. 24, the observable integrated small-angle X-ray intensity predicted by the present theory may be compared with that predicted by Gerold. Dropping the first two terms of Eq. 25, making substitutions for F_1 and F_2, and using the relationship describing the conservation of A atoms, we find

$$V_a Q_0 \text{(observed)} = \frac{P}{N}(m_1 - m_2)^2 (f_A - f_B)^2 - \frac{N}{N}\left(\frac{P}{N}\right)^2 (F_1 - F_2)^2$$
$$= (m_1 - m_A)(m_A - m_2)(f_A - f_B)^2 \qquad 28.$$

* In high-angle diffuse scattering experiments, where the incident intensity is much greater than in small-angle experiments, the Laue monotonic scattering may be observable. Gerold (10) has successfully made such measurements on quenched Al-Ag alloys.

Comparison with Eq. 2 shows that this result is identical to Gerold's expression. The reason for including the negative delta function term was given in the previous section. On the basis of these arguments, it may be concluded that, although in theory the integrated structure-sensitive intensity is independent of the local atomic arrangements, in many physical situations the observable integrated small-angle X-ray scattering is indeed related to and may be used to find the boundaries of the metastable miscibility gap.

A final result of the present work is related to the use of the integrated small-angle scattering data to convert the observed intensity to absolute units. Following Cowley's analysis (5), Walker and Guinier (13) and Freise et al. (20) used a Fourier inversion of Eq. 3 to obtain the probability of finding Ag-Ag pairs of various lengths. In both of these papers, the X-ray intensity was converted to absolute units by noting (Eq. 5) that in this representation the integrated order intensity was a constant which could be evaluated both experimentally and from theory. The arguments given in this paper suggest that in practice the integrated low-angle intensity does not measure the total order intensity and hence all of their values of $P^{Ag-Ag}(r)$ are too large [except $P^{Ag-Ag}(0)$, which must be unity]. That such is the case was found by Freise et al. (20) when they also converted their data to absolute units by measuring the intensity of the incident beam with a series of foils, and found that the values of $P^{Ag-Ag}(r)$ computed in this way were all considerably lower than the previous values.

SUMMARY

We have shown that in solid solutions where the solute atoms are clustered into zones, the integrated order intensity, described by Cowley's order series, consists of four physically distinguishable terms. Each of these terms is dependent on the way the atoms are distributed between the zones and the matrix, but their sum is not. Often, two of these terms (the modulated Laue monotonic intensity scattered by both the zones and the matrix) may not be observable with standard small-angle X-ray scattering apparatus.

In order to verify the assumption of neglecting these terms in the calculation of the integrated small-angle X-ray intensity, we believe that it is necessary to have additional experimental evidence, such as high-angle diffuse intensity measurements. In those cases where such an assumption is valid, the integral of the observed scattering data is indeed related to the metastable miscibility gap, as originally derived by Gerold (4). Finally, it is shown that, due to these unobservable portions of the order intensity, the normalization procedure involving the integrated intensity as used by Walker and Guinier (3) and Freise, Kelly and Nicholson (20) is not valid.

ACKNOWLEDGEMENTS

The authors wish to express their thanks to Drs. C. J. Sparks and H. L. Yakel for their most valuable discussion and comments during the progress of this work.

BIBLIOGRAPHY

1. Hardy, H. K. and Heal, T. J., Progr. Metal Phys., **5,** 143 (1954).
2. Guinier, A., Solid State Phys., **9,** 293 (1959).
3. Kelly, A. and Nicholson, R. B., Progr. Mater. Sci., **10,** 151 (1963).
4. Gerold, V., Phys. Stat. Sol., **1,** 37 (1961).
5. Cowley, J. M., J. Appl. Phys., **21,** 24 (1950).
6. Borelius, G. and Larsson, L. E., Arkiv Fysik, **11,** 137 (1956).
7. Gerold, V. and Schweizer, W., Z. Metallk., **52,** 76 (1961).
8. Baur, R. and Gerold, V., ibid., **52,** 671 (1961).
9. Baur, R. and Gerold, V., Acta Met., **10,** 637 (1962).
10. Gerold, V., Z. Metallk., **46,** 623 (1955).
11. Hendricks, R. W., *On the Mechanism and Kinetics of Precipitation in NaCl-AgCl Solid Solutions*, Ph.D. thesis, Cornell University, Ithaca, N.Y., June 1964.
12. Guinier, A., J. Phys. Radium, **8,** 124 (1942).
13. Walker, C. B. and Guinier, A., Acta Met., **1,** 568 (1953).
14. Guinier, A., et al., *Small-Angle Scattering of X-Rays*, J. Wiley and Sons, New York, 1955.
15. Warren, B. E. and Averbach, B. L., *Modern Research Techniques in Physical Metallurgy*, ASM, Cleveland, 1953; pp. 95–130.
16. Sparks, C. J. and Borie, B. S., to be published by the AIME in the Proceedings of the Conference on *"Local Atomic Arrangements by X-ray Diffraction"* held in Chicago on February 15, 1965.
17. Fournet, G., Bull. Soc. Franc. Mineral. Crist., **74,** 39 (1951).
18. Auer, H. and Gerold, V., Z. Metallk., **56,** 240 (1965).
19. Starke, E. A., Gerold, V. and Guy, A. G., Acta Met., **13,** 957 (1965).
20. Freise, E. J., Kelly, A. and Nicholson, R. B., ibid., **9,** 250 (1961).

Local Atomic Configurations in a Gold-Nickel Alloy

S. C. MOSS AND B. L. AVERBACH

Department of Metallurgy, Massachusetts Institute of Technology, Cambridge, Mass.

INTRODUCTION

This work is concerned with the local atomic arrangements in a single crystal of the alloy, 0.6 Au—0.4 Ni. Earlier X-ray studies of this crystal by means of a reflection technique (1) had indicated a tendency for clustering i.e. a preference for like nearest neighbors. The disorder scattering at larger diffraction angles is dominated by the size effect scattering which arises from the large difference in atomic sizes of the two species, and it was thus difficult to separate clearly the modulation of the diffuse scattering which arises from the disorder alone. The current measurements were made in transmission at small scattering angles. Since the size effect scattering is nil at small angles it was hoped that an unambiguous indication of the local order would thus be provided by these experiments. It should be emphasized that the small-angle scattering reported here represents only part of the total picture in the study of local atomic configurations. The general disorder scattering from a solid solution includes a combination of pure disorder scattering (in this case, the low-angle data) and the combined disorder and size contributions which are observed at larger diffraction angles.

SCATTERING THEORY

The scattering theory has been discussed elsewhere [see, for example, Warren and Averbach (2), Guinier (3), and Sparks and Borie (4)], and will only be summarized here. We define the diffraction vector, $\mathbf{h}$, by

$$\mathbf{h} = \frac{4\pi \sin \theta}{\lambda} = 2\pi(h_1 \mathbf{b}_1 + h_2 \mathbf{b}_2 + h_3 \mathbf{b}_3) \qquad 1.$$

where h_1, h_2 and h_3 are continuous variables in a reciprocal lattice with unit cell dimensions, $|\mathbf{b}_1|$, $|\mathbf{b}_2|$ and $|\mathbf{b}_3|$. For a cubic crystal

$|\mathbf{b}_1| = |\mathbf{b}_2| = |\mathbf{b}_3| = 1/a$, where a is the lattice constant and h_1, h_2 and h_3 have one-half of the integral values of the usual Miller indices at the reciprocal lattice sites. Thus, the (100) reflection is denoted by $\frac{1}{2}00$, the (111) by $\frac{1}{2}\frac{1}{2}\frac{1}{2}$. With this notation the pure disorder scattering, which is concerned only with the identity of the local neighbors, is represented as a triply periodic Fourier cosine series with coefficients which are the Warren short-range order parameters, α_{lmn}. This disorder scattering is given by:

$$I_d = Nm_A m_B (f_A - f_B)^2 \sum_{lmn} \alpha_{lmn} \cos 2\pi(h_1 l + h_2 m + h_3 n) \qquad 2.$$

The intensity, I_d, is in electron units, m_A and m_B are the atomic fractions of A and B atoms in the binary alloy with scattering factors f_A and f_B, and α_{lmn} is defined by:

$$\alpha_{lmn} = 1 - \frac{P_{lmn}^A}{m_A} \qquad 3.$$

P_{lmn}^A is the probability of finding an A atom as an lmn neighbor of a B atom at some origin. The indices, lmn, are defined by the lattice vector, $\mathbf{r}_{lmn}$, where

$$\mathbf{r}_{lmn} = l\frac{\mathbf{a}_1}{2} + m\frac{\mathbf{a}_2}{2} + n\frac{\mathbf{a}_3}{2} \qquad 4.$$

Each shell of atoms about the origin atom may be defined either as the ith shell with an order parameter, α_i, or, for a given shell in a cubic crystal, by the indices, lmn, where $\alpha_{lmn} = \alpha_{mln} = \alpha_{\overline{lmn}}$. The coordination number of a shell is the number of permutations possible for that particular lmn. For the case of an f.c.c. lattice, the first shell about a corner atom has an order parameter $\alpha_1 = \alpha_{110}$ with 12 possible permutations of (110). If all $\alpha_{lmn} = 0$, save α_{000}, which by definition is unity, we have a random solution, and the pure disorder intensity is given by the Laue monotonic term, $Nm_A m_B (f_A - f_B)^2$.

Since the series in Eq. 2 takes on the periodicity of the reciprocal lattice, the disorder scattering may be measured in any cell or Brillouin zone of reciprocal lattice. If only pure disorder scattering is present, the disorder intensity, appropriately corrected for the angular variation of the scattering factors and polarization, is exactly repeated in each zone, including the one centered about the origin. The Fourier transform of this intensity then yields all of the correlation functions of the alloy, and the construction of a model of the solid solution can be attempted. In special cases, some features of a model can be constructed without solving quantitatively for the correlation functions. A particular example is the case where the diffuse scattering appears as side bands about some position in the reciprocal lattice, as with 0.68 Cu−0.316 Au (5), where a satellite pattern about the superstructure reflections has been taken as evidence of a nearly periodic anti-phase domain structure. In the case of the pseudo binary CuNiFe (6), a satellite pattern about the fundamental reflections has been interpreted as arising from a periodic modulation in the local composition.

The diffuse scattering from alloys above their ordering or solution temperatures can exhibit more subtle effects. There may be only local deviations from randomness, and even these may not be regular enough or well enough defined to permit the exact description by a specific model. If, however, the disorder scattering can be described by Eq. 2, which requires that the sites of the locally arranged regions be denoted by vectors from the origin sites in the matrix, we are justified in treating the entire disorder portion of the scattering as a whole. If not, and such is the case when structural coherence between matrix and precipitate is lost, we must resort to more careful analyses of the separate contributions, as outlined by Hendricks and Borie in another paper in this volume.

An additional complication arises from the component of diffuse intensity attributable to the differences in atomic sizes of the atoms in the solid solution. There are generally both local and long-range size effect contributions, but Borie (4, 7) has shown that for the completely random elastically isotropic solid solution, which is characterized by atomic displacements which vary as $1/r^2$ from the origin of distortion, the *total* size effect scattering can be written in closed form. For 0.6 Au−0.4 Ni, the diffuse scattering for a random alloy with size effect becomes

$$I_{d\&s} = 0.24N[f_{Au} - f_{Ni}]^2 \\ \times \left\{ 1.0 + (0.4f_{Ni} + 0.6f_{Au})/(f_{Au} - f_{Ni}) \right. \\ \left. \times 0.22\, C_{Ni} \sum_{lmn} \frac{2\pi(lh_1 + mh_2 + nh_3)}{(l^2 + m^2 + n^2)^{3/2}} \right. \\ \left. \times \sin 2\pi(lh_1 + mh_2 + nh_3) \right\}^2 \quad 5.$$

where C_{Ni}, in units of Å³, is a measure of the strength of the distortion field and is generally negative for the smaller atom and positive for the larger.

Although this alloy is certainly not elastically isotropic $[(C_{11} - C_{12})/2C_{44} = 0.34$ (8)] and can be far from random, depending on the details of the heat treatment and quench, the expression in Eq. 5 does allow us to discuss qualitatively several aspects of the size effect contribution. The usual separation of this scattering into local and Huang diffuse parts has also been extended by Borie and Sparks (9) who collected the static and thermal diffuse scattering (TDS) terms together in such a way as to permit the separation of the diffuse intensity into a pure disorder part and a TDS and static displacement scattering.

Turning to Eq. 5, $C_{Ni} < 0$ because the Ni atom is the smaller of the two. The lattice parameters are: $a(Ni) = 3.564$ Å and $a(Au) = 4.078$ Å. Since f_{Au} is also greater than f_{Ni}, the amplitude-modulated sine series takes on negative values between the origin of reciprocal space and the point ($\frac{1}{2}$ 00), positive values between ($\frac{1}{2}$ 00) and (100), negative between (100) and ($\frac{3}{2}$ 00), etc. The size effect scattering is a nonperiodic function in reciprocal space which alternately subtracts from and adds to the pure disorder scattering.

In Eq. 5, the solution is assumed to be random, and this accounts for the term 1.0 within the bracket. If the solution is not random, the disorder and size effect scattering must be treated separately, but the principal conclusions are qualitatively the same. In this case, however, the size effect contributions are separated into a local modulation series and a Huang diffuse scattering. The local size effect modulation for a crystal in which the distortions show the same symmetry as the lattice itself can be written as:

$$I_s = Nm_A m_B (f_A - f_B)^2 \sum_{lmn} \beta_{lmn}$$
$$\times 2\pi(h_1 l + h_2 m + h_3 n) \sin 2\pi(h_1 l + h_2 m + h_3 n) \quad 6.$$

The coefficients β_{lmn} are not constants but depend on the scattering factors and local order parameters, α_{lmn}, as well as on the local displacements of both kinds of atoms. The oscillations in sign of I_s are the same as those predicted by the series in Eq. 5.

If we have clustering, the value of $P_{lmn}{}^A$ is less than m_A, and $\alpha_{lmn} > 0$. Thus the Fourier series in Eq. 2 generally takes on the shape of Fig. 1a with the sharpness of the peak being determined by the range or the extent of the correlations. If the size effect modulations of Eq. 6 are included, we may observe scattering similar to either Fig. 1b or 1c. Figure 1b characterizes the gold-nickel system, where both f_{Au} and $a(Au)$ are larger than f_{Ni} and $a(Ni)$. For a system such as aluminum-copper, $f_{Cu} > f_{Al}$ but $a(Al) > a(Cu)$, and we would expect a redistribution of the intensity in the manner indicated by Fig. 1c. If there is some definite structural aspect to the clustering, i.e. some specific composition modulation, then the profiles

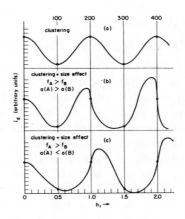

Fig. 1. Idealized diffuse scattering along [100] for a binary single crystal with clustering and size effect.

of Fig. 1 are still further altered, and the general procedure for analyzing these effects has recently been treated by deFontaine (10) for the one-dimensional case.

The plots in Fig. 1 are idealized diffuse scattering profiles along a [100] axis in the crystal, and all angular variation in the ideal disorder scattering has been removed. The intensity units are arbitrary. The contributions of the fundamental diffraction peaks, the (200) and (400) reflections for the f.c.c. structure, and their associated TDS, have also been omitted. In addition, the Huang scattering, which peaks about the fundamentals, is not represented. Thus it appears that, as h_1 becomes large, the intensity becomes negative. If all of these diffraction effects are included, the resulting profiles in 1a, 1b and 1c are partly obliterated, and the total intensity is always positive. However, if we wish to retrieve the profile of curve 1a, we must resort either to a small-angle scattering experiment or to a method of separating out the pure disorder scattering, such as the one developed by Borie and Sparks. It should be noted that the portions of 1a, 1b and 1c below $h_1 = 0.5$ are quite similar, being only slightly depressed (1b), or enhanced (1c), by the size effect scattering.

In the present discussion, the disorder scattering from the 0.6 Au−0.4 Ni crystal will be treated as a whole, and the small-angle part will be presented as the undistorted disorder intensity. Finally, from the observed shape of the high- and low-angle data, we will try to derive a solid solution model without recourse to the quantitative short-range order parameters.

EXPERIMENTAL RESULTS

The Au-Ni phase diagram (11) is shown in Fig. 2. The major characteristics are the complete miscibility of the two f.c.c. metals at higher temperatures and the miscibility gap. The decomposition process at temperatures within the gap is heterogeneous in nature, originating mainly at grain boundaries or at the surface in a single crystal. The coherent spinodal, or locus points within the gap where the second derivative of the chemical partial molar free energy with respect to composition is zero, appears to be suppressed several hundred degrees in this system (12) because of the elastic energy associated with the formation of coherent gold-rich and nickel-rich clusters (13). The heterogeneous transformation is easily suppressed on quenching from temperatures above the miscibility gap, but the spinodal decomposition is apparently not so easily avoided.

The crystal which was used for the small-angle measurements reported here is the same one used in the earlier work (1) and contained about 40 at.% nickel. A (100) slice was cut adjacent to the sample used for the reflection measurements. This slice was polished, etched and solution-treated for one week at 890°C and water-quenched in about one second to room temperature. The crystal was then mounted on a polishing block with a small central hole, and the thickness reduced to about 0.0009 in. using a sequence of polishing papers and diamond grit. The thickness in the final stages was obtained from measurements of the transmission of CuKα X-rays. The crystal was quite hard and polished easily without smearing. This is shown by a typical transmission Laue photograph taken at a thickness of 0.0012 in. (Fig. 3). There is a pronounced splitting of the spots from the initial mosaic structure, but the individual spots are quite sharp. The striking four-fold streaking at the center is not caused by small-angle scattering, but by the intersection of a short wavelength Ewald shell from the white spectrum, with the intense diffuse scattering in the vicinity of the {200} peaks. The diffuse scattering in this region of reciprocal space has been discussed earlier (1).

At a thickness of 0.0009 in. the polishing became difficult, but it was still not possible to obtain small-angle transmission photographs

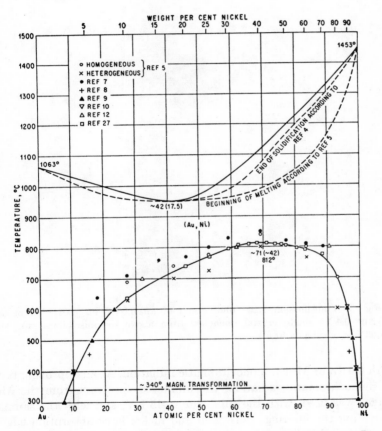

Fig. 2. The gold-nickel phase diagram. [From *The Constitution of Binary Alloys*, Second Edition, p. 220, by M. Hansen, Copyright © 1958 by the McGraw-Hill Book Company, Inc. Used by permission of McGraw-Hill Book Company.]

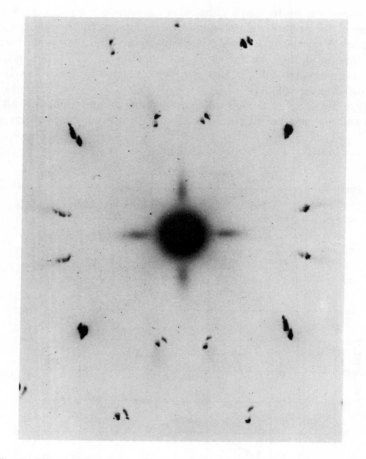

Fig. 3. Transmission Laue photograph of an 0.0012 in. (001) foil of an 0.6 Au − 0.4 Ni single crystal quenched from above the miscibility gap. Mo target X-ray tube at 40 KV.

with CuKα or MoKα monochromatic radiation. As an approximation to monochromatic radiation, the unfiltered spectrum from the Mo target was used, with the tube run at 45 KV, and with the sample providing the filtering. Both gold and nickel have absorption edges at considerably longer wavelengths than the short wavelength limit of the tube ($\lambda_{SWL} = 0.275$ Å). Consequently, the gold-nickel foil cuts off the longer wavelengths, and there is a resultant passband

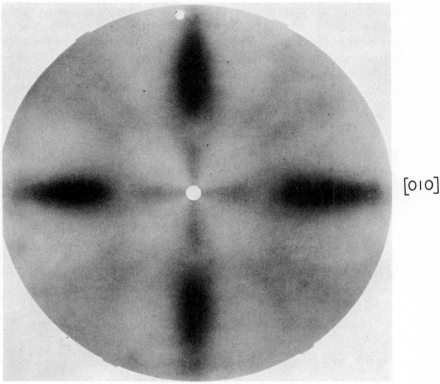

Fig. 4. Small-angle X-ray transmission photograph from the same crystal as in Fig. 3, but with a foil thickness of 0.0009 in.

which peaks at about 0.36 Å and falls off more slowly toward the longer wavelengths. Because of the diffraction from a wavelength band, the resultant scattering was smeared along radial lines from the center of the pattern, but the data are still instructive.

The small-angle photo in Fig. 4 was taken with an extended vacuum chamber Kiessig Camera with medium resolution, a sample-to-film distance of 176 mm, and an exposure time of 112 hours. The four dark outer diffuse streaks along ⟨100⟩ directions are the same as those shown in Fig. 3 and arise from the diffuse scattering in the vicinity of the (200) reflections. The diffuse scattering in reciprocal space for this crystal is shown in Fig. 5, which is taken from the previous study (1). If we assume that the center of the outer dark streaks in Fig. 4 corresponds to the diffraction vector extending to the approximate maximum in the diffuse intensity near (200) in Fig. 5, the resultant wavelength is about 0.36 Å, corresponding to the peak of the wavelength distribution used in these experiments.

The increased resolution of the small-angle scattering experiment makes it possible to observe four diffuse streaks in ⟨100⟩ directions

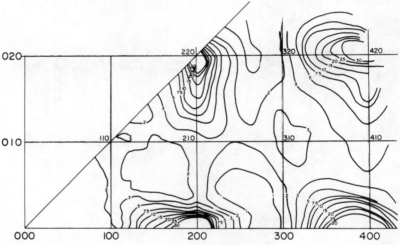

Fig. 5. Equi-intensity contours, in Laue monotonic units, in reciprocal space of the diffuse scattering from single crystal 0.6 Au−0.4 Ni. The crystal was quenched in about one second from above the miscibility gap and the data were obtained in reflection [Ref. (1)].

within the outer streaks of Fig. 4. These inner diffuse streaks are assumed to arise from the pure disorder effects from the region of reciprocal space close to the origin, where the size effect makes very little contribution. Microphotometer traces along each of these lines showed that the intensity rises from the origin, falls off slightly and rises again rapidly as the outer streaks are approached.

Figure 4 also shows pairs of very weak streaks in the four ⟨110⟩ directions. It appears that these may be due to the diffuse intensity in the vicinity of (110), and these streaks arise from the intersection of the Ewald shell with the cluster and size effect scattering in this region of reciprocal space, denoted by the contour $I' = 3.0$, at (110) in Fig. 5.

Figure 4 appears to confirm the earlier conclusion that there are alternate gold-rich and nickel-rich regions in the quenched alloy. The earlier conclusions were based on the reflection data, which were greatly influenced by the size-effect scattering. It was also suggested that the development of the clustering was accentuated by the relatively slow quench through the spinodal region of the miscibility gap. In order to investigate the influence of the quench, we have performed additional diffuse reflection scattering measurements on an identical crystal, which was solution treated for a week at 890°C and quenched into ice-water in about $\frac{1}{50}$ sec. It was hoped that the more rapid quench would prevent spinodal decomposition, and that the resulting diffraction pattern would be more characteristic of the equilibrium solid solution above the miscibility gap. Figure 6 shows the comparison of the [100] diffractometer tracings from the slowly (6a), and rapidly (6b), quenched crystals. The radiation employed was CuKα monochromatized by a doubly bent LiF crystal (15), and the full-scale intensity was 10,000 counts/min. These reflection data could not be extended below $2\theta = 8°$ because of the divergence of the beam at the crystal. The entire diffractometer was enclosed by a helium-filled bag to eliminate air scattering. The upper curve, 6a, clearly shows both the clustering and size effect scattering in these unreduced and uncorrected data. The very broad and weak peak near $2\theta = 14°$ corresponds to the center of the inner small-angle streak shown in Fig. 4 and is attributed to the cluster scattering. The strongly asymmetric scattering about (200) arises from the large size-effect diffuse scattering; f_{Au} and $a(Au)$ are larger than f_{Ni} and $a(Ni)$, and this corresponds to case (b) in Fig. 1. The fundamental reflections are about 10^5 times greater than the diffuse background intensity.

The lower curve in Fig. 6b shows no clear evidence of the cluster scattering in the rapidly quenched specimen, at least on this scale. The asymmetry about (200) is still noticeable but much less pronounced, and the rapidly-quenched crystal is apparently characterized by a more random arrangement of the constituent atoms. In fact, the standardized and reduced intensity of Fig. 6b can be almost entirely accounted for by Eq. 5 with a value of $C_{Ni} = -0.24$ Å^3. It appears that the rapidly-quenched crystal is probably quite close to being random, with the major diffraction effects arising from the difference in atom sizes.

The position of the diffuse scattering in the vicinity of (100) for both crystals appears to eliminate the possibility of short-range order, i.e. a preference for unlike pairs, as reported by Flinn, Averbach and Cohen (14) on the basis of powder data on samples quenched even more slowly. This discrepancy points out the inherent limitation of the powder technique, since a very large size effect contribution

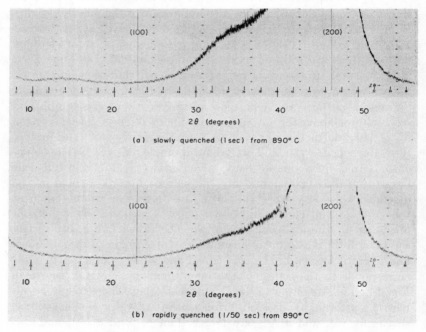

Fig. 6. Diffractometer recordings from two 0.6 Au—0.4 Ni crystals quenched at different rates, in reflection.

averaged over the powder pattern sphere resembles short-range order scattering quite closely.

DISCUSSION

The combination of the small-angle transmission data and the results in reflection indicate that the slowly quenched crystal probably exhibits a composition modulation along ⟨100⟩ axes in the crystal.

On the other hand, the rapidly quenched crystal appears to be more random, with an indication of a strong size-effect scattering, suggesting that the atoms are not precisely at the average lattice sites. In principle, Borie and Sparks (9, 16) have shown that the reflec-

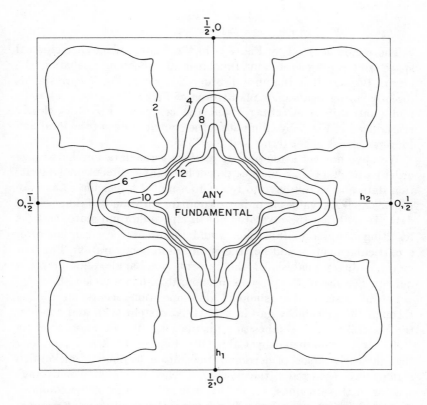

Fig. 7. Equi-intensity contours of the pure disorder scattering in Laue monotonic units, obtained from Fig. 5 by Borie and Sparks [Ref. (16)]. Used by permission of Dr. B. S. Borie, Oak Ridge National Laboratory.

tion data themselves can yield much the same information as the low-angle scattering experiment, if the undistorted disorder scattering is separated from all of the other contributions. They have applied their separation technique to the data in Fig. 5 to obtain the equi-intensity contour map of Fig. 7, in which contours are labelled in Laue monotonic units, $I'(h_1h_20)$ where:

$$I'(h_1h_20) = \frac{I_d}{Nm_Am_B(f_A - f_B)^2} = \sum_{lm} A_{lm} \cos 2\pi(h_1l + h_2m) \qquad 7.$$

and

$$A_{lm} = \sum_n \alpha_{lmn} \qquad 8.$$

The region covered by Fig. 7 is in the $\{h_1h_20\}$ plane of reciprocal space and centered on any fundamental reflection, including the origin. Because it is the pure disorder scattering, Fig. 7 corresponds directly to the small-angle photograph of Fig. 4. Thus the weak fourfold $\langle 100 \rangle$ diffuse streaking about the center of Fig. 4 is closely reproduced in Fig. 7 by eliminating the contributions of the local and long-range size effect terms and the TDS.

We have not yet attempted to analyze these data for short-range order parameters, α_{lmn}, since we do not have three-dimensional diffraction data and would have to resort to a model to calculate the two-dimensional A_{lm} values in order to compare with the two-dimensional data. In addition the crystal is not in an equilibrium state, and the resulting correlation functions would represent the situation after a particular quench and not the equilibrium configuration. Furthermore, a direct analysis of the present small-angle photograph is difficult because of the presence of more than one wavelength.

Let us assume that spinodal decomposition occurs on cooling through the miscibility gap in the Au-Ni system (11), and consider the possibility that composition fluctuations, or waves, occur along $\langle 100 \rangle$ axes, as predicted by Cahn (13). Figures 4, 6a and 7 suggest that the early stages of decomposition into gold-rich and nickel-rich regions have occurred in the slowly quenched specimen. The average spacing of these regions, i.e. the average wavelength of the composition modulation, may be deduced from the position of the inner diffuse streaks in Figs. 4 and 6a. A broad peak is observed in the vicinity of $h_1 = \frac{1}{4}$, and this corresponds to a wavelength of about two unit cells, or approximately 8 Å. The streaking of the diffuse scattering along $\langle 100 \rangle$ directions also suggests that the composition fluctuation occurs in the form of $\{100\}$ platelets with this average thickness.

The most recent study of the decomposition in this system by means of transmission electron microscopy (17) has revealed a periodicity of the order of 40 Å in a 0.64 Au−0.36 Ni alloy which has been quenched and annealed at 150°C for 100 hours. The longer wavelength is evidently the result of a period of growth and is quite consistent with our result of a much shorter wavelength for a crystal which has been slowly quenched through the decomposition range.

SUMMARY

We may summarize our observations on the local order in a single crystal of 0.6 Au−0.4 Ni alloy as follows:

1. A quenched single crystal appears to contain a modulation of the local composition into gold-rich and nickel-rich regions. The extent of the modulation depends on the rapidity of the quench. On the other hand, there is no indication of short-range order, i.e. a preference for unlike neighbors, in these quenched crystals.

2. A very rapid quench appears to suppress the decomposition process, and the Au and Ni atoms seem to be arranged randomly in the solution. This suggests that the equilibrium arrangement above the miscibility gap may be essentially random.

3. The transmission small-angle scattering reveals a strong $\langle 100 \rangle$ directionality to the clustering in the slowly quenched sample. This directionality is reproduced quite well by Borie and Sparks (16), using our earlier high-angle reflection data and an appropriate technique for separating out the size effect from the pure cluster scattering.

4. The $\langle 100 \rangle$ scattering is to be expected from a sample undergoing spinodal decomposition. An average wavelength of about 8 Å has been deduced for the weakly predominating composition fluctuation in this particular slowly quenched crystal.

ACKNOWLEDGEMENTS

We wish to thank Dr. David Keating of the Brookhaven National Laboratory for his kind permission to use, and his assistance with, the Kiessig small-angle camera; Professor J. W. Cahn for several stimulating discussions on the spinodal mechanism; and Dr. B. S. Borie and Dr. C. J. Sparks for their permission to use the contour map in Fig. 7. This research was supported by a grant from the U.S. Atomic Energy Commission to whom we have had continuing occasion to express our deep appreciation.

BIBLIOGRAPHY

1. Moss, S. C., Local Order in Solid Alloys—I, in *Local Atomic Arrangements Studied by X-Ray Diffraction*, Cohen, J. B. and Hilliard, J. E., editors. To be published by Gordon and Breach, New York.
2. Warren, B. E. and Averbach, B. L., *Modern Research Techniques in Physical Metallurgy*, ASM, Cleveland, 1953; pp. 95–130.
3. Guinier, A., *X-Ray Diffraction in Crystals, Imperfect Crystals and Amorphous Bodies*, W. H. Freeman and Co., San Francisco, 1963.
4. Sparks, C. J. and Borie, B., "Method of Analysis for Diffuse X-ray Scattering Modulated by Local Order and Atomic Displacements," in *Local Atomic Arrangements Studied by X-Ray Diffraction*, Cohen, J. B. and Hilliard, J. E., editors. To be published by Gordon and Breach, New York.
5. Scott, R. E., J. Appl. Phys., **31**, 2112 (1960).
6. Hillert, M., Cohen, M. and Averbach, B. L., Acta. Met., **9**, 536 (1961).
7. Borie, B., Acta Cryst., **12**, 280 (1959).
8. Golding, B., preliminary result from thesis in progress for the Sc.D. degree, Massachusetts Institute of Technology.
9. Borie, B. and Sparks, C. J., Metals and Ceramics Div. Ann. Progr. Rept., June 30, 1965, ORNL-3870.
10. deFontaine, D., "A Theoretical and Analogue Study of Diffraction from One-Dimensional Modulated Structures," in *Local Atomic Arrangements Studied by X-Ray Diffraction*, Cohen, J. B. and Hilliard, J. E., editors, to be published by Gordon and Breach, New York.
11. Hansen, M., *Constitution of Binary Alloys*, Second Edition, McGraw-Hill, New York, 1958; p. 220. [The most reliable data on the miscibility gap are the points of Ref. 27—A. Muenster and K. Sagel, Z. physik. Chem. (Frankfurt), **14**, 296 (1958).]
12. Fisher, R. M. and Embury, J. D., Proceedings, Third Regional Meeting on Electron Microscopy—Prague, Czechoslovakia, August (1964).
13. Cahn, J. W., Acta Met., **10**, 179 (1962).
14. Flinn, P. A., Averbach, B. L. and Cohen, M., ibid., **1**, 664 (1953).
15. Chipman, D. R., Rev. Sci. Instr., **27**, 164 (1956).
16. Borie, B. and Sparks, C. J., to be published.
17. Woodilla, J., preliminary results from thesis in progress for the Sc.D. degree, Massachusetts Institute of Technology.

X-Ray Scattering by Point-Defects

A. M. LEVELUT AND A. GUINIER

Laboratoire de Physique des Solides, Faculté des Sciences,
Orsay (S.-et-O.), France

INTRODUCTION

Up to now, it has not been possible to observe directly the point defects in a crystal; we have only more or less indirect experiments at our disposal for their detection: that is, the variation of a physical property which is sensitive to the presence of point defects.

To interpret such a measurement, one must

(a) assume the type of defect which is present and

(b) know theoretically the influence of one isolated defect. Thus for instance, the increase of resistivity after quenching may give the number of frozen-in vacancies. But the difficulties come from the uncertainty of the theoretical calculation of the resistivity increment per vacancy. Another method, the comparison of density and lattice parameter measurements, is more certain because the theory is simpler, but far less sensitive.

An effect of point defects is the production of X-ray diffuse scattering, provided that these defects are associated with a local variation of the electron density. Furthermore the X-ray scattering may easily be calculated from this variation. Therefore, it seemed worthwhile to investigate the possibilities of the measurement of X-ray scattering for the study of some point defects. A similar method had been previously used with neutrons (1) but not with X-rays to our knowledge.

The chief difficulty is the low value of the scattered intensity which may be predicted in typical cases of crystals containing a reasonable proportion of point defects as vacancies, interstitials or impurity atoms. It is too low to be detectable with a conventional X-ray diffractometer. Therefore we have designed a new device especially for this purpose. In this paper we shall describe the apparatus, the tests of its sensitivity and its application to some problems concerning point defects in crystals.

PRINCIPLE OF THE EXPERIMENT

Let us consider a continuous medium of uniform electron density. It gives no X-ray scattering: We shall always here disregard the domain of the very small angles (say $\theta < 1'$), where the external shape of the diffracting grain intervenes. If the electron density is a rigorously periodic function, as in a perfect crystal, the object produces diffractions at large angles, but still no small-angle scattering. Such a scattering appears when the crystal is no longer perfectly homogeneous. Suppose that it contains domains of diameter d and density ρ_1, different from the normal value ρ_0. The small-angle scattering will be roughly limited to an angle of the order of λ/d; its maximum value at 0 angle is proportional to the product of the total volume of heterogeneities, the volume of a single domain and the square of the density contrast $(\rho_1 - \rho_0)^2$. If the size of the individual heterogeneity domains decreases, their total volume being constant, the small-angle scattering becomes broader and its intensity decreases. The limiting case for crystals corresponds to isolated B atoms substituted at the nodes of the otherwise unperturbed crystalline lattice of A atoms. This is the *Laue monotonic scattering* (2):

$$I = c_A c_B (f_A - f_B)^2 \qquad 1.$$

where c_A, c_B are the atomic concentrations and f_A and f_B the scattering factors. I is precisely the scattering power per atom, i.e., the ratio of the scattered intensity per atom of the sample to the intensity scattered by a free electron in the same experimental conditions. If the substituted atoms are in small proportion, $c_A \simeq 1$, and

$$I = c_B (f_A - f_B)^2 \qquad 2.$$

In particular, if vacancies are randomly dispersed with a small concentration c in a lattice *having no other defects*, the scattering will be

$$I = c f_A^2 \qquad 3.$$

The formulae 1 or 3 show that the scattering decreases slowly with increasing angles; this decrease is due to the variation of the scattering factor: it is well known and the study of the scattering as a function of angle provides no information. The two important points are the verification of the decrease proportional to f^2 and the absolute value of the intensity for a given angle, for instance, for very small angles.

The situation is thus quite different from an ordinary small-angle scattering experiment, where the shape of the scattering curve is determined, in order to find the size of the defect. Therefore, in the usual small-angle scattering apparatus, it is necessary to know accurately the angle for which the scattering is measured and the solid angle of the measured scattered beam must be small. But this basic condition does not hold if one is interested only in point defects. The measurement of the scattered intensity in a large solid angle, *if on an absolute scale*, is sufficient, provided that one verifies that the variation of scattering with angle is small. Nevertheless, the measurement may only be done at a small angle for various reasons: (1) It is necessary to avoid any Bragg reflection in the scattered beam: this is certainly realized for any orientation of the crystal if the maximum scattering angle is smaller than the smallest Bragg angle (2) The atomic scattering factors are maximum at zero angle (3) The other causes of scattering which cannot be eliminated are at their minimum at zero angle: these are the scattering due to thermal agitation, the Compton scattering and the scattering due to lattice distortions.

The principle of the experimental device is illustrated by Figs. 1 and 2. The primary beam is monochromatic and convergent on a small focus F. It is obtained by reflection from a LiF crystal with double curvature. A thin (<0.2 mm thickness) LiF lamella is pressed at room temperature against a stainless block with a toroidal surface of radii R_1 and R_2. If θ is the Bragg angle [$\theta = 20°$ for CuKα and (200) LiF], the condition for point focussing is $R_2/R_1 = \sin^2 \theta$. We have used $R_1 = 350$ mm and $R_2 = 51.3$ mm, so the distances of crystal to source or focus are equal to 130 mm.

The focus which we have effectively obtained with the best monochromator is a spot of 0.5×1 mm, the source being a fine-focus Philips tube.

Because of the imperfections of the LiF crystal, the cross-section of the reflected beam is not of uniform intensity but that has no importance for the planned experiments. The beam passes through the specimen which must be a lamella, the thickness of which is made equal to the optimum thickness to maximize the scattered intensity. Its surface must be greater than the section of the beam (1 mm²).

The scattered radiation emitted by the specimen is limited by two screens in the plane passing through the focal point of the mono-

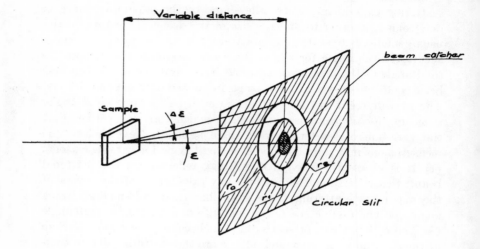

Fig. 1. Principle of the camera.

chromatic beam, so that the scattered beam is confined between two coaxial cones of revolution. The minimum aperture for the inner cone is determined by the beam trap. Thus the radius of the inner screen, r_1, has a minimum value r_0. By changing the screens, the radii r_1 and r_2 may be changed. So it is possible to select the domain of scattering angles of the beam defined by the screens. In our

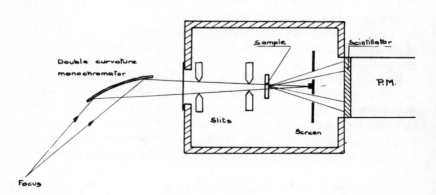

Fig. 2. Scattering chamber.

apparatus, ϵ and $\epsilon + \Delta\epsilon$ being the semi-apex angles of the scattered beam, ϵ may vary from 2°30′ to 6° and $\Delta\epsilon$ has a minimum value of 30′.

The scattered beam is received behind the screens on an X-ray detector which must count the photons with the same efficiency for all possible scattered rays. A large scintillating crystal is associated with a photomultiplier: the sensitivity of such a detector is practically uniform for a surface of 40 mm diameter.

The most important condition for the success of the experiment is the protection of the detector against parasitic radiation which could surpass the radiation to be measured. The main sources of this secondary radiation include the scattering by the air irradiated by the primary beam, the monochromator lamella itself and the edges of the slits touched by the primary beam. This is a general problem encountered in every small-angle scattering device and we used similar solutions: the camera is evacuated, the slits have tantalum edges very carefully polished and adjustable (3) with high accuracy movements so that they can be placed just at the limits of the monochromatic beam without touching it.

DESCRIPTION OF THE APPARATUS

The main part is a box of bronze solid enough to be rigorously undeformed when it is evacuated. The bottom of the box supports the slits, the sample, the screens and the beam trap. The primary beam enters through an aluminum window and the scattered beam leaves the box through a thin beryllium window. The scintillator and photomultiplier are fixed behind this window.

The sample can be moved along the beam from outside so that it is possible without breaking the vacuum to change the distance between sample and screens and hence the mean scattering angle.

An important feature of our apparatus is the possibility of carrying out the measurements at temperatures varying from 100°K to 800°K. Furthermore for the experiments at low temperature, the specimen may be taken in a reservoir of liquid nitrogen and transferred into the position of measurement *without being reheated* to room temperature. These conditions necessitate a somewhat complicated cryostat attached to the removable top of the box. The cooling system has been recently described (4). The procedure to perform an experiment on a cold sample is the following (Fig. 4): The sample holder is a copper block A; this block is plunged into the liquid nitrogen bath where the sample is kept. By a manipulation inside the liquid, the

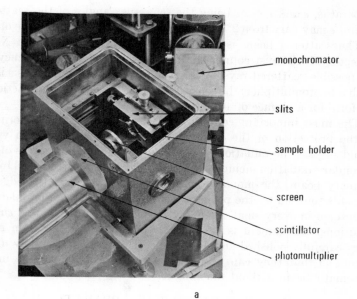

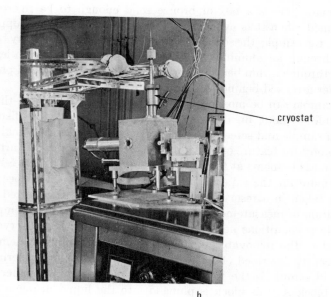

Fig. 3. (a) Scattering chamber (open) (b) Chamber with cryostat.

X-RAY SCATTERING BY POINT-DEFECTS

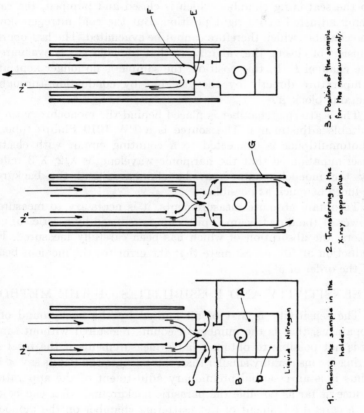

Fig. 4. Scheme of the cryostat.

sample B is fixed into the block A. The cold gas flow is then put into operation and the sample holder slides into the hollow cane G. Thus the sample is kept at low temperature and protected against icing, because the gas flow and the insulating piece D keep it away from atmospheric moisture. The whole cooling system is then transferred to the scattering chamber which is closed and pumped, the sample being adjusted at its right position. But the cold nitrogen flows in the chamber which therefore cannot be evacuated. The last operation consists of closing the valve E, so that the cold gas is evacuated by the channel F. In this position, the sample is no longer kept at low temperature directly by the gas, but by conduction through the metallic block A.

The scattering chamber is placed behind the monochromator with suitable adjustments. The source is a PW 1010 Philips tube. The photomultiplier is connected to a counting circuit with electronic discrimination so that the harmonic wavelengths $\lambda/2$, $\lambda/3$ reflected by the monochromator may be eliminated and the background reduced as much as possible.

To obtain absolute measurements, it is necessary to measure the power of the direct beam. This beam is reduced by a series of nickel filters, the absorption of which has been carefully measured. For a reduction of 10^4, we estimate that the error for the incident beam is of the order of 5%.

SENSITIVITY AND POSSIBILITIES OF THE METHOD

The sensitivity is first of all limited by the background of the apparatus, that is the number of counts registered without sample. It is not possible to obtain reliable measurements if the total scattering is considerably smaller than this background (let us say $<1/5$). Thus it is an essential preliminary adjustment of the apparatus to reduce as far as possible the parasitic background: that can be done by a careful alignment of the scattering chamber on the monochromator beam and by precise adjustments of the various slits. Figure 5 gives as a function of the average scattering angle the ratio of the energy received by the counter per unit solid angle to the total energy of the direct beam in a blank experiment for the best conditions we have been able to obtain. This background decreases when the scattering angle increases from 2° to 6°. Below 3°, the measurements are practically not possible. Furthermore various causes of scattering by the sample itself are superposed on the scattering by point defects (for instance Compton scattering, thermal scattering) and their intensity increases with the scattering angle. Thus it is advantageous to remain in the region of the small angles. Practically, the optimum angular domain for the detection of point defects seems to be between 3° and 6°.

To give a concrete idea of the performance of our apparatus, let us take an example. Consider a perfect lattice of an element of mass A, scattering factor f, specific mass ρ and mass absorption coefficient

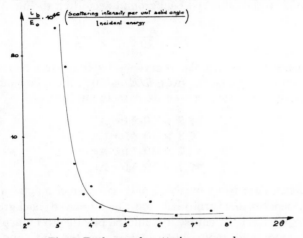

Fig. 5. Background scattering vs. angle.

μ. This crystal contains a concentration $c \ll 1$ of vacancies. Formula 3 gives the scattering power per atom: $I = cf^2 = cZ^2$. The sample is irradiated on a surface s and its thickness is supposed to be the optimum value $1/\mu\rho$. Thus the number of irradiated atoms is

$$n = \frac{s}{\mu\rho} \cdot \rho \cdot \frac{N}{A} \qquad 4.$$

The scattered intensity is

$$i = InI_e \cdot \frac{1}{e} \qquad 5.$$

$1/e$ is the absorption factor corresponding to the optimum thickness, and I_e is the intensity scattered per unit solid angle by a free electron at very small angles:

$$I_e = 7.9 \times 10^{-26} \frac{E_0}{s} \qquad 6.$$

E_0 is the total intensity of the direct beam. Finally formulas 3, 4, 5, and 6 give

$$i = cZ^2 \frac{s}{\mu\rho} \rho \frac{N}{A} \cdot 7.9 \times 10^{-26} \frac{E_0}{s} \frac{1}{e} \qquad 6a.$$

$$= cZ^2 \frac{N}{A\mu} \frac{7.9 \times 10^{-26}}{e} E_0$$

We can calculate the concentration of vacancies giving a scattering just equal to the observed background according to Fig. 5:

$$c = \frac{e}{7.9 \times 10^{-26} N} \frac{A\mu}{Z^2} \cdot \frac{i_b}{E_0} \qquad 7.$$

The nature of the sample intervenes by the factor $A\mu/Z^2$. For the CuKα wavelength and the ratio $i_b/E_0 = 10^{-5}$, the following minimum vacancy concentrations are easily detectable:

2.7×10^{-3} for Al
1.3×10^{-3} for Cu
3.7×10^{-3} for Ag
2.4×10^{-3} for Au.

These figures are to be regarded only as order of magnitude because they depend rather arbitrarily on the observed background in a blank experiment. Of course it would be possible to reduce this background more and more, but, as was pointed out, another more fundamental limitation of the sensitivity would appear, namely the thermal agitation scattering of the sample itself. It corresponds roughly to a concentration of vacancies of a few thousandths at room temperature. Even at liquid nitrogen temperature, the thermal scattering will render very delicate and probably uncertain the detection of isolated vacancies at a concentration less than 10^{-3}. Although we do not as yet have any positive experimental result on this problem, it seems improbable that our method could be used for the study of quenched-in vacancies in metals. But, on the other hand, the method can be tried with confidence in any problem where the expected concentration of point defects exceeds 1%, if this point defect produces a local change of the number of electrons of the order of the atomic number of the atoms in the lattice.

Even if the expected number of vacancies is small, there are cases where it is worthwhile to try the experiment because, if the vacancies are *not* really isolated, a detectable scattering may be found. Consider a cluster of vacancies (or foreign atoms) in a lattice. The resulting scattering is no longer simply proportional to f^2, but decreases from the origin more and more rapidly as the cluster size increases. But for very small clusters, it would be difficult to use the shape of the scattering curve; but at the same time the absolute intensity at zero angle is very sensitive to the cluster size. For instance if vacancies with concentration c are clustered into groups of n vacancies, the concentration of clusters is c/n, but the intensity scattered by an individual cluster is n^2f^2: therefore the scattering power is $I = cnf^2$, it is *n times* the scattering power of the same concentration of isolated vacancies. Thus one can get information about clusters or aggregates, which are too small to be studied by the conventional small-angle scattering method.

SCATTERING BY A PERFECT CRYSTAL

Before studying defects in crystals, it was essential to test our method with a simple crystal as perfect as possible: The measured scattering should then be fully explained by the Compton scattering and the thermal scattering, which may be theoretically evaluated in simple cases.

We have chosen as an example a cleaved lamella of a good LiF crystal: its scattering has been measured at temperatures ranging from 80°K to 700°K. On the other hand, the thermal scattering has been calculated (5) from the elastic constants of lithium fluoride, and the Compton scattering by extrapolation towards zero angle of the data of International Tables of Crystallography (6). Figure 6 shows the good agreement between theoretical and experimental values. These results prove that there are no unexpected sources of small-angle scattering and they prove the validity of our absolute measurements of intensity. Thus, when a difference is found between special LiF crystals and pure perfect samples, this scattering, even weak, may be considered as significant. We have studied, for instance, LiF doped with bivalent cations (Ni) and LiF irradiated with high energy electrons. The presence of point defects has been proved: they are equivalent to a concentration of a few thousandths of F^- vacancies. These results will be published elsewhere because their interpretation requires a comparison with data of other methods (optical, electrical, etc. . . .).

A single crystal of very pure aluminum gives the same result: (Fig. 7) thermal and Compton scattering account for the total

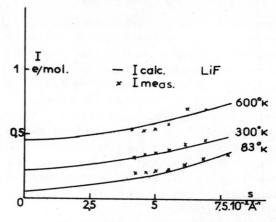

Fig. 6. Scattering by LiF single crystal.

observed scattering but when a polycrystalline sample of the same aluminum is used, the scattering is slightly larger; the extra scattering (about 10% of the thermal scattering at 300°K) is nearly independent of temperature and of scattering angle, except towards the mininum angle (3°) where it increases. As the sample was very carefully annealed and coarse-grained, this scattering is probably not due to lattice defects or grain boundaries. We believe that it can be attributed to double diffraction effects.

STRUCTURE OF DILUTE SOLID SOLUTIONS

This is a typical example of point defects which can be studied by X-rays, provided that the atomic numbers of the two constitu-

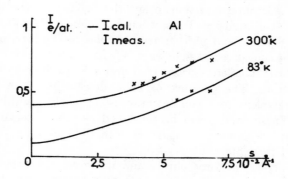

Fig. 7. Scattering by pure Al single crystal.

ents are different enough. Furthermore for concentrations higher than 1%, our sensitivity, as shown in the preceding paragraph, is sufficient.

We have at first tried to verify the Laue formula in the cases of solutions where the distribution of atoms is generally considered as perfectly random. But it seems that perfect randomness is rather an exception. However, we have found at least one good example, Al-Ga.

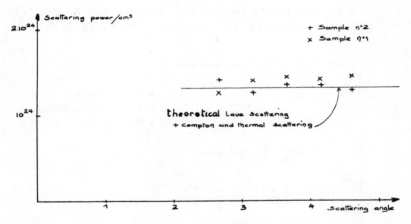

Fig. 8. Scattering by the alloy Al-6.4 at.% Ga.

Al-Ga Alloys

The solubility of gallium in aluminum is 15 at.% at room temperature. We have studied an alloy with only 6.4%; the alloy was homogenized at high temperature and very slowly cooled. The two metals have nearly equal atomic radii because the parameter of the solid solution does not vary much with the gallium concentrations, and they are in the same column of the periodic table. These conditions are favorable to a large solubility limit and a weak influence of the nature of the atoms on the energy of pairs. The scattered intensity has been measured between 2°30' and 4°30' and has been found *constant*, the variation being smaller than the differences between similar samples. Figure 8 shows the good agreement found between the experimental and theoretical values calculated by the Laue formula increased by the Compton scattering (6) and the thermal scattering (supposed equal to that of pure aluminum). Thus the experiment confirms the perfect randomness of the solid solution.

Al-Cu Alloy

We have studied an alloy with a low concentration of copper, 0.45 at.%. According to the solubility curve, it is homogeneous at high temperature and, in the equilibrium state at room temperature, a mixture of a precipitate and a solid solution with less than 0.05% Cu. Two states of the same alloy have been compared: quenched after a long annealing at 400°C, and slowly cooled (15 days from 400°C to 200°C). The second alloy gives practically the scattering of pure Al. The difference between the first and second alloys corresponds to the Laue scattering of the solid solution 0.45% Cu in Al (Fig. 9).

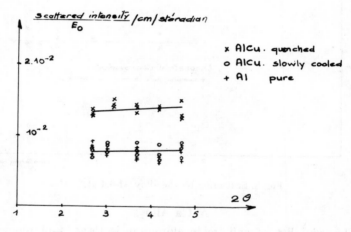

Fig. 9. Scattering by an Al-0.45 at.% Cu alloy.

One may conclude that after quenching the alloy remains homogeneous: the pre-precipitation of Cu is very slow for this low concentration. The slowly cooled alloy is precipitated, but the precipitates are large enough to give no small-angle scattering and the matrix is so poor in Cu that it gives practically the same scattering as pure Al.

Dilute Al-Zn Alloys

The solubility limit at room temperature is approximately 2 at.% (5% weight). Five compositions below this limit have been studied from 0.3 at.% to 1.78 at.%. To evaluate the scattering characteristic of an alloy, we have subtracted from the observed value the scattering measured for the pure aluminum polycrystalline specimen.

(1) Within the experimental error, for each alloy the scattering is constant from 3° to 7°.

(2) The measured intensity is generally larger than the theoretical Laue scattering, the difference being significant, although the experimental error is rather high (10%).

The five alloys have been annealed at 400°C, then slowly cooled to room temperature, in order to reach for each one the equilibrium state at this temperature. The ratio of the experimental scattering to the Laue scattering increases linearly with concentration (Fig. 10).

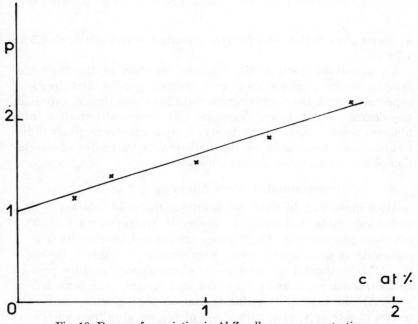

Fig. 10. Degree of association in Al-Zn alloys vs. concentration.

This "degree of association," n, represents the number of atoms per cluster, if one supposes that all Zn atoms are distributed in clusters of uniform size.

Another method of interpretation, more realistic, is to use the calculation of the diffuse scattering from the coefficients of short-range order (7). For the ith shell of neighbors,

$$\alpha_i = I - \frac{n_{ZnAl}}{c_{Al}} \qquad 8.$$

n_{ZnAl} being the proportion of pairs ZnAl and c_{Al} the atomic concentration of Al. If the solution is completely disordered, $\alpha_i = 0$ and $\alpha_i > 0$ if there is a tendency towards clustering of Zn. In our case, the degree of order is very small, and it is negligible except for the nearest neighbors (12 in the f.c.c. lattice). Thus the scattered intensity at small angles can be written (8):

$$I = I_{Laue}(1 + 12\alpha_1) \qquad 9.$$

or

$$\alpha_1 = \frac{n-1}{12}$$

α_1 varies from 0.01 to 0.1 for concentrations increasing from 0.3 to 1.78%.

An important point is the possible variation of the degree of association for a given alloy with temperature. A first series of experiments with the most concentrated alloy shows that, as expected, the degree of association decreases with temperature; these preliminary results must be confirmed and, with a better accuracy of the measurement, it would be possible to determine the energy of association of two Zn atoms in the aluminum lattice.

Supersaturated Al-Zn Alloys (c > 2 at.%)

After quenching to room temperature, the solid solution is in a metastable state and aging at moderate temperatures (<150°C) produces precipitation. G.-P. zones are formed characterized by a scattering at small angles with a maximum at an angle of the order of 1–2°. The angular resolving power of our apparatus is too poor to locate this maximum and it is not adapted to study such large defects (~50 Å). But it was interesting to compare for a given specimen the results of our apparatus with those of the usual diffractometric or photographic methods. Figure 11 shows that the scattering strongly increases with angle and is about 100 times larger than the scattering found with the preceding alloys.

The interesting point is the study of this alloy in a state where the clustering effects had *not* yet been found, for instance after *reversion*. The reversion of an age-hardening alloy is a short heat treatment at a moderate temperature (~200°C) such that the zones are dissolved but no precipitation occurs. It is generally thought that the solid solution is homogeneous after reversion. A second aging produces pre-precipitation but at a much slower rate than after quench. According to previous X-ray measurements, the first traces of the formation of zones are found after about 1 month's aging at room temperature (9).

We have tried to measure the degree of order just after reversion and during the first stages of the second aging. The following results have been obtained:

(1) As was already shown by Bonfiglioli (10), during the reversion treatment the small-angle scattering characterizing the zones progressively decreases but its shape does not change. That means

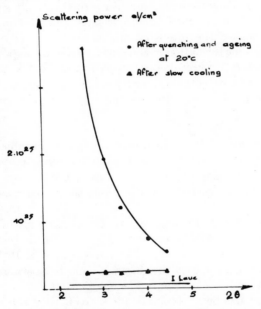

Fig. 11. Scattering by an Al-4 at.% Zn alloy

that the zones are disappearing without progressively changing their size. This evolution is more rapid for higher reversion temperatures (from 120°C to 250°C).

(2) When the reversion is completed, the scattering is almost constant in the range 2–6° for the 4.3 and 6.5% alloy and always higher than the theoretical Laue scattering. The ratio, which can be interpreted as an average number of Zn atoms in the clusters, increases with the Zn content and is comparable to the values found with

alloys below the solubility limit. For the more concentrated alloy, the scattering is not independent of angle and a maximum of scattering is apparent which indicates that the alloy contains zones of the same nature as in the first aging after quench but less numerous and of smaller size. The formation of the zones of the pre-precipitation state begins just after the reversion treatment.

(3) Even in the less concentrated alloys, the second aging at room temperature is *already detected after a few minutes*. The scattered intensity increases with time and a maximum of scattering appears which is more pronounced after longer aging times and for higher concentrations. Thus the evolution seems qualitatively similar to the evolution after quench: zones are progressively formed and grow (this is obvious because the intensity maximum is shifted towards the small angles). This formation of zones at that stage escapes completely when an ordinary diffractometer is used: in the 4.3 at.% alloy, the zones begin to be detectable after 3 months of aging after reversion, while in our present experiments the scattering maximum is well marked after only 6 days.

For concentrated alloys, this evolution is considerable during the time necessary for the establishment of one scattering curve; therefore we are not able to describe the structure of these alloys before any aging: it would be necessary to carry out the measurements at low temperature.

It is only possible to make a very rough estimate of the location of the maximum of the scattering curve (Fig. 12). But it is evident that it occurs at a larger angle ($s = 2\lambda^{-1} \sin \theta \simeq 3.16 \times 10^{-2}$ Å^{-1}) than during the aging after quench; this was expected because the zones are likely smaller.

Another fact has been observed: for a given alloy the rate of zone formation during aging after reversion is very sensitive to the conditions of the reversion treatment. The aging is accelerated when the reversion temperature is increased and when the annealing time at the reversion temperature is increased. The rate of pre-precipitation is dependent on the number of vacancies which aid the diffusion: this number increases with temperature and, on the other hand, they are probably not dispersed uniformly after too short a reversion treatment.

We have also found a considerable effect of a slight coldworking of the metal (bending of the sample). The vacancies or defects introduced by the deformation increase the rate of short-range diffusion of Zn and the scattering measurements clearly show the influence of this increase upon the heterogenization of the solid solution.

Au-Cu Alloys

This alloy was chosen, because, in contrast to Al-Zn, there is a tendency to ordering instead of segregation.

The proportion of Au was smaller than 10 at.% in order to reduce absorption and to remain well outside of the domain where superlattices have been observed.

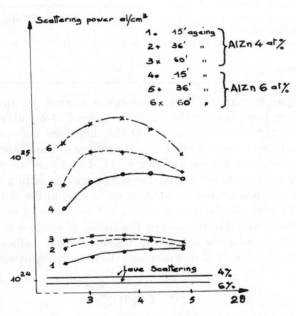

Fig. 12. Scattering by Al-Zn alloys after reversion treatment and aging at room temperature.

After subtraction of the Compton scattering and thermal scattering, calculated for pure copper, the scattering is found to be constant in the range 2–6° and smaller than the Laue scattering.

That could be explained by an incomplete disorder, namely by a short-range order parameter of formula 9 which is negative: around an Au atom, the number of Au neighbors in the first shell would be less than the number corresponding to the average concentration. But this degree of order, in the equilibrium state, should decrease with temperature. Therefore, one expects that the experimental degree of order decreases in samples quenched from increased temperatures. We did not, however, observe any variation. It seems therefore that this explanation is not the correct one.

TABLE I

	Concentration c_{Au} (at.%)			
	7.5	5.4	3.3	2.0
Laue scattering power (10^{24} electrons/cm^3)	12.8	10.6	6.0	3.8
Observed scattering power at small angles (10^{24} electrons/cm^3)	6.9	6.0	4.3	2.8

It is possible that the lattice distortions around Au atoms intervene.* They have been neglected in the case of Al-Zn alloys because the atomic radii are quite close, but the difference between the Au and Cu radii is of the order of 15%. The effects of the distortion have been calculated by Huang and Borie (11, 12): there is a scattering concentrated around the nodes of the lattice and, according to Huang, this scattering has a non-zero limiting value around the center of the reciprocal lattice. In the special case of the Au-Cu solid solution, the Laue intensity would be reduced by the Huang effect, and in a proportion similar to what is effectively observed. But, before drawing definite conclusions, it seems necessary to have verifications in other alloys.

CONCLUSION

We have shown some applications of our apparatus to the study of the distribution of atoms in binary solid solutions.

Small deviations from perfect randomness may be determined at different temperatures or for different heat treatments. Thus this is a new experimental method for investigating the real structure of solid solutions.

Generally speaking, our apparatus is able to detect point defects

* We are indebted to Dr. D. T. Keating for having raised this important point.

or very small defects involving a few atoms only which were not detectable by conventional small-angle apparatus.

It is likely that small heterogeneities could be found not only in crystals but in liquids in spite of the much higher scattering at zero angle for pure liquids. Thus one can think of studies of the distribution of heavy ions in solutions, of the nuclei in supersaturated or supercooled solutions, the demixtion of liquid mixtures, etc.

BIBLIOGRAPHY

1. Dienes, G. J. and Vineyard, G. H., *Radiation Effects in Solids*, Interscience, New York, 1954; p. 67.
2. Guinier, A., *X-Ray Diffraction in Crystals, Imperfect Crystals and Amorphous Bodies*, W. H. Freeman and Co., San Francisco, 1963; p. 264.
3. Luzzati, V. and Baro, R., J. Phys. Radium, **22,** 186A (1961).
4. Bonfiglioli, A. and Testard, O., Acta Cryst., **17,** 668 (1964).
5. de Launay, J., Solid State Phys., **2,** 269 (1956).
6. *International Tables for Crystallography*, vol. III, Kynoch Press, Birmingham, 1962; p. 247.
7. Green, H. S. and Hurst, C. A., *Order-Disorder Phenomena*, J. Wiley and Sons, New York, 1964.
8. Guinier, A., op. cit., p. 266.
9. Graf, R., Recherche Aéronautique, no. 60, **47** (1957).
10. Bonfiglioli, A., in *Metallic Solid Solutions*, J. Friedel and A. Guinier, editors, W. A. Benjamin and Co., New York, 1963; XLI.
11. Borie, B., Acta Cryst., **12,** 280 (1959).
12. Huang, K., Proc. Roy. Soc., **190A,** 102 (1947).

A New Source of Small-Angle X-Ray Scattering*

R. G. PERRET AND D. T. KEATING

Brookhaven National Laboratory, Upton, New York

INTRODUCTION

In the course of a photographic investigation of the small-angle scattering of X-rays from neutron-irradiated diamond single crystals, an anomalous scattering was observed. The anomalous scattering most frequently appears as streaks within a range of several degrees of the direct beam. In turning the crystal the streaks are observed to turn through the same angle and the intensity of the streaks depends upon the crystal orientation and neutron exposure. When the crystal is oriented so that a streak intercepts the direct beam, an intense diffuse scattering about the direct beam is also observed. This orientation is such that a Bragg reflection occurs.

The X-ray effects observed from diamond which has received a neutron exposure of less than 3×10^{20} neutrons/cm² ($E \geq 1$ MeV) are similar to those predicted for crystals containing point defects (1). The point defects in this case are the interstitials and vacancies produced by the neutron irradiation. These effects are: sharp Bragg reflections which are shifted in position corresponding to the change in density, a reduction in the intensity of the reflections by a factor analogous to a temperature factor, and a diffuse scattering localized around the Bragg reflections somewhat analogous to a temperature diffuse scattering. It is this diffuse scattering that produces the anomalous small-angle scattering.

The anomalous scattering is explained as a multiple scattering involving the Bragg and diffuse scattering within the crystal. Two cases can be distinguished depending on the scattering sequence. The first case, which results in streaks, occurs when the direct beam is first diffusely scattered in such a direction as to be subsequently Bragg scattered. The intensity of the resulting reflection conics is determined by the distribution of the diffuse scattering and crystal

* Work performed under the auspices of the U.S. Atomic Energy Commission.

orientation. The second case, which produces an apparent small-angle scattering, occurs when the direct beam is first Bragg scattered and then subsequently is diffusely scattered into the region about the direct beam.

EXPERIMENTAL CONDITIONS

The anomalous scattering was observed under a number of experimental conditions including crystal monochromatized radiation.

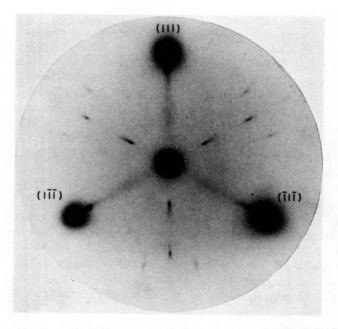

Fig. 1. Transmission Laue photograph of neutron-irradiated diamond. The crystal-to-film distance was 39 mm and exposure time 90 min. The other exposure conditions are given in the text. The direct beam is along [$\bar{1}$11]. The intense diffuse scattering near the (111), (1$\bar{1}\bar{1}$) and ($\bar{1}$1$\bar{1}$) Laue spots and the sharpness of the Laue spots should be noted.

However, all the photographs in the figures were made in an evacuated Kiessig small-angle camera using 0.00035 in. nickel filtered copper radiation from a Philips fine-focus X-ray tube operated in full rectification at 40 KV and 20 ma. The pictures were recorded on "Ilford" industrial G film 100 mm in diameter developed for 6 min at 20°C.

The crystals were mounted on a small goniometer inside the Kiessig camera 17 mm behind the normal sample position so the crystal could be rotated about an axis from outside the camera. The pinhole slit system, K-15, which produces a beam about 1 mm in diameter, was used for all exposures.

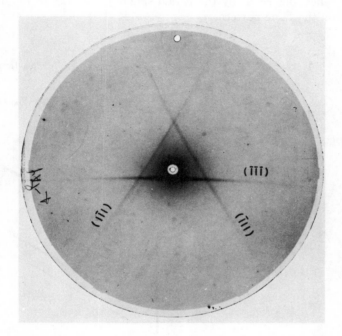

Fig. 2. Small-angle scattering from neutron-irradiated diamond. The crystal-to-film distance was 183 mm and the exposure time 20 hrs. The prominent streaks are reflection conics from the $(\bar{1}\bar{1}\bar{1})$, $(\bar{1}11)$ and $(1\bar{1}1)$ planes reflecting the diffuse scattering near the (111), $(1\bar{1}\bar{1})$ and $(\bar{1}1\bar{1})$ Laue spots of Fig. 1. The closer a point on the streak is to the direct beam the stronger its intensity ($2\alpha_0 = -0.933°$).

CASE I: STREAKS

In order to observe the effects it is essential that the crystal have a large diffuse scattering near the Bragg reflections. Figure 1 is a transmission Laue photograph of the diamond crystal used in this investigation. The crystal was exposed to 2.47×10^{20} neutrons/cm^2

(fast), and has a density 4.9% less than unirradiated diamond. In the photograph the direct beam is along the [$\bar{1}11$] direction. The salient features of the photograph are the intense diffuse spots outside the (111), (1$\bar{1}\bar{1}$) and ($\bar{1}$11) Laue spots and the sharpness of the Laue spots. The three (111) reflections lie just outside the Ewald sphere of copper Kα radiation, and the diffuse scattering associated with these reflections is cut by the Ewald sphere giving rise to the diffuse spots on the Laue pattern.

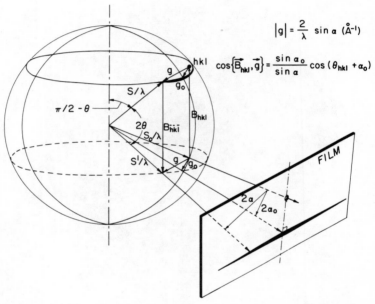

Fig. 3. Reciprocal space construction for streaks. The construction represents the condition of Fig. 2 with $\mathbf{B}_{hkl}$ outside the Ewald sphere. The direct beam, $\mathbf{S}_0/\lambda$, is diffusely scattered in the direction $\mathbf{S}/\lambda$ which is Bragg scattered into the direction $\mathbf{S}'/\lambda$ to produce a point on the streak at the angle 2α from the direct beam.

Figure 2 is a photograph of the small-angle scattering of the same crystal. The direct beam is directed slightly more towards [111] than in Fig. 1. Three streaks are clearly evident and are in fact partial reflection conics from the ($\bar{1}\bar{1}\bar{1}$), ($\bar{1}$11), and (1$\bar{1}$1) planes which are reflecting part of the diffuse scattering near the (111), (1$\bar{1}\bar{1}$) and ($\bar{1}$1$\bar{1}$) Laue spots of Fig. 1. These streaks are similar to the reflection conics observed in diamond by Grenville-Wells (2) and Norman (3) produced by the Compton scattering acting as an internal radiation source. In the present observations, the diffuse source is of limited distribution and strongly dependent upon crystal orientation, so that the closer a point on the streak is to the direct beam, the stronger its intensity.

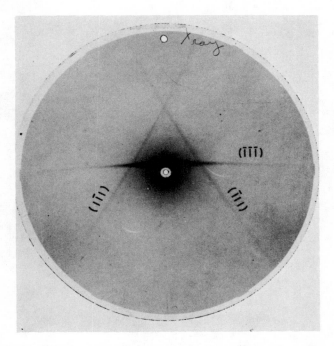

Fig. 4. Small-angle scattering from neutron-irradiated diamond. The (111) reflection lies inside the Ewald sphere ($2\alpha_0 = 0.883°$). The ($\bar{1}\bar{1}\bar{1}$) streak is more intense than in Fig. 2 where the (111) reflection is outside the sphere. All other conditions are the same as in Fig. 2.

This can be better understood with the aid of Fig. 3. $\mathbf{S}_0/\lambda$ is the direct beam which is diffusely scattered into the direction $\mathbf{S}/\lambda$ from a point $\mathbf{g}$ distant from the reflection $\mathbf{B}_{hkl}$. The ray $\mathbf{S}/\lambda$ is traveling in a direction so as to satisfy the Bragg angle for the ($\bar{h}\bar{k}\bar{l}$) reflection and is Bragg scattered into the direction $\mathbf{S}'/\lambda$. The rays $\mathbf{S}/\lambda$ which satisfy the Bragg condition generate a cone whose axis is parallel to

$\mathbf{B}_{hkl}$ with semi-cone angle $\pi/2 - \theta_{hkl}$ and whose apex is the center of the Ewald sphere. The locus of points, $\mathbf{g}$, which contribute to the streak is the intersection of the cone with the Ewald sphere, and $\mathbf{g} = \mathbf{S}'/\lambda - \mathbf{S}_0/\lambda$. Apart from polarization factors (4) the intensity scattered into the streak is proportional to the interference function at $\mathbf{B}_{hkl} + \mathbf{g}$. If 2α is the angle between the direct beam and a point

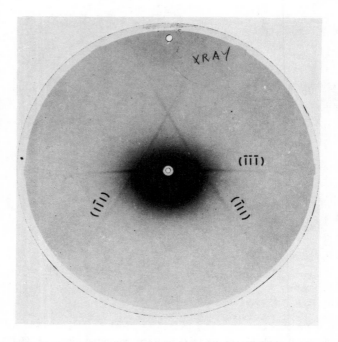

Fig. 5. Small-angle scattering from neutron-irradiated diamond. The (111) reflection lies on the Ewald sphere ($2\alpha_0 = 0$). In addition to the streaks there is an apparent large small-angle scattering which is more intense below the direct beam than above. The other conditions are the same in Figs. 2, 4 and 5.

on the streak $g = (2/\lambda) \sin \alpha$. Taking $2\alpha_0$ as the smallest angle on the streak with the positive direction towards $\mathbf{B}_{hkl}$ we can write $\cos \{B_{hkl}, g\} = [(\sin \alpha_0)/(\sin \alpha)] \cos (\theta_{hkl} + \alpha_0)$. These two quantities enter into the expression for the diffuse scattering from point imperfections.

For a diamond crystal of N unit cells containing a fraction p of the atoms as Frenkel defects and assuming the displacements about an interstitial and vacancy are respectively* $C_i(\mathbf{r}/|r|^3)$ and $C_v(\mathbf{r}/|r|^3)$, the diffuse scattering is given approximately as**

$$I_D = 8p(1-p)f^2 N$$
$$\times \left\{ \left[1 + aC_v \left(\frac{B \cos (B,g)}{g} + 1 \right) \frac{\sin 2\pi rg}{2\pi rg} \right]^2 \right.$$
$$\left. + \left[1 + bC_i \left(\frac{B \cos (B,g)}{g} + 1 \right) \frac{\sin 2\pi rg}{2\pi rg} \right]^2 \right\} \quad 1.$$

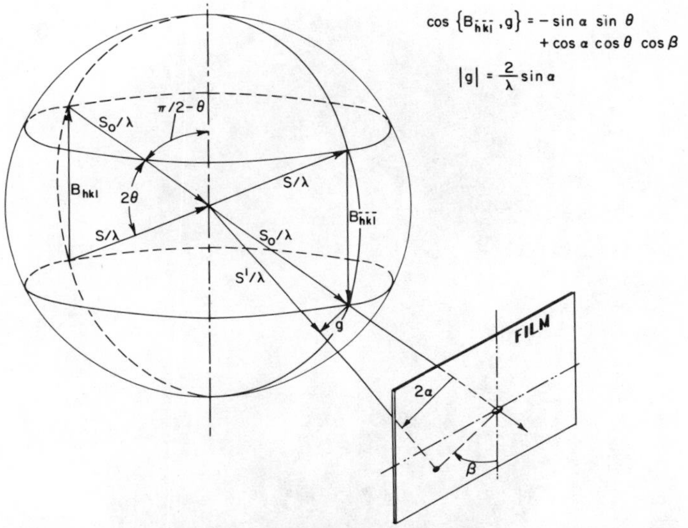

$\cos \{B_{h\bar{k}i}, g\} = -\sin\alpha \sin\theta + \cos\alpha \cos\theta \cos\beta$

$|g| = \frac{2}{\lambda} \sin\alpha$

Fig. 6. Reciprocal space construction for the apparent small-angle scattering. The construction represents the conditions of Fig. 5. The direct beam, $\mathbf{S}_0/\lambda$, is Bragg scattered into the direction $\mathbf{S}/\lambda$ which is diffusely scattered into the direction $\mathbf{S}'/\lambda$.

where $a = -b = 4\pi(2/a_0)^3$ for reflections with even Miller indices and $a = b = 2\pi(2/a_0)^3(1-2p)$ for reflections with odd Miller indices. The cell edge is a_0, and r is the distance of the closest atom to an interstitial or vacancy. Qualitatively Eq. 1 accounts for many of the

* This type of displacement is expected in an elastically isotropic medium (5) and C_i and C_v are the strength of an interstitial and vacancy. The anisotropy ratio, $(C_{11} - C_{12})/2C_{44}$, is unity for a cubic isotropic medium. The elastic constant data for diamond give 0.628 (6), 0.667 (7) and 0.875 (8) for this ratio.

** The expression was derived using the results of Smirnov and Tikhonova (9) grouping terms as Borie (10) has done, and using Ekstein's (11, 12) approximation for the lattice sums involved.

observed features of the streaks. It correctly predicts that the most intense point on the streak is that nearest the direct beam for which g is smallest. Since irradiated diamond expands, $C_v + C_i$ is positive (5) so that the most intense (111) streaks should be found when $\cos\{B,g\}$ is positive. This is indeed found to be the case as can be seen by comparing the $(\bar{1}\bar{1}\bar{1})$ streaks in Figs. 2 and 4. In Fig. 4 the (111) reflection lies outside the Ewald sphere so that $\cos\{B,g\}$ is

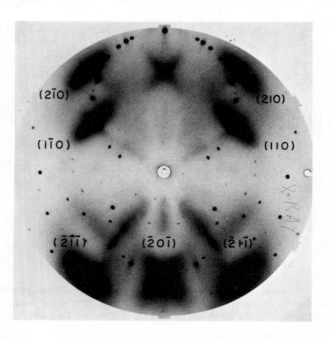

Fig. 7. Transmission Laue photograph of phenanthrene. The crystal-to-film distance was 83 mm and the exposure time was 15.8 hrs. The direct beam is 6.9° from [001] towards [$\bar{2}0\bar{1}$]. The other conditions are given in the text. The most intense sources of thermal diffuse scattering are identified.

positive and the $(\bar{1}\bar{1}\bar{1})$ streak is more intense than in Fig. 2 where the cosine is negative. The values of $2\alpha_0$ for Figs. 2 and 4 are $-0.933°$ and $0.883°$ so that the only major difference is the cosine factor.

CASE II: APPARENT SMALL-ANGLE SCATTERING

When the crystal is oriented so that the streak and main beam intersect, a large anomalous scattering about the direct beam is observed as in Fig. 5. This corresponds to the (111) reflection being on the Ewald sphere and while this condition is to be avoided in practice the anomalous scattering is not double Bragg scattering (13). In Fig. 6 the direct beam, $\mathbf{S}_0/\lambda$, is Bragg scattered into the direction $\mathbf{S}/\lambda$. $\mathbf{S}/\lambda$, playing the role of an incident beam, is diffusely scattered into the direction $\mathbf{S}'/\lambda$ from a point of $\mathbf{g}$ distant from $\mathbf{B}_{\overline{hkl}}$. All the points $\mathbf{g}$ which contribute to this diffuse scattering must lie on the Ewald sphere. The expression for g is the same as in the case of the streaks and $\cos\{B_{\overline{hkl}},g\} = -\sin\alpha\sin\theta + \cos\alpha\cos\theta\cos\beta$. β is the

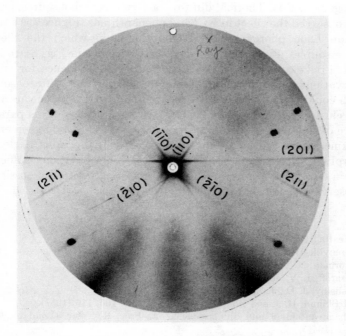

Fig. 8. Small-angle scattering from phenanthrene. The crystal-to-film distance was 183 mm and the exposure time 39.7 hrs. The reflection conics associated with the diffuse sources of Fig. 7 are identified [For the (201) streak $2\alpha_0 = -0.62°$].

angle between the traces of the planes containing $\mathbf{S}'/\lambda$ and $\mathbf{S}_0/\lambda$ and $\mathbf{S}/\lambda$ and $\mathbf{S}_0/\lambda$ on the film. The $\cos\{B_{\overline{hkl}},g\}$ is generally positive in the region below the direct beam and the scattering in Fig. 5 is more intense in this region in qualitative agreement with Eq. 1.

SUMMARY

A parasitic source of small-angle scattering in single crystals, not previously reported, has been observed. It is a multiple scattering effect involving the diffuse scattering from point imperfections and the Bragg scattering. Two cases can occur depending on the scattering sequence. The first case occurs when the direct beam is first diffusely scattered in such a direction as to be subsequently Bragg scattered to produce reflection conics. The second case occurs when the direct beam is first Bragg scattered and subsequently diffusely scattered about the direct beam to produce an apparent small-angle scattering. Similar effects should be observed in crystals with a diffuse scattering analogous to this. Thermal diffuse scattering is such a source, and we have observed the effect in KCl and phenanthrene. Figure 7 is a transmission Laue photograph of phenanthrene in which the thermal diffuse sources giving rise to the reflection conics in Fig. 8 are identified. Fricke and Gerold have reported strong evidence for similar observations in copper and aluminum (14). Other diffuse sources which could produce similar effects are found in heterogeneous alloys with clustering or in crystals with stacking disorders. This parasitic source of small-angle scattering was discussed using only geometrical and kinematical considerations for the scattering processes, neglecting absorption. However, we feel that these considerations suffice to explain our observations, and to alert investigators to the possibility of observing similar effects.

BIBLIOGRAPHY

1. Keating, D. T., Acta Cryst., **16,** A113 (1963).
2. Grenville-Wells, H. J., Nature, **168,** 291 (1951).
3. Norman, N., Acta Cryst., **11,** 1 (1958).
4. Azaroff, L. V., ibid., **8,** 701 (1955).
5. Eshelby, J. D., J. Appl. Phys., **25,** 255 (1954).
6. McSkimin, H. J. and Bond, W. L., Phys. Rev., **105,** 116 (1957).
7. Bhagavantam, S. and Bhimasenachar, J., Proc. Roy. Soc. (London), **A187,** 381 (1946).
8. Prince, E. and Wooster, W. A., Acta Cryst., **6,** 450 (1953).
9. Smirnov, A. A. and Tikhonova, E. A., Soviet Phys.-Solid State, **3,** 899 (1960).
10. Borie, B., Acta Cryst., **12,** 280 (1959).
11. Ekstein, H., Phys. Rev., **68,** 120 (1945).
12. Cochran, W., Acta Cryst., **9,** 259 (1956).
13. Guinier, A. and Guyon, E., J. Appl. Phys., **30,** 622 (1959).
14. Fricke, H. and Gerold, V., ibid., **30,** 661 (1959).

Small-Angle Scattering from Dislocation Rings in Crystals, and Its Relation to the Depleted Zones in Neutron-Irradiated Metals

A. SEEGER AND P. BRAND

Max-Planck-Institut für Metallforschung and Institut für theoretische und angewandte Physik der Technischen Hochschule Stuttgart, Germany

INTRODUCTION

Both theoretical and experimental evidence exists [the latter coming mainly from radiation-hardening experiments (1, 2)] that neutron irradiation of metals creates extended defects containing a net lack of atoms, the so-called depleted zones (3–5). The depleted zones should give rise to small-angle scattering of X-rays or long wavelength neutrons. Small-angle X-ray scattering experiments have demonstrated that vacancies in quenched copper may cluster into voids, presumably of oblate shape (6–8). The experimental evidence presented by Galligan and Washburn (8) indicates that these clusters are responsible for the hardening of quenched copper crystals. It appears that the small-angle scattering of X-rays or subthermal neutrons is a powerful tool for studying the vacancy agglomerates occurring in quenched or irradiated metals, particularly if these are too small to be investigated by transmission electron microscopy.

MODELS FOR VACANCY AGGLOMERATES

We consider two possible models for the depleted zones in neutron-irradiated metals: (a) spheres of radius R_0 with an electron (or nuclear) density reduced by a "factor of dilution" p (b) circular edge dislocation rings of radius ρ_0 and dislocation strength b (9, 10).

Approximating the scattered intensity by Guinier's law (11)

$$i(h) = i(0) \exp(-R_s^2 h^2/3) \qquad 1.$$

we have

(a) $\quad i(0) = Nf^2 \left(\dfrac{4\pi}{3}\right)^2 p^2 R_0^6/\Omega^2; \qquad R_s = R_0\sqrt{3/5}$ \hfill 2a.

(b) $\quad i(0) = Nf^2 \left(\dfrac{1-2\nu}{1-\nu}\right)^2 \dfrac{8\pi^2}{15} b^2 \rho_0^4/\Omega^2; \qquad R_s = \rho_0\sqrt{9/14}$ \hfill 2b.

Here N = density of scattering centers, Ω = atomic volume, ν = Poisson's ratio. Equation 2b holds for $h\rho_0 \leq 2.7$ for an isotropic medium with an isotropic distribution of rings. The influence of the anisotropy of the elastic medium and of the anisotropy of the ring distribution have been studied; for cubic crystals the results are not very different from Eq. 2a (10).

For a given radius of gyration R_s and within the range of validity of Eq. 1, the two kinds of defects lead to the same scattering, if

$$R_s p/b = \frac{1-2\nu}{1-\nu} \cdot \frac{14}{25\sqrt{2}} \approx \frac{1}{5} \qquad 3.$$

In principle they can be distinguished through their asymptotic small-angle scattering laws, namely $i(h) \sim h^{-4}$ for spheres (11), and $i(h) \sim h^{-3}$ for rings (9). However, since the determination of the asymptotic scattering law requires a rather high experimental accuracy, it may be convenient to consider instead the shape factor

$$w = \left(\frac{3}{5}\right)^{\frac{3}{2}} \cdot \frac{3}{4\pi} \cdot \frac{i(0)}{R_s^3 Q(0)} \qquad 4.$$

where

$$Q(0) = \frac{4\pi}{(2\pi)^3} \int_0^\infty i(h) h^2\, dh \qquad 5.$$

is the integrated small-angle intensity. The shape factor w has been defined in such a way that for spheres $w = 1$. For spheroids with axial ratios $v:1:1$ we have

$$w = v\left(\frac{3}{2+v^2}\right)^{\frac{3}{2}} \leq 1 \qquad 6.$$

The shape factor for edge dislocation rings of radius ρ_0 is given by (10)

$$w_d = \frac{4\pi}{5}\left(\frac{14}{15}\right)^{\frac{3}{2}} \left[\ln\left(\frac{4\rho_0}{\rho_i}\right) - \frac{11}{6}\right]^{-1} \qquad 7.$$

Here $\rho_i(\approx 1$ Å) denotes the inner cut-off radius for the strain-fields of the dislocation rings. According to Eq. 7, $w_d < 1$ for $\rho_0 > 15\rho_i$ and $w_d > 1$ for $\rho_0 < 15\rho_i$. In quenched and annealed copper single crystals radii of gyration R_s between 5 and 14 Å and form factors w between 0.3 and 0.5 were observed (7). In agreement with other evidence (6) it is concluded that the quenched-in vacancies have agglomerated into cavities of oblate shape.

Schilling and Schmatz (12) have shown that neutron irradiation $(1.1 \times 10^{19}$ n/cm² with energies $E > 1$ MeV) of polycrystalline copper gives rise to a measurable small-angle scattering of sub-thermal neutrons of wavelengths too long for double Bragg reflections to occur. The scattering shows the annealing behavior to be expected from the depleted zones. Only an integrated intensity (average over both scattering angle and wavelength) could be determined. Making the assumption that each primary knock-on atom creates one depleted zone of spherical shape (model a), Schilling and Schmatz could account for the observed intensity if they assumed that each depleted zone contained on the average 15 vacant lattice sites. (This result is virtually independent of the distribution of the vacant sites, provided the radius of gyration is smaller than about 8 Å. If a larger value of R_s and a correspondingly smaller factor of dilution p is assumed, the number of vacant sites per zone comes out somewhat larger.) If we employ model b and assume that the depleted zones scatter sub-thermal neutrons as edge dislocation rings consisting of about 15 vacant lattice sites (diameter approximately 10 Å), it is found that 8 such rings produced per primary knock-on would account for the observed intensity. The evidence from radiation hardening and from radiation damage theory (5) is in favor of the production of more than one depleted zone per primary knock-on in Cu. It appears that the best description of the depleted zones is intermediate between models a and b, but closer to the dislocation ring model.

THEORY OF SMALL-ANGLE SCATTERING FROM EDGE-TYPE DISLOCATION RINGS

General Remarks

We must emphasize that the formulae for the scattering from dislocation rings employed in the previous section were based on the linearized theory of elasticity. It is well known (13–15) that the volume expansion of plastically deformed crystals due to dislocations is caused by non-linear effects, specifically by the quadratic terms in the stress-strain relations. For a quantitative treatment of the small-angle scattering from dislocation rings it therefore appears essential to investigate the effect of these quadratic terms. We shall do this on the basis of the Kröner-Seeger second-order theory of internal stresses (16, 17). Our starting point is the following relation between the small-angle scattering amplitude $a(\mathbf{h})$ and the Fourier transform $\tilde{\Theta}$ of the volume dilatation $\Theta(\mathbf{r})$ in the deformed state

$$a(\mathbf{h}) = \frac{f}{\Omega} \tilde{\Theta}(\mathbf{h}). \qquad 8.$$

Theory of Internal Stresses (16, 17)

The basic equations are (for further explanations see Appendix 1):
(a) The fundamental geometrical equation, expressing the coherence of the medium in the presence of dislocations

$$(\text{Ink } \mathcal{E})^{ij} = \eta^{ij} + \tfrac{1}{2}\epsilon^{jnm}\epsilon^{ilk}(-2\mathcal{E}_{nkq} + h_{nkq})(-2\mathcal{E}_{mlp} + h_{mlp})_{(ij)} \qquad 9.$$

(b) The equilibrium condition ($\check{\sigma}$ = stress tensor in the deformed state)

$$\text{Div } \check{\sigma} \equiv \nabla_i \sigma_{ij} = 0 \qquad 10.$$

(c) The strain-stress relation (to second order)

$$\mathcal{E}_{ij} = \frac{1}{2\mu}\left(\sigma_{ij} - \frac{\nu}{1+\nu}\sigma_I \delta_{ij}\right) + C_2(\sigma_I)^2 \delta_{ij} + C_3 \sigma_{II} \delta_{ij} \\ + C_5 \sigma_I \sigma_{ij} + C_7 \sigma_{III}(\sigma^{-1})_{ij} \qquad 11.$$

C_2 to C_7 are coefficients depending on the second-order and third-order elastic constants (see Appendix 2).

In what follows we restrict ourselves to an isotropic medium.

First-Order Approximation (9)

Within the linear approximation the equations of the previous subsection reduce to (α = dislocation density tensor)

$$\text{Ink } \mathcal{E} \equiv \mathfrak{n} = -\text{Sym Rot } \alpha \qquad 9\text{a.}$$

$$\text{Div } \check{\sigma} = 0 \qquad 10\text{a.}$$

$$\mathcal{E}_{ij} = \frac{1}{2\mu}\left(\sigma_{ij} - \frac{\nu}{1+\nu}\sigma_I \delta_{ij}\right) \qquad 11\text{a.}$$

Inserting (11a) into (9a) using (10a), we obtain the inhomogeneous Beltrami equation

$$\Delta \sigma_{ij} + \frac{1}{1+\nu}(\nabla_i \nabla_j \sigma_{kk} - \Delta \sigma_{kk} \delta_{ij}) = 2\mu \eta_{ij} \qquad 12.$$

or

$$\Delta \sigma_{\mathrm{I}} = -2\mu \frac{1+\nu}{1-\nu} \eta_{\mathrm{I}} \qquad 12\mathrm{a}.$$

and, with

$$\Theta(\mathbf{r}) = \frac{1}{2\mu} \frac{1-2\nu}{1+\nu} \sigma_{\mathrm{I}} \qquad 12\mathrm{b}.$$

$$\Delta \Theta(\mathbf{r}) = -\frac{1-2\nu}{1-\nu} \eta_{\mathrm{I}}(\mathbf{r}) \qquad 13.$$

Fourier transformation of 13 gives us the desired result

$$\tilde{\Theta}(\mathbf{h}) = (2\pi)^{\frac{3}{2}} \frac{1-2\nu}{1-\nu} \frac{1}{h^2} \tilde{\eta}_{\mathrm{I}}(\mathbf{h}) \qquad 14.$$

which allows us to express the scattering amplitude a in terms of the dislocation distribution $\boldsymbol{\alpha}$.

Second-Order Approximation

In the quadratic theory we wish to retain the advantages of the method developed for the linear case. This can be done as follows:

(a) We expand stress and strain in a power series of a small parameter, starting with the solution of the linearized problem, and subdivide the second-order term of the strain into two parts, such that for each of them an equation analogous to 9a of the linear theory holds.

(b) We write the fundamental equations as Fourier transforms.

Modifying a proposal of Stojanovich (18), we obtain from the expansion of stresses,

$$\boldsymbol{\sigma} = \boldsymbol{\sigma}^{(1)} + \boldsymbol{\sigma}^{(2)} + \cdots \qquad 15.$$

the expansion for the strains,

$$\boldsymbol{\varepsilon} = \boldsymbol{\varepsilon}^{(1)} + \boldsymbol{\varepsilon}^{(2)} + \cdots \qquad 16.$$

by applying the strain-stress relation. We have

$$\varepsilon_{ij}^{(1)} = \frac{1}{2\mu}\left(\sigma_{ij}^{(1)} - \frac{\nu}{1+\nu}\sigma_{\mathrm{I}}^{(1)}\delta_{ij}\right) \qquad 16\mathrm{a}.$$

$$\varepsilon_{ij}^{(2)} = \frac{1}{2\mu}\left(\sigma_{ij}^{(2)} - \frac{\nu}{1+\nu}\sigma_{\mathrm{I}}^{(2)}\delta_{ij}\right) + C_2(\sigma_{\mathrm{I}}^{(1)})^2\delta_{ij}$$
$$+ C_3\sigma_{\mathrm{II}}^{(1)}\delta_{ij} + C_5\sigma_{\mathrm{I}}^{(1)}\sigma_{ij}^{(1)} + C_7\sigma_{\mathrm{III}}^{(1)}(\sigma^{-1})_{ij}^{(1)} \qquad 16\mathrm{b}.$$

Going back to the basic equations 9 to 11, we see that 11 is already satisfied by our Eq. 16. Equation 10 takes the form

$$\nabla_i(\sigma_{ij}^{(1)} + \sigma_{ij}^{(2)}) = 0$$

or
$$\nabla_i \sigma_{ij}^{(2)} = 0 \qquad 10\mathrm{b}.$$

Inserting 16 into 9 we get

$$\mathrm{Ink}\,\boldsymbol{\varepsilon}^{(1)} = \mathbf{n}^{(1)}$$
$$\mathrm{Ink}\,\boldsymbol{\varepsilon}^{(2)} = \mathbf{Q}^{(2)} \qquad 17.$$

Here **Q** stands for the second term on the right-hand side of 9.

By splitting the expression 16b for $\varepsilon_{ij}^{(2)}$ into two terms

$$\boldsymbol{\varepsilon}^{(2)} = \boldsymbol{\varepsilon}^{(2,1)} + \boldsymbol{\varepsilon}^{(2,2)} \qquad 18.$$

with

$$\varepsilon_{ij}^{(2,1)} = [C_2(\sigma_{\mathrm{I}}^{(1)})^2 + C_3\sigma_{\mathrm{II}}^{(1)}]\delta_{ij} + C_5\sigma_{\mathrm{I}}^{(1)}\sigma_{ij}^{(1)} + C_7\sigma_{\mathrm{III}}^{(1)}(\sigma^{-1})_{ij}^{(1)} \qquad 18\mathrm{a}.$$

$$\varepsilon_{ij}^{(2,2)} = \frac{1}{2\mu}\left(\sigma_{ij}^{(2)} - \frac{\nu}{1+\nu}\sigma_{\mathrm{I}}^{(2)}\delta_{ij}\right) \qquad 18\mathrm{b}.$$

we may modify 9 to read

$$\mathrm{Ink}\,\boldsymbol{\varepsilon}^{(2,2)} = \mathbf{Q}^{(2)} + \mathbf{P}^{(2)} = \mathbf{R}^{(2)} \qquad 19.$$

where $\mathbf{P} = -\mathrm{Ink}\,\boldsymbol{\varepsilon}^{(2,1)}$.

A comparison between 19, 10b, 18b and 9a to 11a according to the scheme

linear	*quadratic*
$\mathrm{Ink}\,\boldsymbol{\varepsilon}^{(1)} = \mathbf{n}$	$\mathrm{Ink}\,\boldsymbol{\varepsilon}^{(2,2)} = \mathbf{R}$
$\mathrm{Div}\,\boldsymbol{\sigma}^{(1)} = 0$	$\mathrm{Div}\,\boldsymbol{\sigma}^{(2)} = 0$
$\varepsilon_{ij}^{(1)} = \frac{1}{2\mu}\left(\sigma_{ij}^{(1)} - \frac{\nu}{1+\nu}\sigma_{\mathrm{I}}^{(1)}\delta_{ij}\right)$	$\varepsilon_{ij}^{(2,2)} = \frac{1}{2\mu}\left(\sigma_{ij}^{(2)} - \frac{\nu}{1+\nu}\sigma_{\mathrm{I}}^{(2)}\delta_{ij}\right)$

shows that we may calculate the second-order contribution by the same procedure as the first-order approximation.

We may therefore immediately write down the result for $\tilde{\mathcal{E}}_I{}^{(2,2)}$ in analogy to 14:

$$\tilde{\mathcal{E}}_I{}^{(2,2)} = \frac{1-2\nu}{1-\nu} \frac{1}{h^2} \tilde{R}_I(\mathbf{h}) \qquad 20.$$

In the expression for the second-order scattering amplitude

$$a^{(2)} = \frac{f}{\Omega} (2\pi)^{\frac{3}{2}} \left[\tilde{\mathcal{E}}_I{}^{(2,1)} + \tilde{\mathcal{E}}_I{}^{(2,2)} + \frac{1}{2} \widetilde{(\mathcal{E}_I{}^{(1)})^2} - 2\tilde{\mathcal{E}}_{II}{}^{(1)} \right]. \qquad 21.$$

20 is the only genuinely quadratic term. The other terms are already determined from the linear approximation and can be calculated by means of Parseval's theorem, using the Beltrami equation, 12, to obtain the Fourier transforms $\tilde{\sigma}_{ij}{}^{(1)}$ and $\tilde{\mathcal{E}}_{ij}{}^{(1)}$.

The main ideas of the preceding treatment may be summarized as follows:

(a) Not to work out the volume dilatation and then Fourier transform it, but to subject the basic equations to a Fourier transformation, going thereby from a nonlinear system of coupled differential equations to a system of nonlinear algebraic equations.

(b) To expand and subdivide stress and strain suitably in order to employ as far as possible the elegant linear method of solution.

Some Details of the Calculation

The application of the theory to specific dislocation arrangements is not quite straightforward, but may lead to certain difficulties which we shall now discuss briefly.

For an edge-type dislocation loop of circular shape the linear approximation gives (9) (ψ is the angle between $\mathbf{h}$ and $\mathbf{b}$)

$$a(\mathbf{h}) = \frac{f}{\Omega} 2\pi \left(\frac{1-2\nu}{1-\nu} \right) b\rho_0 \frac{\sin \psi}{h} \cdot J_1(h\rho_0 \sin \psi) \qquad 22.$$

The total intensity

$$\frac{1}{8\pi^3} \int_0^\infty a^2(\mathbf{h}) \, d\tau_h$$

diverges. So do the convolution integrals introduced in formula 21 for $a^{(2)}(\mathbf{h})$ by the application of Parseval's theorem.

The reason for these divergences is that the dislocation core causes a singularity in the volume dilatation if the dislocation density is represented by a Dirac δ-function.

On physical grounds, however, it is obvious that the scattering by the dislocation core cannot be very important. Unfortunately, the usual procedure to introduce a suitably chosen cut-off radius fails here, since the cut-off radius would appear in the final result in various powers (up to the second) and since it would be hard to see how the additional interferences affect the scattering intensity. We must therefore resort to using an extended dislocation core instead of a singular one.

Another difficulty arises from the ambiguity of the scattering amplitude at $h = 0$ (i.e. $\theta = 0$), which is evident from 22. This ambiguity arises from the transition from 13 to 14 if we disregard the influence of boundaries. In order to get the correct behavior at $h = 0$, we must employ Green's formula and evaluate the resulting surface integral.

A detailed calculation, however, shows that in general the scattering amplitude as computed from 14 must be corrected only for extremely small angles at which observations cannot be made anyway.

The amplitude $a(0)$, which is related to the total volume expansion, depends on the boundary conditions in such a way that it vanishes for a finite medium with free surfaces, whereas for an infinite medium it assumes the unique value

$$a^{(1)}(0) = \frac{2\pi}{3}\left(\frac{1-2\nu}{1-\nu}\right)b\rho_0^2 \qquad 23.$$

in agreement with the results of Eshelby derived for a spherical center of dilatation (19).

Carrying the expansion of Eq. 15 to higher terms shows that whereas the structures of the linear and quadratic terms are different, higher-order terms would not bring in any new features. This, together with the fact that the numerical contributions of the third- and higher-order terms are expected to be small, justifies the neglect of higher terms.

RESULTS OF THE SECOND-ORDER THEORY

We shall now present a few selected results for a circular edge-dislocation loop.

We give first the formula for the quadratic intensity $i_f^{(2)}(0)$ at zero angle for a medium with free surfaces. In this case we are only concerned with the first and the two last terms of 21, since according to Colonnetti's theorem $\tilde{\mathcal{E}}_I^{(2,2)}(0)$ vanishes. The result reads

$$i_f^{(2)}(0) = N \frac{f^2(0)}{\Omega^2} \frac{\pi^2}{16} \frac{b^4 \rho_0^2}{(1-\nu)^4}$$
$$\times S^2(\rho_0)\{1 + 2(2\mu)^2[3(3C_2 + C_5)(1+\nu)^2 + (3C_3 + C_7)\nu]\}^2 \quad 24.$$

Here $S(\rho_0)$ is a function which, compared to ρ_0^2, varies only slowly with the radius ρ_0 of the dislocation loop.

From the amplitude corresponding to 24 we can calculate the total volume change of the specimen if a dislocation loop is introduced. In order to create an edge-dislocation ring in a homogeneous elastic medium we have to take out of the interior of the body a piece of matter of the shape of a flat cylinder with radius ρ_0 and height b (physical realization by condensation of vacant sites). Then the top and the bottom of the hole are joined together. The volume of the disappearing hole is called the plastic volume change $\Delta V_{pl} = -\pi b \rho_0^2 < 0$. From Colonnetti's theorem (20) it follows that within the range of validity of the linearized theory of elasticity the elastic strains cause no further volume expansions. However, the non-linear elastic properties cause an additional elastic volume expansion $\Delta V_{el} > 0$. We obtain $V_2 - V_1 = \Delta V_{el} + \Delta V_{pl}$, where V_1 and V_2 are the total volumes of the body when containing the hole and after welding, respectively. For the ratio of elastic to plastic volume change we have the formula

$$\frac{\Delta V_{el}}{\Delta V_{pl}} = \frac{V_2 - V_1 - \Delta V_{pl}}{\Delta V_{pl}} = \frac{V_2 - V_1}{-\pi b \rho_0^2} - 1 = -\frac{a_f^{(2)}(0)}{\pi b \rho_0^2} \quad 25.$$

Numerical results are given in Table I (fifth column).

The intensity at zero angle for a single loop in an infinite specimen as obtained by the extrapolation of $a_i^{(2)}(\mathbf{h})$ to $h = 0$, neglecting surface effects, is given by

$$i_i^{(2)}(0) = \frac{f^2(0)}{\Omega^2} \pi^2 b^4 \rho_0^2$$
$$\times \left[\frac{\rho_0}{b} D_4 \sin^2 \psi + S(\rho_0)[D_1 + D_3 + (D_2 - D_3)\cos^2 \psi]\right]^2 \quad 26.$$

[ψ is the angle between $\mathbf{h}$ and $\mathbf{b}$, D_1 to D_4 are second- and third-order elastic constants (21)]. (See Appendix 2.)

TABLE I

$\frac{\rho_o}{b}$	$\dfrac{i^{(2)}(\psi=90°, h=0)}{i^{(1)}(\psi=90°, h=0)}$	$\dfrac{i^{(2)}(\psi=0°, h=0)}{i^{(2)}(\psi=90°, h=0)}$	$\dfrac{\langle i^{(2)}(0)\rangle}{\langle i^{(1)}(0)\rangle}$	$\dfrac{\Delta V_{el}}{\Delta V_{pl}}$	
2,5	–	–	1,31	–0,20	
5	–0,143	0,296	1,15	–0,13	quadratic theory
10	–0,017	0,039	1,07	–0,07	
15	–0,0125	0,019	1,05	–0,05	
limit $\rho_o \to \infty$	0	0	1	0	linear theory

The influence of the second-order effects on the small-angle scattering from an edge-dislocation ring for copper.

First column: Ratio of ring radius to modulus of Burgers vector.
Second column: Quadratic contribution to scattering amplitude divided by linear amplitude, both for angle $\psi = 90°$ between scattering vector and Burgers vector, extrapolated to zero scattering angle ($h = 0$).
Third column: Ratio of second-order intensities for $\psi = 0°$ and $\psi = 90°$, extrapolated to $h = 0$.
Fourth column: Ratio of averaged quadratic intensity to averaged linear intensity for $h = 0$. The average is over the angle ψ assuming an isotropic distribution of ring orientations.
Fifth column: Ratio of elastic to plastic volume change.

For a polycrystal we have

$$\langle i^{(2)}(0)\rangle = N \frac{f^2(0)}{\Omega^2} \pi^2 b^4 \rho_0^2$$

$$\times \left[\frac{8}{15} D_4^2 \left(\frac{\rho_0}{b}\right)^2 + \frac{S(\rho_0)\cdot\rho_0}{b} D_4 \left(\frac{4}{3} D_1 + \frac{4}{15} D_2 + \frac{16}{15} D_3\right) \right.$$
$$\left. + S^2(\rho_0)\left(D_1^2 + \frac{1}{5} D_2^2 + \frac{8}{15} D_3^2 + \frac{2}{3} D_1 D_2 + \frac{4}{3} D_1 D_3 + \frac{4}{15} D_2 D_3\right) \right] \quad 27.$$

if we assume that all loop orientations were equally probable.

Inserting the elastic constants of copper (22, 23) into 27 gives:

$$\langle i_i^{(2)}(0)\rangle = N \frac{f^2(0)}{\Omega^2} \pi^2 b^4 \rho_0^2$$

$$\times \left[0.116 \left(\frac{\rho_0}{b}\right)^2 + 0.044 \left(\frac{\rho_0}{b}\right) S(\rho_0) + 0.249 S^2(\rho_0) \right] \quad 27\text{a}.$$

The first term in the bracket is the averaged linear intensity $\langle i_i^{(1)}(0)\rangle$. The influence of the second-order terms can be seen from Table I (fourth column).

The formula for the scattered intensity for finite **h** cannot be cast into a simple analytic expression. We shall therefore give some curves representing results for Cu and $\rho_0 = 5b$ obtained from a computer (Figs. 1 to 3).

Figure 1 shows the linear and the quadratic contributions to the scattering amplitude for four orientations of the scattering vector **h** with respect to **b** as a function of the scattering angle. The quadratic

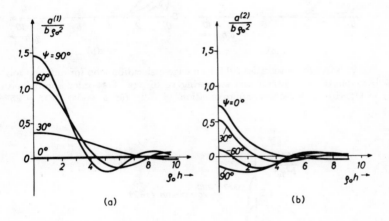

Fig. 1. Scattering amplitudes $a^{(j)}$ of an edge-dislocation loop for different angles ψ between Burgers vector and scattering vector (in units of $b\rho_0^2$); (a) first-order approximation ($j = 1$); (b) second-order terms ($j = 2$) for a loop of radius $\rho_0 = 5b$, i.e. $b\rho_0^2 = 25b^3$.

contribution is seen to counteract the marked anisotropy of the linear approximation.

Figure 2 gives the intensities $i^{(1)}(h)$ (linear approximation) and $i^{(2)}(h)$, demonstrating the stronger concentration of the intensity towards zero angle in the second-order approximation. This can also be seen from Fig. 3, showing the averaged linear and quadratic intensities $\langle i^{(1)}\rangle$ and $\langle i^{(2)}\rangle$ for an isotropic polycrystalline sample.

In Fig. 4 the polar diagram of the scattered intensity is plotted as a function of the angle ψ between **h** and **b**. The different curves correspond to different radii ρ_0. It can be seen that with increasing ring radius the anisotropy increases, the limit for large radii being

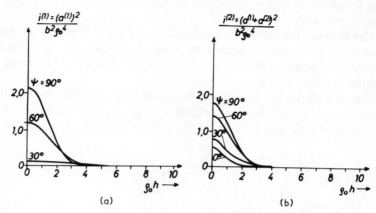

Fig. 2. Scattered intensities $i^{(j)}$ for an edge-dislocation loop for different angles ψ between Burgers vector and scattering vector; (a) first-order approximation ($j = 1$); (b) second-order approximation ($j = 2$) for $\rho_0 = 5b$, ($b^2\rho_0^4 = 25^2 b^6$).

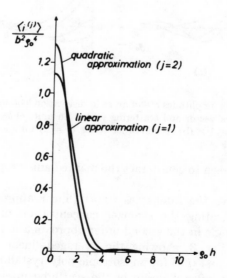

Fig. 3. Linear and quadratic approximation to the scattered intensity for an edge-dislocation loop of radius $\rho_0 = 5b$, averaged over an isotropic distribution of loop orientations. Units as in Fig. 2.

the linear approximation. (The intensities are given in units of $\langle i_i^{(1)}(0)\rangle$ in order to be immediately comparable.)

The principal information contained in plots such as Fig. 3 can be summarized by giving the ratio of the radius of gyration R_s to the ring radius ρ_0, and the ratio of the averaged intensity at zero angle $\langle i^{(2)}(0)\rangle$ calculated by the quadratic approximation to the averaged intensity at zero angle $\langle i^{(1)}(0)\rangle$ calculated by the linear theory (Fig. 5). We see that the stronger concentration of the scat-

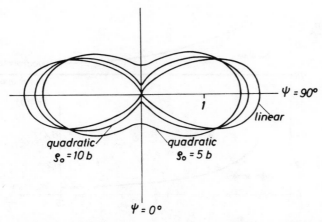

Fig. 4. Polar diagram of the ratio $i_i^{(2)}(0)/\langle i_i^{(1)}(0)\rangle$ of the intensity at angle ψ for a single loop to the averaged linear intensity for an isotropic loop distribution, both extrapolated to scattering angle $2\theta = 0$. Ring radii $\rho_0 = 5b$, $10b$, ∞ (denoted by "linear").

tered intensity towards smaller scattering angles resulting from the inclusion of the second-order terms has the consequence that for small enough ring radii the radius of gyration may become *larger* than the ring radius.

It is evident from Figs. 4 and 5 as well as Table I that the quadratic effects become more and more important the smaller the ring radius ρ_0. We illustrate these effects for the example of $\rho_0 = 5b$. Given measured values of R_s and of $i(0)$, the application of the first-order approximation instead of the second-order theory would lead to an apparent ring radius $\rho_0^{(1)} = 1.3\rho_0$ and to a reduction of the number of rings by a factor 0.41. Compared with linear theory the more complete theory thus reduces the number of vacancies per ring by a factor 0.58 and enhances the total number of vacancies by a factor 1.5. The numerical statements in the section on Models for Vacancy Agglomerates must be modified accordingly.

THE DILATATION FIELD

As mentioned in the previous section, the volume integral of the dilatation, i.e. the total elastic volume change, is zero in the linear approximation and positive in the quadratic approximation. As a

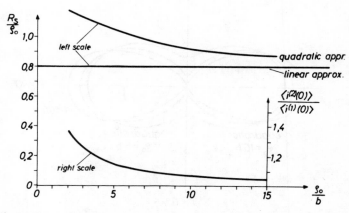

Fig. 5. Ratio of radius of gyration R_s to ring radius ρ_0 in the linear and the quadratic approximation and intensity $\langle i^{(2)}(0)\rangle/\langle i^{(1)}(0)\rangle$ extrapolated to zero scattering angle as a function of ρ_0 (numerical results for Cu).

by-product of our calculations, we obtain the result that the dilatation terms dominating at large distances r from the ring (disregarding surface effects) have the same form in both approximations, namely $r^{-3}P_2(\vartheta)$, where $P_2(\vartheta)$ is the second Legendre polynomial and ϑ the angle between Burgers vector and radius vector. The at first sight surprising result that both the first- and second-order strains vary asymptotically as r^{-3} appears to be general for closed dislocation loops of any shape.

APPENDIX 1

For a detailed explanation of the meaning of the fundamental Eq. 9 we must refer the reader to the original papers (16, 23). This appendix serves to clarify the definitions of the symbols used in this equation.

$\mathcal{E}_{ij}$ is Euler's strain tensor defined by $\mathcal{E}_{ij} = \frac{1}{2}(a_{ij} - g_{ij})$, where a_{ij} and g_{ij} are the metric tensors of the deformed and the so-called natural state respectively. [For the notion of the natural state see Ref. (23)]. From $\mathcal{E}_{ij}$ one derives

$$\mathcal{E}_{nkq} = \tfrac{1}{2}[\nabla_n \mathcal{E}_{qk} + \nabla_k \mathcal{E}_{nq} - \nabla_q \mathcal{E}_{kn}].$$

ϵ^{jnm} is the totally antisymmetric 3rd-rank Levi-Civita tensor, which in the case of Cartesian coordinates has the value 0 if two or all three of its indices are equal, and $+1$ or -1 otherwise, the $+$ sign corresponding to an even permutation of the indices.

From the dislocation density tensor $\alpha^i{}_j$ one derives

$$h_{mlp} = -\tfrac{1}{2}[\epsilon_{mls}\alpha^s{}_p + \epsilon_{pms}\alpha^s{}_l - \epsilon_{lps}\alpha^s{}_m]$$

The indices (ij) in parentheses signify that the symmetric part of the expression with respect to ij is to be taken.

The incompatibility tensor $\eta^i{}_j$ is given by

$$\eta^i{}_j = -(\epsilon_{jkl}\nabla_k \alpha^i{}_l)_{(ij)}.$$

It may alternatively be represented in the form

$$\mathbf{n} = \text{Sym}\,(\boldsymbol{\alpha} \times \boldsymbol{\nabla}) \equiv -\text{Sym Rot}\,\boldsymbol{\alpha}$$

'Ink" is a second-order linear symmetric differential operator

$$\text{Ink}\,\mathcal{E} \equiv \boldsymbol{\nabla} \times \mathcal{E} \times \boldsymbol{\nabla}.$$

APPENDIX 2

According to Murnaghan (24) the elastic energy density of an isotropic medium up to the third order in $\mathcal{E}_{ij}$ can be expressed by

$$\Psi(\mathcal{E}) = \frac{\lambda + 2\mu}{2}\mathcal{E}_\text{I}{}^2 - 2\mu\mathcal{E}_\text{II} + \frac{l + 2m}{3}\mathcal{E}_\text{I}{}^3 - 2m\mathcal{E}_\text{I}\mathcal{E}_\text{II} + n\mathcal{E}_\text{III},$$

where $\mathcal{E}_\text{I}$, $\mathcal{E}_\text{II}$, $\mathcal{E}_\text{III}$ are the three invariants of the strain tensor, λ, μ the second-order Lamé constants, and l, m, n, the third-order Murnaghan elastic constants.

The third-order elastic constants L, M, N defined by Pfleiderer, Seeger and Kröner (17) are related to the l, m, n, by

$$N = -\frac{1}{8\mu^2}\left[6 + \frac{n}{\mu}\right]$$

$$M = \frac{1}{2\mu^2}\left[\frac{1}{2} - \frac{2-3\nu}{1+\nu} + \frac{1}{18\kappa}\left(12\lambda + 24\mu + 6m + n\left(\frac{3\kappa}{2\mu} - 1\right)\right)\right]$$

$$L = -\frac{l}{81\kappa^3} - \frac{m}{18\kappa\mu^2} + \frac{n}{27\mu^3\kappa^3}\left[\frac{\kappa^3}{4}(\mu - \kappa) - \frac{\mu^3}{27}\right]$$

$$+ \frac{1}{81\kappa^2}\left[4 - \frac{1}{3\kappa}(6\lambda + 4\mu)\right] + \frac{1}{6\mu^2}\left[-\frac{1}{3} + \frac{2-3\nu}{1+\nu} - \frac{1}{3\kappa}(2\lambda + 4\mu)\right]$$

where $\kappa = E/3(1 - 2\nu)$, $E = 2\mu(1 + \nu)$ and ν = Poisson's ratio.

The elastic constants C_2 to C_7 appearing in the strain-stress relation 11 are defined as follows (17):

$$C_2 = \frac{1}{E^2}[3\nu(1 - \nu) - 1] + 3L + M$$

$$C_3 = \frac{3(1-\nu)}{2\mu E} + M$$

$$C_5 = \frac{3\nu - L}{2\mu E} - M$$

$$C_7 = -\frac{1}{4\mu^2} + N$$

In order to get more compact final formulae we introduce the abbreviations

$$D_1 = \frac{(2\mu)^2}{16(1-\nu)^3}\left[8(1+\nu)^3 C_2 + 8\nu(1+\nu)C_3 + 2(1+4\nu)(1+\nu)C_5 \right.$$
$$\left. + (-1 + 5\nu + 2\nu^2)C_7 + \frac{95 - 188\nu + 80\nu^2 + 16\nu^3}{16(1-\nu)^3}\right]$$

$$D_2 = \frac{(2\mu)^2}{16(1-\nu)^2}[4(1+\nu)(1-2\nu)C_5 + (1-2\nu)C_7]$$

$$D_3 = \frac{(2\mu)^2}{16(1-\nu)^3}[(1-2\nu)C_7 - (1+\nu)(1-2\nu)C_5]$$

$$D_4 = \frac{1-2\nu}{1-\nu}$$

For Cu we took from the literature (22) the following third-order constants: $l = -2 \times 10^4$ kp/mm^2, $m = -6 \times 10^4$ kp/mm^2, $n = -15.5 \times 10^4$ kp/mm^2.

BIBLIOGRAPHY

1. Seeger, A. and Essmann, U., *Rendiconti della Scuola Internationale di Fisica* "E. Fermi," XVIII Corso, Academic Press, New York, 1962; p. 717.
2. Diehl, J., *Radiation Damage in Solids*, **I,** Int. Atomic Energy Agency, Vienna, 1962; p. 129.
3. Seeger, A., *Proc. Second Int. Conf. Peaceful Uses Atomic Energy (Geneva 1958)*, **6,** United Nations, New York, 1958; p. 250.
4. Seeger, A., *Radiation Damage in Solids*, **I,** Int. Atomic Energy Agency, Vienna, 1962; p. 101.
5. v. Jan, R., Phys. Stat. Sol., **8,** 331 (1965).
6. Chik, K.-P., Seeger, A. and Rühle, M., *Vth Int. Congress for Electron Microscopy*, paper J-11, Academic Press, New York, 1962.
7. Seeger, A., Gerold, V. and Rühle, M., Z. Metallk., **54,** 493 (1963).
8. Galligan, J. and Washburn, J., Phil. Mag., **8,** 1455 (1963).
9. Seeger, A. and Kröner, E., Z. Naturforschg., **14a,** 74 (1959).
10. Seeger, A. and Rühle, M., Ann. Physik, **11,** 216 (1963).
11. Guinier, A., et al., *Small-Angle Scattering of X-Rays*, J. Wiley and Sons, New York, 1955.
12. Schilling, W. and Schmatz, W., Phys. Stat. Sol., **4,** 95 (1964).
13. Zener, C., Trans. Amer. Inst. Min. Met. Engrs., **147,** 361 (1942).
14. Seeger, A., and Haasen, P., Phil. Mag., **3,** 470 (1958).
15. Seeger, A., Suppl. Nuovo Cimento, **VII [X]**, 632 (1957).
16. Kröner, E. and Seeger, A., Arch. Rational Mech. Anal., **3,** 97 (1959).
17. Pfleiderer, H., Seeger, A. and Kröner, E., Z. Naturforschg., **15a,** 758 (1960).
18. Stojanovich, R., *Proceedings of the Int. Symposium on Second-Order Effects in Elasticity, Plasticity and Fluid Dynamics*, Haifa, 1962; Macmillan Co., New York, 1964; p. 142.
19. Eshelby, J. D., J. Appl. Phys., **25,** 255 (1954).
20. Colonnetti, G., Atti Accad. Naz. Lincei, Rend., Cl. Sci. Fis. Mat. Natur., (5), **272,** 155 (1918).
21. Brand, P., Diplomarbeit, Technische Hochschule Stuttgart (1964).
22. Seeger, A. and Buck, O., Z. Naturforschg., **15a,** 1056 (1960).
23. Kröner, E., Arch. Rational Mech. Anal., **4,** 273 (1960).
24. Murnaghan, F. D., *Finite Deformation of an Elastic Solid*, J. Wiley and Sons, New York, 1951.

Critical Scattering of X-Rays from Binary Alloys

A. MÜNSTER

Institute of Theoretical Physical Chemistry, University of Frankfurt (Main)

INTRODUCTION

A few years after the discovery of the critical point (1869) it was found by Avenarius (1) that a condensing gas immediately above the critical point exhibits a strong increase of light-scattering, especially in the forward direction. As first shown by Guthrie (2), the same effect can be observed in binary liquid mixtures near the critical point of demixtion. This phenomenon is called critical opalescence. Careful experimental studies performed by Friedländer (3) and Travers and Usher (4) have established that the appearance of critical opalescence is uniquely determined by the thermodynamic state variables and therefore must be considered as a true equilibrium property.

The first step in the direction of a theoretical understanding was made by Smoluchowski (5), who suggested that critical opalescence might be explained by a strong increase of local density fluctuations when approaching the critical point. This idea was worked out by Einstein (6), but his result fails to describe the angular dependence of the scattered intensity which is the most characteristic feature of the critical opalescence. Now the Smoluchowski-Einstein theory is based on the assumption that density fluctuations in different volume elements are statistically independent of each other. Ornstein and Zernike (7) pointed out that this is not permissible in the immediate neighborhood of the critical point. They were able to show that the critical opalescence can be explained by an enormous increase of the correlation range which tends to infinity at the critical point itself. This general idea can be proved to be correct quite rigorously. On the other hand, the validity of the special equation given by Ornstein and Zernike (which henceforth will be referred to as O.-Z. theory) is still an open question.

Till the middle of this century the study of critical opalescence was confined to fluid systems. Nothing was known about analogous phenomena in the solid state and the general opinion was that either they did not exist or, at least, could not be observed. In the meantime, however, the situation has basically changed, and nowadays the study of critical fluctuations plays a similar role in solid state physics and in the physics of fluids. [For a review of the field we refer to Münster (8).] In this paper we are concerned with one of these effects discovered during the last decade; this effect may be considered indeed as the true analog of the critical opalescence occurring in binary liquid mixtures near the critical point of demixtion.

The theoretical investigation of the diffuse scattering of X-rays from binary alloys led Krivoglaz (9) and independently Münster and Sagel (10) to the conclusion that in the neighborhood of the critical point of demixtion a critical opalescence of X-rays is to be expected which means that the intensity of small-angle scattering should greatly increase when approaching the critical point. By measuring the small-angle scattering of the alloy aluminum-zinc, Münster and Sagel (11, 12) showed that this effect indeed exists, in substantial agreement with theoretical expectation.

In the following, we shall first outline* the theory of the effect whereas in the last section new experimental results will be reported and discussed.

LOCAL FLUCTUATIONS. CORRELATION FUNCTIONS

From our introductory remarks, it appears that the correlation between local density fluctuations plays a central role in the theory of critical opalescence. Although these concepts are quite familiar in the theory of fluids it must be admitted that their physical meaning is less obvious in the problem which we are concerned with. First of all one may ask how to define density fluctuations in a binary alloy where the diffusion constant is of the order $10^{-8} - 10^{-12}$ cm^2 sec^{-1}. The answer is that the temporal regression of fluctuations does not play any role at all in our problem and therefore the term "local density fluctuations" means nothing else but the spatial inhomogeneities of the system or, in other words, the local deviations from the average composition. Secondly, in the theory of light-scattering one is used to imagining local density fluctuations as defined for small

---

* For the details the reader is referred to references (8) and (13).

volume elements which still contain comparatively large numbers of particles. It is obvious that this is not permissible in the case of X-rays where indeed density fluctuations at a given point of the physical space must be considered. Therefore a given configuration of the system is to be specified by means of Dirac δ-functions. Since critical opalescence, as mentioned above, is an equilibrium property and the critical point itself is defined in terms of thermodynamic state variables, we have to use ensemble averages of the pertinent quantities as defined in statistical mechanics. In carrying out this program we meet with another significant difference between fluids and solids. As is well known, scattering experiments yield information about interatomic distances only. Now it turns out that the pair distribution function of a fluid, the so-called radial distribution function, depends only on the scalar distance r between two particles and therefore is an experimentally accessible quantity. On the contrary, the pair distribution functions of a binary alloy depend on two position vectors and the true analog of the radial distribution has to be defined more carefully. Finally we have to examine to what extent the aforementioned concepts remain meaningful in the case of a polycrystalline sample.

Let us now consider a binary alloy made up of N_1 atoms of the species 1 and N_2 atoms of the species 2. The system is supposed to be a single crystal. We shall assume that the so-called adiabatic approximation holds good, which means that with sufficient accuracy a Hamiltonian may be constructed which depends on the nuclear coordinates only. For our present purpose we may then consider the atoms as structureless mass points obeying semi-classical statistics. Long-range interaction forces, however, will be excluded. Let $\mathbf{q}_{ik}$ denote the position vector of the ith atom of the species k. Then for a given configuration of the system the number density of the species k is given by

$$\nu_k^{(1)}(\mathbf{q}) = \sum_{i=1}^{N_k} \delta(\mathbf{q} - \mathbf{q}_{ik}) \qquad 1.$$

where $\delta(\mathbf{q} - \mathbf{q}_k)$ is the three-dimensional δ-function. We now ask for the conditional probability density of finding a particle of the species l at $\mathbf{q}' = \mathbf{q} + \mathbf{r}$ if a particle of the species k is known to be at $\mathbf{q}$. Obviously this quantity will depend on the vectors $\mathbf{q}$ and $\mathbf{r}$. Therefore, taking the weighted average over all possible positions of the k-atom we obtain a function which depends only on the vectorial distance **r** between the two particles. This quantity, which will be denoted by $\bar{N}_k \langle \nu_{kl}^{(1,1)} \rangle$ turns out to be the convolution of two singlet densities.* Thus we have

$$\bar{N}_k \langle \nu_{kl}^{(1,1)} \rangle = \nu_k^{(1)} \star \nu_l^{(1)} = \sum_{i=1}^{N_k} \sum_{j=1}^{N_l} \delta(\mathbf{r} + \mathbf{q}_{ik} - \mathbf{q}_{jl}) \qquad 2.$$

where the star is a symbolic notation for the "convolution product." Furthermore we have

$$\bar{N}_k \int \langle \nu_{kl}^{(1,1)} \rangle d\mathbf{r} = N_k N_l - N_k \delta_{kl} \qquad 3.$$

where δ_{kl} is the Kronecker delta. To obtain equilibrium properties, we now take the grand canonical averages of the preceding equations. Then the single-particle distribution function is defined as

$$\rho_k^{(1)}(\mathbf{q}) = \overline{\nu_k^{(1)}(\mathbf{q})} \qquad 4.$$

The conditional probability density of finding an l-particle at a distance **r** from a k-particle, averaged over all positions of the k-particle, is given by

$$\rho_{kl}^{(1,1)} = \overline{\langle \nu_{kl}^{(1,1)} \rangle} \qquad 5.$$

The normalization of this function reads

$$\bar{N}_k \int \langle \rho_{kl}^{(1,1)} \rangle \, d\mathbf{r} = \overline{N_k N_l} - \bar{N}_k \delta_{kl} \qquad 6.$$

Let us now put

$$\rho_k^{(1)}(\mathbf{q}) = \rho_k g_k^{(1)}(\mathbf{q}), \qquad \langle \rho_{kl}^{(1,1)} \rangle = \bar{\rho}_l g_{kl}^{(1,1)} \qquad 7.$$

Then we have the normalization

$$\lim_{V \to \infty} V^{-1} \int g_k^{(1)}(\mathbf{q}) \, d\mathbf{q} = 1, \qquad \lim_{V \to \infty} V^{-1} \int \langle g_{kl}^{(1,1)} \rangle \, d\mathbf{r} = 1 \qquad 8.$$

It is readily shown that the function $\langle g_{kl}^{(1,1)} \rangle$ reduces to the radial distribution function in the case of fluid systems and therefore it must be considered as its true analog. For large absolute values of the distance **r** the probability density for finding an l-particle will be independent of the presence of the k-particle. Hence

$$\lim_{r \to \infty} \langle g_{kl}^{(1,1)} \rangle = g_l^{(1)} \qquad 9.$$

In our applications the quantity of interest is indeed the difference between $\langle g_{kl}^{(1,1)} \rangle$ and its asymptotic value. Thus we define

* The horizontal bar denotes the grand canonical average.

$$g_{kl}(\mathbf{r}) = \langle g_{kl}^{(1,1)} \rangle - g_l^{(1)} \qquad 10.$$

with the normalization

$$\lim_{V \to \infty} V^{-1} \int g_{kl}(\mathbf{r}) \, d\mathbf{r} = 0 \qquad 11.$$

The functions g_{kl} are called correlation functions. They will play a central role in our theory since the correlation between local density fluctuations as well as the density fluctuations of the system as a whole are determined by the correlation functions.

So far we have confined ourselves to single crystals. The experiments to be reported in the last section of this paper, however, have been performed using polycrystalline samples and the question arises how to define molecular distribution functions in this case. From a molecular point of view a polycrystalline sample must be considered indeed as a heterogeneous system since the molecular structure is neither coherent nor reproducible. Therefore it must be concluded that molecular distribution functions in the rigorous sense do not exist. On the other hand a polycrystalline sample has not only well defined thermodynamic properties but certain features of the molecular structure are likewise well defined and reproducible as shown by the X-ray patterns which are of the Debye-Scherrer type whereas Bragg or Laue diagrams do not exist. Hence we may hope to find an appropriate definition of molecular distribution functions by retaining only those features which all macroscopically identical polycrystalline samples have in common. This can be done on the basis of the following assumptions (14):

(a) Any crystallite can be considered as a thermodynamic system for which molecular distribution functions obeying the above normalization relations exists.

(b) The orientation of the crystallites can be mapped on the unit sphere by a continuous density function which is assumed to be a constant.

(c) Correlations between molecular distributions in different crystallites are negligible.

Then averaging over all orientations leads to a new set of correlation functions $g_{kl}(r)$ which depend on the scalar distance r only and therefore are spherically symmetric. It is readily shown that these averaged functions obey the relations 6, 8 and 11. Hence, for the sake of simplicity, they will be denoted by the same symbols as the functions defined for a single crystal. It is obvious, however, that the averaged functions are physically meaningful only for distances which are smaller than the linear dimension of the crystallites.

Let us now consider the local density fluctuations. Denoting by $\Delta\rho_k$ and $\Delta\rho'_l$ the local deviations from the average densities at $\mathbf{q}$ and $\mathbf{q'} = \mathbf{q} - \mathbf{r}$, respectively, we have

$$\langle \overline{\Delta\rho_k \Delta\rho'_l} \rangle = \rho_k \rho_l g_{kl}(\mathbf{r}) \qquad (r \neq 0) \qquad 12.$$

where the left-hand member is the nonweighted average of the covariance over all positions of the k-particle. As to the density fluctuations of the system as a whole, we obtain

$$\frac{\overline{(N_k - \bar{N}_k)(N_l - \bar{N}_l)}}{\bar{N}_k \bar{N}_l} = V^{-1} \int g_{kl} \, d\mathbf{r} + \bar{N}_k^{-1} \delta_{kl} \qquad 13.$$

Thus the total density fluctuations may be conceived as a superposition of the local fluctuations given by Eq. 12. To conclude this section, we make a few further remarks on the correlation functions. So far we have not yet proved that the limit defined by Eq. 11 exists. This question can be settled by using the theory of the grand canonical ensemble. It turns out that the limit in question indeed exists almost everywhere, the only exceptions being the singularities corresponding to phase transitions and critical points. Hence we may infer that, apart from these special cases, g_{kl} is at most of $O(r^{-n})$ with $n > 3$ if $r \to \infty$. Experimental studies on the alloy aluminum-zinc show that practically g_{kl} has assumed its asymptotic value for $r > 10$ Å. This suggests rather an exponential decay found theoretically and experimentally for the radial distribution function of simple liquids. In the range $r < 10$ Å, g_{kl} has an oscillating character reflecting the molecular structure of the system. In the case of spherical symmetry this corresponds to the so-called coordination shells surrounding the central particle. Now, as already mentioned in the introduction, in the study of critical fluctuations one is mainly interested in the asymptotic behavior of the correlation function. Hence we shall find it convenient to smear out the molecular details by a local averaging process. This leads to the definition of semifine correlation functions which, assuming spherical symmetry, may be written

$$\hat{g}_{kl}(r) = g_{kl} \star \Theta(s) \qquad 14.$$

with

$$\Theta(s) = (a/\pi)^{\frac{3}{2}} \exp(-as^2) \qquad 15.$$

where a is a positive constant to be chosen in such a way that the molecular details are completely smeared out. It follows immediately that

$$g_{kl}(r) = \hat{g}_{kl}(r) \quad \text{for} \quad r \to \infty \qquad 16.$$

Moreover using Fourier transforms and the convolution theorem we obtain

$$\lim_{V \to \infty} \int g_{kl}(r) \, d\mathbf{r} = \lim_{V \to \infty} \int \hat{g}_{kl}(r) \, d\mathbf{r} \qquad 17.$$

The physical meaning of the semifine correlation functions becomes particularly transparent in the Fourier representation. Let $\eta_k(\mathbf{u})$ be the Fourier transform of the singlet density as defined by Eq. 1. Then the spectral distribution of the local fluctuations is given by the function

$$\phi_{kl}(\mathbf{u}) = \overline{\eta\eta^*} - \bar{N}_k \bar{\eta}_l \qquad 18.$$

Obviously the molecular structure will be reflected only in the domain of short wavelengths. Therefore we cut off this part of the spectrum, putting

$$\hat{\phi}_{kl}(\mathbf{u}) = \phi_{kl}(\mathbf{u}) \exp(-u^2/4a) \qquad 19.$$

Since

$$\mathfrak{F}^{-1} \exp(-u^2/4a) = \Theta(s) \qquad 20.$$

(where $\mathfrak{F}$ is the Fourier operator) we obtain with the aid of the convolution theorem

$$\mathfrak{F}^{-1}\hat{\phi}_{kl}(\mathbf{u}) = \bar{N}_k \bar{\rho}_k \hat{g}_{kl}(\mathbf{r}) \qquad (\text{for } a^{-\frac{1}{2}} < r) \qquad 21.$$

Thus the introduction of the semifine correlation function amounts to a cut-off of the short-wavelength part of the spectrum. Moreover we have from 19, as $u \to 0$

$$\hat{\phi}(u) = \phi(u) \quad \text{for} \quad u \to 0 \qquad 22.$$

Consequently the asymptotic behavior of the correlation function can be determined unambiguously from the (experimental or theoretical) knowledge of the spectrum near the origin. This important result plays a fundamental role in the study of critical opalescence.

Finally we shall give an important relation between the thermodynamic quantity defining the critical point of demixtion and the correlation functions of a binary alloy. Let us denote by Π the osmotic pressure* and by G_{kl} the limiting value of the space integral over the correlation function g_{kl} for $V \to \infty$. Then combining Eq. 13 with the theory of the grand canonical ensemble one obtains (14, 15):

$$\left(\frac{\partial \Pi}{\partial c_2}\right)_T = \frac{kT}{1 + G_{22}c_2} \qquad 23.$$

with $c_2 \equiv \bar{\rho}_2$.

DIFFUSE SCATTERING OF X-RAYS FROM BINARY ALLOYS

It was predicted by von Laue (16) that a binary alloy will exhibit, besides the well known Debye-Scherrer reflections, a diffuse scattering of X-rays in the small-angle region. This effect, which is due to the different scattering powers of the constituent atoms, was first observed by Wilchinsky (17) and by Guinier and Griffoul (18). Refined theoretical treatments were given subsequently by Wilchinsky (17), Cowley (19), Warren et al. (20). In the following we shall briefly describe the theory developed by Münster and Sagel (10, 11), which is most appropriate for our purposes.

As in the preceding section, we consider a single crystal of a binary alloy. The vectors of the primary beam and of the scattered radiation will be denoted by $\mathbf{k}_0$ and $\mathbf{k}$, respectively. Defining the scattering vector

$$\mathbf{s} = \mathbf{k}_0 - \mathbf{k}, \qquad k = 1/\lambda \qquad 24.$$

(where λ is the wavelength of the X-rays) we have

$$s = 2k \sin \theta = 2/\lambda \sin \theta \qquad 25.$$

where 2θ is the scattering angle. The electron density will be denoted by $\rho^{(e)}(\mathbf{q})$. Its Fourier transform is given by

$$\mathcal{F}\rho^{(e)}(\mathbf{q}) = \int \rho^{(e)}(\mathbf{q}) e^{-2\pi i \mathbf{s} \cdot \mathbf{q}} \, d\mathbf{q} \equiv P(\mathbf{s}) \qquad 26.$$

For convenience we define a reduced intensity I by the equation

$$I = \frac{I(\theta) R^2}{I_0 a^2 p} \qquad 27.$$

---

* The osmotic pressure is to be considered here simply as a thermodynamic function.

Here is I_0 the intensity of the primary beam, $I(\theta)$ the scattered intensity as measured at the angle θ and the distance R from the sample,

$$a = e^2/mc^2 \qquad 28.$$

the classical radius of the electron and p the polarization factor. Our derivation will be based on the kinematic diffraction theory. For a discussion of this approach we refer to Hosemann and Bagchi (21). Using the above definitions we may write the basic equation

$$I = PP^* \qquad 29.$$

where P^* is the complex conjugate of P. Defining a function

$$Q(\mathbf{q}) = \mathfrak{F}^{-1}I(\mathbf{s}) \qquad 30.$$

we obtain from 26 and 29, with the aid of the convolution theorem,

$$Q(\mathbf{q}) = \int \rho^{(e)}(\mathbf{x})\rho^{(e)}(\mathbf{q}+\mathbf{x})\,d\mathbf{x} \equiv \rho^{(e)}(\mathbf{q}) \star \rho^{(e)}(-\mathbf{q}) \qquad 31.$$

Thus the Q-function is the "convolution square" of the electron density. Next we introduce the configuration of the centers of gravity of the atoms which will be specified, as previously, by a set of vectors $\mathbf{q}_k$. Choosing the center of gravity as origin we may describe the electron cloud of a single atom of the species k by a density function $\rho_k^{(e)}(\mathbf{r}^{(e)})$, where

$$\mathbf{r}^{(e)} = \mathbf{q} - \mathbf{q}_k \qquad 32.$$

The contribution of this atom to the total electron density at the point $\mathbf{q}$ will be

$$\rho_k^{(e)}(\mathbf{q}) = \int \delta(\mathbf{q} - \mathbf{q}_{ik} - \mathbf{r}^{(e)})\rho_k^{(e)}(\mathbf{r}^{(e)})\,d\mathbf{r}^{(e)} \qquad 33.$$

Hence the total electron density at $\mathbf{q}$ turns out to be the sum of two convolutions which may be written

$$\rho^{(e)}(\mathbf{q}) = \nu_1^{(1)} \star \rho_1^{(e)} + \nu_2^{(1)} \star \rho_2^{(e)} \qquad 34.$$

From Eqs. 31 and 34, we obtain

$$\begin{aligned}Q(\mathbf{q}) = &[\nu_1^{(1)} \star \rho_1^{(e)}(\mathbf{r}^{(e)})] \star [\nu_1^{(1)} \star \rho_1^{(e)}(-\mathbf{r}^{(e)})] \\ &+ [\nu_2^{(1)} \star \rho_2^{(e)}(\mathbf{r}^{(e)})] \star [\nu_2^{(1)} \star \rho_2^{(e)}(-\mathbf{r}^{(e)})] \\ &+ 2[\nu_1^{(1)} \star \rho_1^{(e)}(\mathbf{r}^{(e)})] \star [\nu_2^{(1)} \star \rho_2^{(e)}(-\mathbf{r}^{(e)})] \quad 35.\end{aligned}$$

Using the definition 2, we may write this equation

$$Q(\mathbf{q}) = \nu_1^{(1)} \star [\rho_1^{(e)}(\mathbf{r}^{(e)}) \star \rho_1^{(e)}(-\mathbf{r}^{(e)})]$$
$$+ \nu_2^{(1)} \star [\rho_2^{(e)}(\mathbf{r}^{(e)}) \star \rho_2^{(e)}(-\mathbf{r}^{(e)})]$$
$$+ \sum_{k,l=1}^{2} \bar{N}_k \langle \nu_{kl}^{(1,1)} \rangle \star [\rho_k^{(e)}(\mathbf{r}^{(e)}) \star \rho_l^{(e)}(-\mathbf{r}^{(e)})] \quad 36.$$

According to Eq. 30, the reduced intensity is obtained from 36 simply by Fourier transformation. So far we have confined ourselves to an arbitrary fixed atomic configuration. In order to get equilibrium properties we have now to take the average of Eq. 36 with the aid of the grand canonical ensemble. In carrying out this process we shall assume that the functions $\rho_1^{(e)}$ and $\rho_2^{(e)}$ do not depend on the atomic configuration. As will be seen immediately this amounts to the statement that the atomic scattering powers have fairly well defined values even for crystals, which is indeed confirmed by experience. Then, using the definitions 4 and 5, we obtain

$$\bar{Q}(\mathbf{q}) = [\rho_1^{(e)}(\mathbf{r}^{(e)}) \star \rho_1^{(e)}(-\mathbf{r}^{(e)})] \int \rho_1^{(1)} \, d\mathbf{q}_1$$
$$+ [\rho_2^{(e)}(\mathbf{r}^{(e)}) \star \rho_2^{(e)}(-\mathbf{r}^{(e)})] \int \rho_2^{(1)} \, d\mathbf{q}_2$$
$$+ \sum_{k,l}^{2} \bar{N}_k \int \langle \rho_{kl}^{(1,1)} \rangle [\rho_k^{(e)}(\mathbf{r}^{(e)}) \star \rho_l^{(e)}(-\mathbf{r}^{(e)})] \, d\mathbf{r}_{kl} \quad 37.$$

The observed reduced intensity is now readily obtained by taking the Fourier transform of Eq. 37. Defining the atomic scattering powers f_k by the equation

$$f_k = \mathfrak{F}\rho_k^{(e)} \qquad 38.$$

and putting $N = \bar{N}_1 + \bar{N}_2$, we get

$$I = N\left[x_1 f_1^2 + x_2 f_2^2 + \sum_{k,l=1}^{2} x_k f_k f_l \int \langle \rho_{kl}^{(1,1)} \rangle e^{2\pi i \mathbf{s} \cdot \mathbf{r}} \, d\mathbf{r} \right] \qquad 39.$$

where x_k is the mole fraction of the species k. By a somewhat tedious rearranging, this equation may be cast into the form

$$I = I_D + I_T + I_B \qquad 40.$$

where the last term of the right-hand side gives the Bragg reflections of the crystalline lattice whereas the second one contains their modification by the thermal motion. The first term depends on the difference of the scattering powers f_1 and f_2 and describes the diffuse small-angle scattering in which we are interested. For a polycrystalline sample, this term becomes

$$I = -N \left[x_1 x_2 (f_1 - f_2)^2 4\pi\rho \int g_{12} \frac{\sin hr}{hr} r^2 \, dr \right] \qquad 41.$$

where the quantities

$$h = 2\pi s = \frac{4\pi}{\lambda} \sin \theta, \qquad \rho = (\bar{N}_1 + \bar{N}_2)/V \qquad 42.$$

have been introduced for convenience. Let us now define a reduced intensity

$$i(h) = \frac{I_D}{N x_1 x_2 (f_1 - f_2)^2} \qquad 43.$$

Then Eq. 41 can be rewritten

$$hi(h) = -4\pi\rho \int_0^\infty r g_{12}(r) \sin (hr) \, dr \qquad 44.$$

which, by Fourier inversion, gives

$$-r g_{12}(r) = \frac{1}{2\pi^2 \rho} \int_0^\infty hi(h) \sin (hr) \, dh \qquad 45.$$

Thus the correlation function g_{12} can be determined experimentally from small-angle scattering. On the other hand, the space integral G_{12} can be calculated from purely thermodynamic data. For the system Al-Zn, a satisfactory agreement between the two sets of data has been found.

CRITICAL SCATTERING

Let us now examine the behavior of the density fluctuations near the critical point of demixtion. First we observe that in a binary alloy we have density fluctuations of two kinds of particles which are to be described by three correlation functions. On the other hand as can be seen from Eq. 44, only the correlation function g_{12} is connected with small-angle X-ray scattering. It can be shown, however, that critical fluctuations and critical scattering are indeed completely specified by a single correlation function (22). First, by means of an appropriate transformation, we may pass from density fluctuations of the constituent particles to total density fluctuations and concentration fluctuations. Then it turns out that total density fluctuations are not affected at all when approaching the critical point of demixtion. Thus the problem reduces to the discussion of concentration fluctuations which are described by the correlation function:

$$g = g_{11} + g_{22} - 2g_{12} \qquad 46.$$

This function indeed plays a dominant role in the theory of critical light scattering. At first sight this result seems to be useless for the theory of X-ray scattering. We shall see, however, that with respect to the critical behavior, all correlation functions are completely equivalent.

Since a calculation of the correlation function from first principles is, at present at least, not possible, we shall start with the thermodynamic definition of the critical point of demixtion which for our purposes may be written

$$\left(\frac{\partial \Pi}{\partial c_2}\right)_T = 0 \qquad 47.$$

Then it follows from Eq. 23 that at the critical point the space integral G_{22} diverges. Since the definition of the "solvent" is arbitrary the same must be true for the space integral G_{11}. Now it can be shown that, when approaching the critical point T_c, we have asymptotically

$$\lim_{T \to T_c} \frac{G_{12}{}^2}{G_{11}G_{22}} = 1 \qquad 48.$$

Thus G_{12} likewise tends to infinity at the critical point. Furthermore we have the asymptotic relations

$$\lim_{T \to T_c} \frac{G_{11}}{G} = v_2{}^2 c_2{}^2, \quad \lim_{T \to T_c} \frac{G_{22}}{G} = v_1{}^2 c_1{}^2, \quad \lim_{T \to T_c} \frac{G_{12}}{G} = -v_1 c_1 v_2 c_2 \quad 49.$$

where G denotes the space integral over the correlation function g defined by Eq. 46 and v_k the partial molecular volume of component k.

With regard to Eq. 44, we shall be especially interested in the quantity G_{12}. Since the correlation function g_{12} necessarily has an upper bound, the divergence of the space integral can be due only to its asymptotic behavior which means that the condition mentioned in the previous section must be violated. Hence, introducing the semifine correlation function, we may write

$$\hat{g}_{12}(r) = A \cdot r^{-n}, \quad (0 < n \leq 3), \quad (T \to T_c, r \to \infty) \qquad 50.$$

This shows, in connection with Eq. 12, that the critical point is characterized by an enormous increase of the correlation range between local density fluctuations. On the other hand the appearance of a long-range tail in the correlation function gives rise to an increase of its Fourier transform near the origin which, by virtue of Eq. 44, means that the intensity of small-angle scattering of X-rays will greatly increase when approaching the critical point.

It is now easy to see that, by virtue of 49, all correlation functions will exhibit the same asymptotic behavior for $r \to \infty$ which leads to the relations

$$\hat{g}_{11}(r) = v_2^2 c_2^2 \hat{g}(r), \qquad \hat{g}_{22}(r) = v_1^2 c_1^2 \hat{g}(r), \qquad \hat{g}_{12}(r) = -v_1 c_1 v_2 c_2 \hat{g}(r) \qquad 51.$$

Thus, with respect to the critical behavior, the functions $\hat{g}_{11}$, $\hat{g}_{22}$, $\hat{g}_{12}$, $\hat{g}$ are indeed equivalent. We may pass from one to another simply by multiplying with the appropriate scale factor.

As we have seen, the essential qualitative features of critical scattering of X-rays can be predicted rigorously if the thermodynamic definition of the critical point is taken for granted. The derivation of an explicit formula for the scattered intensity, however, requires an additional assumption. Confining ourselves to the O.-Z. theory we shall briefly outline the aspects which will be needed in the discussion of the experimental results. As can be seen from Eqs. 21 and 44, we need in our problem the Fourier transform of the semifine correlation function. For definiteness, we shall calculate $\hat{g}_{12}$ but since we are interested only in the asymptotic result we might choose any other correlation function as well. Let us define the functions

$$\alpha(u) = \mathfrak{F}\hat{g}_{12}(r) \qquad 52.$$

and

$$\beta(u) = \frac{\rho \alpha(u)}{1 + \rho \alpha(u)} \qquad 53.$$

From Eqs. 18 and 21 we see that $\alpha(u)$ is essentially the spectral distribution of local fluctuations. From the preceding considerations it follows that at the critical point $\alpha(u)$ tends to infinity as $u \to 0$. On the contrary, the function $\beta(u)$ is only weakly influenced by critical conditions and is shown to be almost proportional to the true fluctuation spectrum far away from the critical point. In the O.-Z. theory the critical fluctuation spectrum is expressed in terms of a reference spectrum $\beta(u)$ corresponding to a "normal" behavior of the system.

The underlying assumption is that the singularity of $\alpha(u)$ characterizing the critical point is a pole. This entails that $\beta(u)$ can be expanded into a Taylor series around the origin. Retaining only second-order terms we have

$$\beta(u) = F - \tfrac{1}{6}\epsilon^2 u^2 \qquad 54.$$

where

$$F = \int \hat{f}(r)\,d\mathbf{r}, \qquad \epsilon^2 = \int r^2 \hat{f}(r)\,d\mathbf{r}, \qquad \hat{f}(r) = \mathfrak{F}^{-1}\beta(u) \qquad 55.$$

Now solving 53 for $\alpha(u)$, we get

$$\rho\alpha(u) = \frac{\beta(u)}{1 - \beta(u)} \qquad 56.$$

Substituting 54 into 56, and putting

$$\kappa_1^2 = \frac{6(1 - F)}{\epsilon^2}, \qquad A = \frac{6F}{4\pi\rho\epsilon^2} \qquad 57.$$

we finally obtain

$$\alpha(u) = 4\pi A \,\frac{1 - u^2/4\pi\rho A}{\kappa_1^2 + u^2} \qquad 58.$$

This gives, for $u \to 0$, the asymptotic formula

$$\alpha(u) = \frac{4\pi A}{\kappa_1^2 + u^2}, \qquad (u \to 0) \qquad 59.$$

and by Fourier inversion

$$\hat{g}_{12}(r) = A\,\frac{\exp(-\kappa_1 r)}{r}, \qquad (r \to \infty) \qquad 60.$$

As to the parameters κ_1 and A, we have, from the critical behavior of G_{12}

$$\beta(0) = 1, \qquad (T = T_c) \qquad 61.$$

Hence, using 57,

$$\kappa_1 \to 0 \quad \text{for} \quad T \to T_c \qquad 62.$$

On the other hand, we obtain from 58, putting $u = 0$

$$\frac{A}{\kappa_1^2} = \frac{G_{12}}{4\pi} \qquad 63.$$

Thus the quantity A/κ_1^2 can be determined from thermodynamic measurements. On the contrary, the quantity A itself must be considered rather as a characteristic parameter of the reference spectrum. We have indeed from Eqs. 55 and 57

$$\frac{3}{2\pi\rho A} = \frac{\int r^2 \hat{f}(r)\, d\mathbf{r}}{\int \hat{f}(r)\, d\mathbf{r}} \equiv l^2 \qquad 64.$$

which means that the quantity $(\rho A)^{-1}$ is essentially the square of the correlation length of the reference spectrum. The correlation length of the spectrum can be defined either as

$$r_c = \kappa_1^{-1} \qquad 65.$$

or by the equation

$$L^2 = \frac{\int r^2 \hat{g}_{12}(r)\, d\mathbf{r}}{\int \hat{g}_{12}(r)\, d\mathbf{r}} = 6 r_c^2 \qquad 66.$$

where Eq. 60 has been used. For $T \to T_c$ both quantities tend to infinity as it should be.

To obtain the explicit formula for the critical scattering of X-rays from a binary alloy, we write Eq. 44

$$i(h) = -\rho \mathfrak{F} \hat{g}_{12}(r) \qquad 67.$$

Then substitution of 58 into 67 gives

$$i(h) = 4\pi\rho A\, \frac{1 - h^2/4\pi\rho A}{\kappa_1^2 + h^2} \qquad 68.$$

For sufficiently small values of h, the plot of $1/i(h)$ vs. h^2 yields a straight line according to the formula

$$\frac{1}{i(h)} = \frac{1}{4\pi\rho A}\, (\kappa_1^2 + h^2) \qquad 69.$$

In the following section, this equation will be compared with experimental results.

EXPERIMENTAL RESULTS

As already mentioned in the introduction, the first experimental studies of the critical scattering of X-rays from binary alloys were performed by Münster and Sagel (10, 11, 12). Although the existence of the effect could be established, the quantitative results were not yet satisfactory. Since no other measurements have been reported in the meantime we decided to start a new research program in this field. Rather quickly it became obvious that for a thorough investigation of the problem the construction of a new apparatus was unavoidable. For external reasons, however, this part of our program had to be postponed and as a first step we were compelled to confine ourselves to a refinement of the technique used by Münster and Sagel. In the following we shall briefly summarize the essential results of this preliminary research on the system Al-Zn (23). The phase diagram of this alloy is shown in Fig. 1. The coexistence curve near the critical

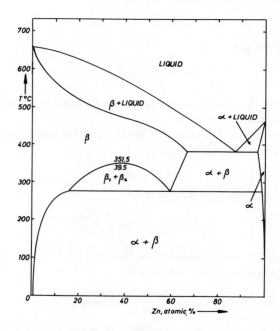

Fig. 1. Phase diagram of the system Al-Zn.

point of demixtion was determined by Münster and Sagel (24) using electrical conductivity measurements. The results are shown in Fig. 2. From these data the following values of the critical temperature T_c and the critical mole fraction x_{Zn} have been calculated:

$$T_c = 351.5° \pm 0.3°\text{C} \qquad 70.$$

$$x_{Zn} = 0.395 \pm 0.002 \qquad 71.$$

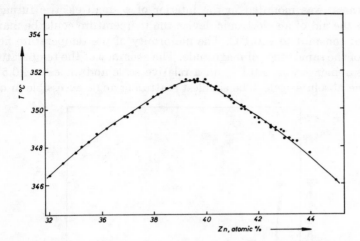

Fig. 2. Coexistence curve of the system Al-Zn near the critical point of demixtion.

Therefore the study of critical opalescence requires measurements between 350° and 400°C and the experimental problem is essentially the adaptation of temperature control and small-angle technique to this temperature range.

As in the earlier work, an X-ray camera of the Guinier type was used which is shown schematically in Fig. 3. The width of the beam was about 0.5 mm at the sample and 0.1 mm at the Geiger-Müller counter. The height of the beam was 14 mm, the intensity in this direction being fairly constant. The sample, a foil of about 25 μ

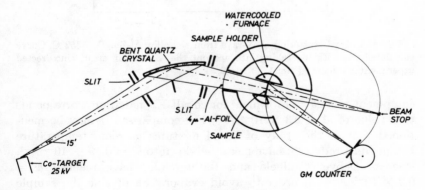

Fig. 3. Guinier camera geometry.

thickness, was mounted in the interior of a small electric furnace. With the aid of an electronic device the temperature could be maintained constant to ±0.1°C. The uniformity of the temperature field was of the same order of magnitude. The accuracy of the temperature measurements was ±0.1°C on a relative scale and at least ±0.5°C on the absolute scale. The smallest scattering angle accessible in our

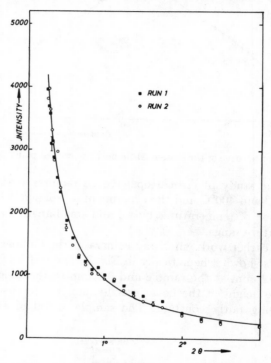

Fig. 4. Critical scattering of X-rays from the alloy Al-Zn. $T = 352°C$. Curve calculated from Eq. 69 taking into account the collimation error. Uncorrected experimental values: ■ run 1, ○ run 2.

equipment was $2\theta = 15'$ which, for cobalt $K\alpha$ radiation, corresponds to a value of $h^2 = 2.3 \times 10^{-4}$ Å^{-2}. For comparison it may be mentioned that in the study of liquid mixtures at room temperature Frisch and Brady (25) have measured down to $h^2 = 0.49 \pm 10^{-4}$ Å^{-2}, whereas Debye, Caulfield and Bashaw (26) have reached $h^2 = 0.2 \times 10^{-4}$ Å^{-2}. In order to avoid evaporation of zinc the sample was held under hydrogen at normal pressure. The intensity scattered from the hydrogen as well as the parasitic scattering (slits etc.) have been subtracted from the measured intensities corrected for absorption and the geometry of the camera. These values must still be corrected for the collimation error; this was done using a formula proposed by Gerold (27). At the beginning of each run the sample was annealed for 2^h at 385°C and subsequently held 4^h at the temperature of the measurement. From the electrical conductivity measurements of Münster and Sagel (24) it may be concluded that under these conditions internal equilibrium can be taken for granted.

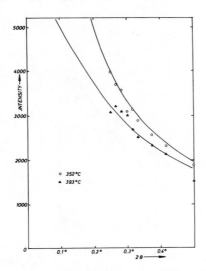

Fig. 5. Critical scattering at 352°C compared with small-angle scattering at 393°C. Curves calculated from Eq. 69 taking into account the collimation error. Uncorrected experimental values.

The reproducibility of the measurements is demonstrated in Fig. 4 which shows two runs at 352°C for the same sample. The vertical bars indicate the statistical error of the counter. These values have not been corrected for the collimation error.

The effect of critical scattering as it appears for these uncorrected values is shown in Fig. 5.

In both diagrams the curves have been calculated by introducing the collimation error into the O.-Z. formula, Eq. 68. It is seen that these curves fit the experimental data satisfactorily in the proper small-angle region (up to $2\theta \approx 30'$). For larger scattering angles small but systematic deviations occur.

For the quantitative comparison between theory and experiment, however, it is more appropriate to plot $1/i(h)$ versus h^2 where the intensity is now corrected for the collimation error. For $T = 352°C$ and a wider range of h^2-values, the results are presented in Fig. 6. We shall not discuss the details of this plot but it becomes rather obvious that for the discussion of the O.-Z. theory only the range up

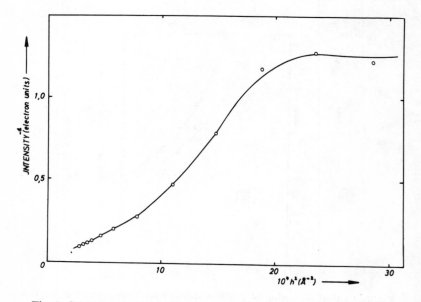

Fig. 6. Critical scattering of X-rays from the alloy Al-Zn at $T = 357°C$ over a wide range of h.

to $h^2 \approx 5 \times 10^{-4}$ Å^{-2} should be considered. Furthermore it is interesting to note that the general shape of the curve is very similar to that found by Debye, Caulfield and Bashaw (26) for the system perfluorotributylamine-isopentane.

The small-angle results at four temperatures above the critical temperature are shown in Fig. 7. It is seen that, within the limits of experimental error, we have indeed a linear behavior up to $h^2 \approx 4 \times 10^{-4}$ Å^{-2}. The subsequent bending upward can be understood on the basis of the O.-Z. theory if Eq. 68 is used instead of Eq. 69. Furthermore the initial slope increases monotonically when approaching the critical point. This is in agreement with the results obtained from light-scattering measurements (26, 28), and it is to be expected theoretically on the basis of Eq. 64 which states that the initial slope is proportional to the square of the correlation length of the

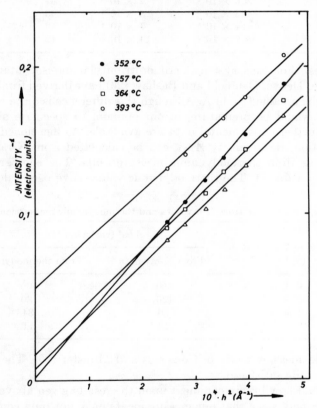

Fig. 7. Critical scattering of X-rays from the alloy Al-Zn at four temperatures. Ornstein-Zernike plot.

reference spectrum. The values of the parameters A and κ_1 used in Fig. 7 as well as the corresponding correlation lengths l and r_c have been listed in Table I. As is to be expected from the considerations in the previous section, r_c increases much more strongly than l when approaching the critical point. The values of l, however, are surprisingly high in comparison with the values obtained for liquid

TABLE I
Values of the parameters A and κ_1. Correlation lengths.

$T - T_c$ (°C)	A (Å · molecule^{-1})	κ_1 (Å^{-1})	l (Å)	r_c (Å)
0.5	3.6×10^{-3}	2×10^{-3}	47	500
5.5	4.3×10^{-3}	5×10^{-3}	43	200
12.5	4.4×10^{-3}	7×10^{-3}	43	140
41.5	4.6×10^{-3}	14×10^{-3}	42	70

mixtures. For the system perfluorotributylamine-isopentane, for instance, Debye, Caulfield and Bashaw (26) have derived from X-ray scattering a value $l = 14$ Å. Although this difference is quite remarkable, it would be premature, in our opinion, to speculate about it before further experimental data are available. As mentioned in the last section, the quantity A/κ_1^2 can be calculated from X-ray data as well as from thermodynamic measurements. The comparison is shown in Table II. The thermodynamic values have been calculated

TABLE II
The quantity A/κ_1^2 from X-ray data and thermodynamic measurements.

$T - T_c$ (°C)	A/κ_1^2 (cm^3 mole^{-1})	
	From X-ray data	From thermodynamics
0.5	600	800
5.5	120	80
12.5	50	34
41.5	16	11

from the measurements of Corsepius and Münster (29). The agreement is quite satisfactory.

We shall now briefly comment upon the results given above. First of all we may state that our measurements have not only confirmed the experimental results of Münster and Sagel (which have not been discussed in this paper) but moreover, by extension to the true small-angle region, have definitely proved the existence of the critical scattering of X-rays from binary alloys. Furthermore it has been shown that our results can be interpreted consistently on the basis of the O.-Z. theory and the existing thermodynamic data. On the other hand we want to stress that the accuracy of our measurements is not yet sufficient to follow the increase of critical scattering in detail. Indeed even the differences between the curves for 352°C, 357°C and 364°C are near the limits of experimental error. As can be seen from Figs. 5 and 7, this is due in part to the fact that, as a consequence of the "crossing over," significant differences are to be expected only for scattering angles below 15'. Similarly, our present results do not yet prove the validity of the O.-Z. theory near the critical point. It is obvious that both problems could be attacked successfully only if the measurements are extended to smaller angles and at the same time the accuracy of X-ray readings and temperature control is considerably improved. This will be the goal of future research in this field.

ACKNOWLEDGEMENT

I want to express my gratitude to Mr. U. Würz who performed the measurements and numerical calculations.

BIBLIOGRAPHY

1. Avenarius, M., Ann. Physik, [2], **151**, 303 (1874).
2. Guthrie, F., Phil. Mag., [5], **18**, 30, 497, 504 (1884).
3. Friedländer, J., Z. physik. Chem., **38**, 385 (1901).
4. Travers, M. W. and Usher, F. L., ibid., **57**, 365 (1907).
5. Smoluchowski, M. v., Ann. Physik, [4], **25**, 205 (1908).
6. Einstein, A., ibid., [4], **33**, 1275 (1910).
7. Ornstein, L. S. and Zernike, F., Proc. Acad. Sci. (Amsterdam), **17**, 793 (1914).
8. Münster, A., "Critical Fluctuations" in *Fluctuation Phenomena in Solids*, R. Burgess, ed., Acad. Press, New York, 1965.
9. Krivoglaz, M. A., Zh. Eksperim. i. Teor. Fiz., **31**, 625 (1656).
10. Münster, A. and Sagel, K., Z. physik. Chem. (Frankfurt), **12**, 145 (1957).
11. Münster, A. and Sagel, K., Naturwiss., **44**, 535 (1957).
12. Münster, A. and Sagel, K., Mol. Phys., **1**, 23 (1958).
13. Münster, A., "Statistische Thermodynamik kondensierter Phasen" in *Encyclopedia of Physics* (S. Flügge, ed.), vol. XIII, Springer-Verlag, Berlin, 1962.
14. Münster, A. and Sagel, K., Z. physik. Chem. (Frankfurt), **22**, 81 (1959).
15. Kirkwood, J. G. and Buff, F., J. Chem. Phys., **19**, 774 (1951).
16. von Laue, M., Ann. Physik, [4], **56**, 497 (1918).
17. Wilchinsky, Z. W., J. Appl. Phys., **15**, 806 (1944).
18. Guinier, A. and Griffoul, R., Compt. Rend., **221**, 555 (1945).
19. Cowley, J. M., J. Appl. Phys., **21**, 24 (1950).
20. Warren, B. E., Averbach, B. L. and Roberts, B. W., ibid., **22**, 1493 (1951).
21. Hosemann, R. and Bagchi, S. N., *Direct Analysis of Diffraction by Matter*, North-Holland Publishing Co., Amsterdam, 1962.
22. Münster, A. and Schneeweiss, Ch., Z. physik. Chem. (Frankfurt), **37**, 353 (1963).

23. Münster, A. and Würz, U. (to be published) (1965).
24. Münster, A. and Sagel, K., Z. physik. Chem. (Frankfurt), **7,** 296 (1956).
25. Frisch, H. and Brady, G. W., J. Chem. Phys., **37,** 1514 (1962).
26. Debye, P., Caulfield, D. and Bashaw, J., ibid., **41,** 3051 (1964).
27. Gerold, V., Acta Cryst., **10,** 290 (1957).
28. Debye, P., Chu, B. and Kaufman, D., J. Chem. Phys., **36,** 3378 (1962).
29. Corsepius, H. and Münster, A., Z. physik. Chem. (Frankfurt), **22,** 1 (1959).

Small-Angle X-Ray Scattering from the Perfluoroheptane-Isooctane System in the Critical Region

GEORGE W. BRADY

Bell Telephone Laboratories, Murray Hill, New Jersey 07971

and

D. McINTYRE, M. E. MYERS, JR., AND A. M. WIMS

National Bureau of Standards, Washington, D. C. 20234

INTRODUCTION

The critical state, which can be defined as a region in which the thermodynamic properties characterizing a system undergo discontinuous changes, has been a challenging subject of study for scientists of all disciplines for many years. In fact, no textbook of statistical mechanics is complete without a thorough treatment of phase transitions and critical point phenomena. But over the last few years, more advanced theoretical and experimental techniques have been developed and have spurred research in this field. The more sophisticated statistical mechanical techniques of cluster algebra (1) and Ising lattice gas models (2) have extended the earlier theories, many of which were either heuristic or classically thermodynamic in nature. At the same time the remarkable improvement in experimental techniques has suggested new studies which clearly distinguish between the different theories. A review article of the recent developments in critical X-ray scattering by Brumberger (3) summarizes the more recent experimental findings, and gives the nonworker in this field an idea of the technical difficulty of the measurements, combining as they do the problems of obtaining accurate temperature control, high purity materials, and low intensity X-ray measurements at very small scattering angles.

In this paper, we will present a brief discussion of the newer theories of the critical transformation and their relation to the scattering of electromagnetic radiation, which is one of the most direct methods for studying the critical phenomena because the correlation functions can be deduced from the intensity curves. We have chosen to study the low-angle X-ray diffraction from the system perfluoroheptane-i-octane in the critical region. This system has a convenient critical consolute temperature (23.78°C), a relatively low absorption ($\mu = 0.31$), and an appreciable difference in electron density, hence scattering power, of the two components.

THEORY

In a two-component system of molecules A and B, the scattering intensity in electron units, I_{eu}, is given by the expression (4)

$$I_{eu} = x_A f_A^2 + x_B f_B^2 + 4\pi\rho \int_0^\infty \left[r^2 \{x_A^2 f_A^2 [g_{AA}(r) - 1] + x_B^2 f_B^2 [g_{BB}(r) - 1] + 2x_A x_B f_A f_B [g_{AB}(r) - 1]\} \frac{\sin sr}{sr} \right] dr \quad 1.$$

where x_A and x_B are the mole fractions of the two species in the irradiated volume V containing N molecules, f is the molecular scattering factor, the $g_{\mu\nu}$ terms, with $\mu, \nu = A, B$, are the three independent radial distribution functions, and $s = 4\pi\lambda^{-1} \sin \theta$, where λ is the wavelength and θ is one-half the scattering angle. This expression assumes spherical symmetry of the molecules A and B, both in their scattering power and their radial distribution. In a single component system, the expression reduces to

$$\frac{I_{eu}}{f^2} = 1 + \rho G(s) = I(s) \quad 2.$$

where

$$G(s) = 4\pi \int_0^\infty r^2 [g(r) - 1] \frac{\sin sr}{sr} dr \quad 3.$$

In certain cases, Eq. 2 can, by Fourier inversion, be used to determine the characteristic correlation function of the system. We are interested in the behavior of $I(s)$ in the neighborhood of $s = 0$, that is, in the behavior of $g(r)$ at large r. In this region, beyond a certain value of r the order of magnitude of the three radial distribution functions is comparable and if the system is chosen such that the scattering power of one of the components predominates, then in the region of low s Eq. 2 is satisfied with $\rho = \rho_A = N_A/V$, $f(s) = f_A(s)$ and $g(r) = g_{AA}(r)$. Because of the much greater scattering power of fluorine compared to hydrogen, our fluorocarbon-hydrocarbon mixture fulfills these requirements, at least for the purposes of outlining the theoretical approach to the phenomenon of scattering.

Thus, Eqs. 2 and 3 illustrate that the problem devolves into a study of the behavior of $g(r)$ in the vicinity of the critical consolute temperature. As an example, we can consider the case of a one-component system, such as a gas at its critical point. It can be shown that for this system the compressibility is related to the scattering at zero angle, $I(0)$, by the relation (5),

$$kT \left(\frac{\partial P}{\partial \rho}\right)_T^{-1} = 1 + \rho \int_0^\infty [g(r) - 1] 4\pi r^2 \, dr \qquad 4.$$

since for small s, $\sin sr/sr \to 1$. At the critical point $(\partial \rho/\partial P)_T$ is infinite and as a result $g(r)$ decays sufficiently slowly to its asymptotic value of unity with increasing r to render the integral in Eq. 4 divergent. The very long range of correlation implied by this is what characterizes the scattering of radiation in the critical region. In a two-component liquid system the analog (6) of the compressibility is the term $(\partial \ln a_A/\partial \phi_B)^{-1}$. a_A is the activity of component A, ϕ_B is the volume fraction of component B, and the derivative is the slope of the activity isotherm. At the critical point, the first derivative becomes zero and the isotherm has a point of inflection which makes the second derivative also vanish. Thus the property of $g(r)$ which is of interest is the r-dependence of its decay as a function of temperature. This dependence is implied in the low-angle scattering measurements through Eqs. 2 and 3.

The first major attempt at a solution of the problem was undertaken by Ornstein and Zernike (O.-Z.) (7). Their original approach was mathematical in nature and involved the postulate that the correlation $G(r)$ between two molecules can be described by a short-range correlation function $C(r)$ which measures the direct influence of molecule 1 on 2, and a long-range indirect contribution involving the influence of molecule 1 on a third one which in turn exerts its influence on molecule 2. $G(r)$ and $C(r)$ are not independent (8) and can be shown to be related at the critical point by the expression

$$1 + \rho G(s) = 1/[1 - \rho C(s)] \qquad 2a.$$

and thus through Eq. 4 to the divergence of the compressibility, $I(0)/kT$ by

$$1 - \rho C(0) = 1 - \rho \int C(r) \, dr = 0 \qquad 4a.$$

We have written down these equations to point out that implicit in the O.-Z. formulation is the fact that $C(r)$ is finite at the critical point as is clearly shown by Eq. 4a, and is shorter-ranged than $G(r)$. Further extension of the theory centers around the existence of the higher moments of $C(r)$. From a more thermodynamic viewpoint, the implication is that the local free energy of the fluid can be expanded in terms of the local density and its spatial gradients in a region close to and including the critical point and therefore that these derivatives exist at the critical point. The treatment is fairly involved and for a fuller discussion the reader is referred to the excellent review article by Fisher (8). We will return later to a critique of the O.-Z. theory.

Debye (9) has applied the O.-Z. approach to regular solution theory to develop an approximate theory of critical opalescence, which leads to an attractive and experimentally tractable formulation of the critical scattering problem. Using his well-known expression for the correlation function $G(r)$, which relates fluctuations at two points A and B a distance r apart, $\langle \delta\eta_A \delta\eta_B \rangle_{Av} = G(r)\langle \delta\eta^2 \rangle_{Av}$ and inserting it in Eq. 2, one can expand for small s to get

$$\frac{I}{I_{max}} = 1 - \frac{s^2}{6} L^2 + \cdots \qquad 5.$$

where L is a persistence length defined by

$$L^2 = \frac{\int r^2 G(r)\, d\tau}{\int G(r)\, d\tau} \qquad 6.$$

with $d\tau$ a volume element in the system.

To derive a relation between Eq. 5 and the actual intermolecular range of forces, Debye expands the potential energy of a central molecule in terms of the density. Thus, if the attraction between two molecules can be written as $-E(r)$, the potential energy of one molecule in the field of its n neighbors is

$$E = -\int nE(r) 4\pi r^2\, dr = -\int nE(r)\, d\tau \qquad 7.$$

The range of integration is taken to be from the point of contact of two molecules to infinity. Expanding in powers of n we have for molecules A

$$E_{AA} = -n_A \int E_{AA}(r)\, d\tau - \frac{\Delta n_A}{6} \int r^2 E_{AA}(r)\, d\tau \qquad 8.$$

where Δ is the Laplace operator, and the coordinate system has its origin at the central molecule. Similar expressions result for interactions of molecules A with B, B with B, and B with A. This leads to an expression for the total potential energy of the system as the sum of four terms E_{AA}, E_{AB}, E_{BA}, and E_{BB}; clearly E_{AB} and E_{BA} are equivalent. In addition, Debye defines a length l_{AA} by the expression

$$\frac{l_{AA}{}^2 W_{AA}}{W_{AA}} = \frac{\int r^2 E_{AA}\, d\tau}{\int E_{AA}\, d\tau} \qquad 9.$$

where each term

$$E_{\mu\nu} = W_{\mu\nu}\left[n_\mu + l_{\mu\mu}{}^2 \frac{\Delta n_\mu}{6}\right] \qquad 10.$$

The excess energy may then be obtained by subtracting from the sum of these terms (integrated over the volume) the energy of the two separate components. The excess entropy of the mixture over that of the separate components can be written in the usual way as an integral over $d\tau$ of the sum of terms $(\Phi_\mu/\omega_\mu)k \ln \Phi_\mu$ $(\mu = A,B)$ where Φ_μ and ω_μ are the volume fraction and molecular volume of component μ. The excess free energy is then obtained from the difference between the two. We can represent the thermal fluctuations in terms of the volume fractions as $\Phi_1 + \eta$ and $\Phi_2 - \eta$, since by definition the sum of the fractions must be unity and the integral of η over the volume is zero. Then on converting n in Eq. 9 into volume fractions and remembering that

$$\int \eta \Delta \eta\, d\tau = -\int \mathrm{grad}^2\eta\, d\tau + \sigma \qquad 11.$$

where σ is a surface term which can be ignored, the excess free energy is

$$\frac{F}{kT} = \int d\tau \left[\frac{\Phi_1}{\omega_1}\ln \Phi_1 + \frac{\Phi_2}{\omega_2} + \frac{\Omega}{2kT}\Phi_1\Phi_2 \right.$$
$$\left. + \left(\frac{1}{\omega_1\Phi_1} + \frac{1}{\omega_2\Phi_2} - \frac{\Omega}{kT}\right)\frac{\eta^2}{2} + \frac{H}{2kT}\mathrm{grad}^2\eta\right] \qquad 12.$$

In this expression Ω and H are

$$\Omega = \frac{W_{AA}}{\omega_A{}^2} + \frac{W_{BB}}{\omega_B{}^2} - 2\frac{W_{AB}}{\omega_A\omega_B}$$

$$H = \frac{1}{6}\left[\frac{W_{AA}}{\omega_A{}^2}l_{AA}{}^2 + \frac{W_{BB}}{\omega_B{}^2}l_{BB}{}^2 - 2\frac{W_{AB}}{\omega_A\omega_B}l_{AB}{}^2\right] \qquad 13.$$

Ω is an averaged cohesive energy density and the ratio H/Ω is a measure of the range of intermolecular forces. The two final steps in the Debye treatment of critical scattering are to introduce the Brillouin concept of fluctuations, which treats them as a group of plane sonic waves whose wavelength Λ can be visualized as the spacing between a series of parallel planes from which the measuring wavelength λ is diffracted at an angle θ, according to Bragg's law. There is thus for each angle only one corresponding periodic fluctuation in density. One can write for the density

$$\eta = \eta_0 \sin \frac{2\pi x}{\Lambda}$$

where $2\pi/\Lambda = s$ and x is a coordinate normal to the wave front. The integral over the volume in Eq. 12 can then be evaluated. Finally taking into account that the first and second derivatives of the activity isotherm are zero at the critical point and the classical equation of Einstein for the angular independent scattered radiation from liquids (10), there results for the intensity of scattering near the critical point the relation:

$$\frac{I}{I_M} = \frac{1}{1 + \frac{1}{(\tau - 1)} \frac{H}{\Omega} s^2} \qquad 14.$$

Putting $H/\Omega = l^2/6$ and expanding for small s we get:

$$\frac{I}{I_M} = 1 - \frac{l^2 s^2}{6(\tau - 1)} + \cdots \qquad 15.$$

which when combined with Eq. 5 leads to the relation

$$L^2 = \frac{l^2}{(\tau - 1)} \qquad 16.$$

between the persistence length and the range of intermolecular forces. We note how Eq. 16 gives an explicit and simple formulation of the relationship between $G(r)$ and the short-range direct correlation distance l.

It should be emphasized again, as Debye does, that this theory is an approximation in two important respects. First it takes into account only the first term in the density expansion, and secondly it assumes that the thermodynamic properties of liquids can be adequately defined in terms of a van der Waals or regular solution approach. Accepting these limitations, its great advantages are the insight it gives into critical phenomena and the stimulus it has given to their experimental investigation. This has resulted because of the simple form of the final equations. According to Eq. 14 a plot of I^{-1} against s^2 should result in a series of parallel lines whose slopes are proportional to l^2 and whose intercepts on the ordinate are directly proportional to ΔT; at $T = T_c$ the line should pass through zero. A considerable amount of data has been accumulated over the past few years from light scattering and small-angle X-ray measurements. The former unfortunately suffer from significant errors due to multiple scattering and the latter from the difficulty of extending the measurements into sufficiently small angles. However, the data show that over a considerable range of s^2, straight lines whose slopes give reasonable values of l are obtained. On the other hand, the lines do not always remain parallel as the temperature is changed, and there has been no definite case where it has been established that the curve for $\Delta T = 0$ passes through the origin. Indeed the X-ray measurements have seemed to indicate a turning down of the curves at the smallest measurable angle (21).

Returning now to a consideration of the O.-Z. theory, some fundamental objections to its exact validity have been raised. Frisch and Stillinger (11) have pointed out that there is an ambiguity in the predicted dimensional dependence of $g(r)$. In d dimensions the O.-Z. theory implies that at the critical point,

$$G(r) = g(r) - 1 \approx A \frac{e^{-\beta r}}{r^{d-2}} \qquad 17.$$

with $d > 2$ and $\beta = \beta(T - T_c) \to 0$ as $T \to T_c$; A is a constant. In two dimensions, however, the result is

$$G(r) \approx A \ln r e^{-\beta r} \qquad 18.$$

which is a physically impossible result. Since there are real two-dimensional systems, such as adsorbed films, which exhibit phase transitions, they should also satisfy the condition that $[g(r) - 1]$ vanishes at infinity, as it does in the three-dimensional case. Thus in a fundamental sense the O.-Z. predictions can be shown to be inadequate since they exclude a large class of known critical phenomena. In view of this obvious defect in the theory it is necessary to examine the alternative methods of determining $G(r)$.

In principle there does in fact exist a rigorous method of calculating

$G(r)$ in terms of the cluster theory of fluids (1), where $G(r)$ formally is shown to be the solution of an integral equation. However, there is in the kernel of these equations an infinite set of cluster diagrams which have no nodes, articulation points or subdiagrams, and thus are not reducible to the more simple series and parallel chains which can be computed. Since this prevents a rigorous solution of the equation, approximate methods have been resorted to. Green (12) adopted the procedure of neglecting them entirely and found that for a three-dimensional fluid

$$G(r) \sim \frac{\text{const}}{r^2} \qquad 19.$$

whose r-dependence is widely at variance with the classical O.-Z. result. The fact that this leaves out a large set of the Mayer (1) cluster bonds $[f_{ij} = e^{-\varphi(r_{ij})kT} - 1]$ and, in consequence, their contribution to the correlation function, introduces an element of uncertainty into the calculation, but it is extremely doubtful that this neglect could account for such a large deviation.

Stillinger and Frisch (11) in an attempt to test the Green result have applied the cluster technique to a two-dimensional lattice gas, where the molecular positions are restricted to periodically located sites of a planar array, which may however assume any number of configurations (square, triangular, etc.). The reason for choosing this approach is that certain exact results are available for comparison. Kaufman and Onsager (13), considering only nearest neighbor interactions corresponding to the two-dimensional Ising model of ferromagnetism have found that for a square planar lattice, $G(r)$ at the critical point is proportional to $r^{-\frac{1}{4}}$. Under equivalent conditions, application of the cluster theory leads to the conclusion that

$$G(r) \sim \frac{\text{const}}{r^{\frac{4}{3}}} \qquad 20.$$

Thus unlike the prediction of the O.-Z. theory, this two-dimensional result is qualitatively correct, although the exponential decay is too rapid. Finally, Stillinger and Frisch show that the wider-sense irreducible graphs (IG) neglected by Green would have to be of the form

$$\text{IG} \sim \frac{\text{const}}{r^{\frac{1}{2}}} \qquad 20.$$

for large r, in order for the cluster theory to produce the exact solution of Kaufman and Onsager.

To summarize, the cluster algebra and lattice gas techniques appear to be qualitatively correct in their predictions about the critical behavior of $G(r)$. Quantitatively the former suffers from the great difficulty of evaluating the contribution of the more complicated diagrams, although some progress in this direction has been made by Domb and Sykes (14) who have attempted a summation for the Ising model in three dimensions of *all* clusters by evaluating each of them up to some finite order then extrapolating to infinite order. Their results appear to indicate that the critical pair correlation function is asymptotically proportional to $r^{-\frac{7}{4}}$, which is in line with the conjectures outlined above. The latter approach, i.e., the lattice gas, has the disadvantage that it is probably too idealized, because of the rigidity of the model which is its basic postulate. It is possible to extend the requirement of a fixed array of lattice points, to take account of the partially random nature of the liquid state by the simple expedient of assuming that some calculable number of sites be vacant. But this extension does not seem to be satisfactory in describing the thermal movement of molecules, since it requires that this complex motion be restricted to jumps from one fixed site to another.

The net result of all these calculations leads to the conclusion that the r-dependence of $g(r)$ at the critical point does not follow the classical form predicted by O.-Z. but should on the basis of the lattice gas calculations be described by the relation

$$G(r) \sim \frac{\text{const}}{r^{1+\epsilon}} \qquad 21.$$

with $1 \geq \epsilon > 0$.

In Fourier transform notation, this leads to the following expression for the small-angle X-ray scattering

$$I(s) \sim \frac{1}{[\beta^2(T) + s^2]^{1-\epsilon/2}} \qquad 22.$$

where $\beta(T) \sim \beta(T - T_c)^\nu$ is a reciprocal correlation length, which vanishes as $T \to T_c$, and ν is related to the nature of the corresponding divergence of the compressibility as $1/(T - T_c)^\gamma$, through the relation $\nu(2 - \epsilon) = \gamma$. [A discussion of the relationship between the exponents is too detailed for this sketchy review and the reader is again referred to the article by Fisher (8).] Equation 22 is experimentally usable, since a plot of $I^{-1/(1-\epsilon/2)}$ against s^2 should be linear and therefore a determination of ϵ is possible. As is evident from the discussion, this in turn should be of great help in the development of more realistic models and the more rigorous theories of fluids.

EXPERIMENTAL DETAILS

The measurements were made with two different experimental arrangements. The first was a slightly modified version of an apparatus previously described (15). The slit system was improved by increasing the target-to-sample distance from 100 mm to 150 mm, and the sample-to-counter distance from 200 mm to 300 mm. With these improved optics it proved possible to make measurements to within 5 minutes of the main beam. The cell and temperature control equipment were designed by J. I. Petz (16). This device uses a Peltier couple which permits both heating and cooling of the sample, and in consequence an easy adjustment of the temperature in the range between 12°C and 50°C is feasible. The temperature is regulated to 0.03°C. Background corrections were made as previously described.

The new experimental results which will be presented and discussed in the following section extend the earlier measurements to much smaller angles (5×10^{-4} radians) with much more accurate temperature control than that attainable with the previous apparatus. These experiments were made at the National Bureau of Standards. For this purpose a Kratky camera, fitted with a proportional counter and associated amplifying and scaling circuits, was used. The temperature control system consisted essentially of a water bath with a capacity of 100 gallons maintained at a temperature constant to within at least 0.003°C, as measured by a platinum resistance thermometer. From this reservoir the water was circulated at a rate of 1 liter per minute through a copper block surrounding the X-ray cell. The block was drilled out laterally to allow insertion of the cell and had an opening in the direction of the incident and scattered beam to allow free passage of the X-rays through it. The cell itself had a path length of 1 mm and was fitted with 1.6 mil mica windows. The space between the faces of the copper block and the windows was enclosed by cementing 0.3 mil "Mylar" film to the block, thus ensuring that any temperature gradients between the assembly and the surroundings would be across the "Mylar" film. This intervening space was filled with air.*

---

* This apparatus will be described in detail in a separate publication.

The intensity data were corrected for the finite slit by the method of Schmidt and Hight (17). This method is based on an analytical solution of the slit correction formula which requires a Gaussian intensity distribution for the incident beam. The trapezoidal profile of the Kratky camera beam was accordingly approximated by a Gaussian curve enclosing the same area.

The perfluoroheptane was subjected to a rigorous purification process which consisted of placing it in sealed tubes with added KOH, heating for 48 hours at 130°C, washing with distilled water and then drying over "Drierite." This procedure was repeated seven times, after which the perfluoroheptane was distilled. We are indebted to J. L. Lundberg and M. Y. Hellman of Bell Telephone Laboratories for their help in carrying out this lengthy procedure. The consolute curve, also measured by Lundberg and Hellman, is in fair agreement with that determined previously by Hildebrand, Benesi and Fisher (18) and indicates that the critical concentration is 41.5 mole % C_7F_{16} and that the critical temperature is 23.78°C.

RESULTS AND DISCUSSION

We will first discuss briefly some earlier results (15, 19, 20, 21, 22) obtained on the C_7F_{16}-i-C_8H_{18} system to provide a background for a presentation of the latest findings from our investigation in the region of much smaller angles than were previously studied. In this way we hope to present a more complete picture of the complex phenomena which occur, and also indicate how our knowledge has evolved during the course of the research.

A very important property of binary liquid systems is measured by the Zimm clustering function, K_{AA}. This function has an exact molecular definition. It is the double volume integral of the pair correlation function of species A, divided by the total volume. It is directly related to the activity isotherm of the system as shown in the following expression,

$$\frac{K_{AA}}{v_A} = -\left[(1 - \Phi_A)\frac{\partial}{\partial a_A}\left(\frac{a_A}{\Phi_A}\right) - 1\right] \qquad 23.$$

where Φ_A and a_A are respectively the volume fraction and activity of component A, and v_A is its molar volume. The quantity $\Phi_A(K_{AA}/v_A)$ is the mean number of molecules A in excess of the random expectation in the neighborhood of a given molecule A, and is thus a direct measure of their clustering tendency. For an ideal solution, the activity is directly proportional to the volume fraction and the clustering function is equal to minus one molecular volume, i.e., the reference molecule itself. If it is positive and large, the molecules tend to form large clusters. Figure 1 shows a plot of this function at 30°C, obtained numerically using the vapor pressure and activity data of Mueller and Lewis (23). We note that even away from the critical temperature clustering exists, increasing considerably in magnitude in a $\sim 25\%$ composition interval around the critical composition, and this is associated with the increase in low-angle scat-

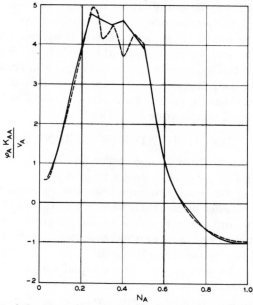

Fig. 1. Plot of the Zimm clustering integral vs. mole fraction (NA) of C_7F_{16} in C_7F_{16}-i-C_8H_{18} solutions at 30°C determined by two methods of numerical integration of the data of Mueller and Lewis [J. Chem. Phys., **37**, 1514 (1962)].

tering observed in this region. Unfortunately, there do not exist enough thermodynamic data as a function of temperature throughout the critical region to evaluate the cluster integral and relate it directly to the X-ray results. The clusters, if independent quasi-particles, would have a mean dimension related to the persistence length L, which can be obtained directly from the scattering curves. However, this dimension is defined more generally as being a correlation distance of two density fluctuations without specifying any particulate character to it. Nevertheless, under certain conditions (21) the curves do indeed furnish strong evidence for the existence of separate aggregates. This can be seen in Fig. 2, where a series of scattering curves of a 50–50 mole % solution is plotted for different temperatures. Curves a, b and c show well-defined interference maxima which can only be accounted for by an "interparticle" type of scattering. (A fuller discussion of this phenomenon can be found in the original publication.) Even more interesting perhaps is the fact that curve d, the one closest to the separation temperature, does

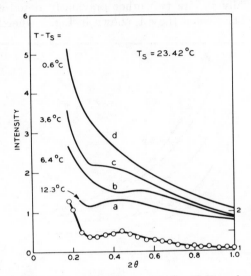

Fig. 2. Small-angle X-ray scattering curves for 50-50 mole % C_7F_{16}-6-C_8H_{18}. The curves have been plotted without experimental points, because background corrections have been applied. To give an idea of the actual data, curve e represents the actual data of curve a. The ordinate for curve e appears on the right [J. Chem. Phys., **40**, 2747 (1964)].

not exhibit any maximum and a plot of I^{-1} (or $I^{-1/2}$) against s^2 gives a straight line, characteristic of a random series of fluctuations. This behavior indicates that there is a qualitative change in the correlation of the molecules somewhere in the critical region. The local fluctuations away from the critical region are cluster-like in character and appear to change to an extended mutually interpenetrating structure of the two components.

Thus we see that the very complex mechanisms that the system undergoes in the critical region have as a precursor a tendency to incipient microscopic segregation, characterized directly by the shape of the scattering curves or, equivalently, by the form of the correlation function. The clustering function predicts this type of behavior well above the critical temperature. In terms of regular solution theory the clustering tendency is associated with the relative magnitude of the two cohesive energy densities, which plays an important role in the Debye development (cf. Eq. 13) of the critical scattering theory.

We wish finally to cite two other previously noted experimental observations. The first of these is shown in Fig. 3 wherein is plotted

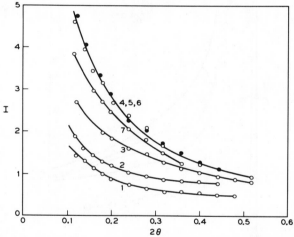

Fig. 3. A typical set of intensity curves measured as the temperature is lowered. The curve labeled 4, 5 and 6 is the superimposition of three curves [J. Chem. Phys., **35**, 2234 (1961)].

a typical set of intensity curves measured as the temperature is lowered. The concentration is about 30 mole % C_7F_{16}. As the consolute temperature is approached the intensity increases rapidly at first (curves 1, 2, 3, 4) to a maximum value, after which point a region of nearly constant intensity is reached. In this region the intensity curves superimpose (curves 4, 5, 6), although a careful examination shows that the intensity probably does increase slightly, within the precision of the measurements. The temperature interval over which this effect is observed is of the order of $\sim$1.0 degrees depending on the composition. Following this rise the intensity curve begins to decrease again (curve 7).

Figure 4 illustrates this effect even more clearly; here the apparent persistence length L, determined from Guinier plots of $\ln I$ against s^2, is plotted as a function of temperature. (It might be pointed out in passing that this is a convenient way to determine L. For small values of s^2, $I/I_M = 1 - s^2\bar{R}^2/3$, and by comparison with the Debye expression in Eq. 5 it is seen that $L^2 = 2\bar{R}^2$. Of course, this is only a mathematical equivalence and not a physical one.) Corresponding to the intensity effect we notice a flat portion on the curve, again extending over an equivalent temperature interval; it is preceded by an increase, and followed by a decrease in the long-

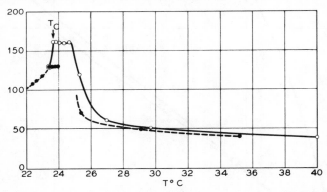

Fig. 4. Persistence lengths L plotted vs. a function of temperature. T_c is the temperature at which two phases appear. Solid line, 40% C_7F_{16}, dashed line, 30% C_7F_{16} [J. Chem. Phys., **35**, 2234 (1961)].

range correlation parameter. The phase separation temperature for the two solutions in each case coincides with the lower temperature end of the flat region. After separation the ordinate of the curve for the higher concentration, 40% C_7F_{16} (23.75°C), decreases rapidly until the separation temperature (23.44°C) of the lower concentration, 30% C_7F_{16}, is reached. After this the curves superimpose. This is to be expected since once the consolute curve is reached the solution separates into two phases whose composition is uniquely determined by the temperature and is independent of the starting composition.

Lastly, Fig. 5 shows an O.-Z. plot of $[I(s) - 1]^{-1}$ against s^2. {In the critical region $[I(s) - 1] \cong I(s)$.} The points represent the reciprocal relative scattered intensity for a 30 mole % C_7F_{16} solution. The curves

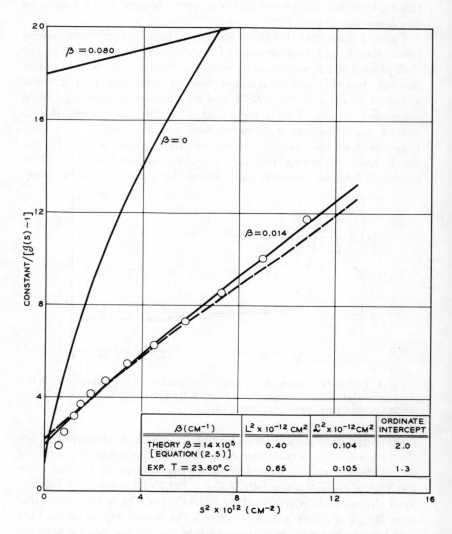

Fig. 5. O.-Z. plots of $I(s) - 1$ vs. s^2 according to Eq. 24 for various values of β, $\epsilon = 0.75$, $b = 10$. The dashed line is a corresponding plot for a more tractable but nearly equivalent form of equation 24, for $\beta = 0.014$, $\epsilon - 0.75$ and $B = 5$ (see text and references) [J. Chem. Phys., **37**, 1514 (1962)].

are calculated from Eqs. 2 and 3 using for the correlation function the expression*

$$G(r;\beta,b) = \frac{A_\epsilon e^{-\beta r}}{r^{1+\epsilon}} (1 + b\beta r)^\epsilon \qquad 24.$$

where b is a constant, and A_ϵ can be considered as constant in the vicinity of $\beta = 0$. We note that the apparent linear plot actually has an S-shaped character bending upwards at large s, probably due to contributions from shorter-range correlations, and curving downwards around $s^2 \sim \beta^2$. Unfortunately, the data are least reliable in this latter region due to a rapidly increasing background correction caused by the proximity of the main beam. However, the experimental validity of this downward trend is attested to by the results of Fig. 3. From Eqs. 2 and 22, and the definition of β, the s-space value of $G(0)$ should be infinite at $T = T_c$ only, and clearly since the curves superimpose over a finite range of ΔT, there must exist a region at very low s-values where the curves separate from one another in such a way that only the reciprocal critical curve has a zero intercept.

In summary, our experiments have established (a) that significant clustering exists in the solution at temperatures well above the critical; this is characterized by a correlation length calculable in certain cases from the scattering curves (b) that there is a change in the nature of the correlation as one approaches very close to the consolute temperature. It is hoped that further experiments will provide sufficient evidence to allow a more exact physical interpretation of these effects; in particular, a quantitative relation between the Zimm cluster integral and the correlation function would be extremely useful. (c) There is a temperature region of apparently constant intensity extending into small values of s, whose O.-Z. plot shows a linear behavior over a considerable range and a characteristic downward bend at lowest s. Our subsequent discussion will be concerned with the scattering measurements in the angular region and temperature interval below this bend, because it is precisely in this region that deviations from classical behavior are to be expected.

The scattering curves as measured by the Kratky camera are

---

* This equation was proposed by Frisch and Brady (21) after consideration of the arguments of Fisher (8) and Green (12). For large b it reduces to an O.-Z. type of dependence on r. For $\epsilon = 1$, and small b or $\beta \to 0$, the Green expression results. For a detailed discussion of the equation and the significance of the quantities in the insert, see reference (22).

shown in Fig. 6. The vertical line at $2\theta = 1.8 \times 10^{-4}$ rad. indicates the lower angular limit attainable with the earlier apparatus, and the discussion up to now has been concerned with the interpretation of the results obtained in the region to the right of this line. Some striking features are at once evident. As ΔT decreases, the intensity at small angles increases, quite rapidly at first and then at a slower rate (for example curves a and b measured at ΔT values of 0.007° and 0.020° differ by only $\sim$4%). The region of constant intensity at

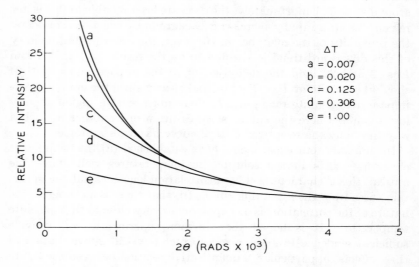

Fig. 6. X-ray scattered intensity curves measured with a Kratky camera. The minimum measurable value of the intensity was determined by the relative magnitude of the background correction. The curves have been corrected for window scattering, electronic background and main beam interference. Slit corrections have not been applied.

relatively larger angle is again present, although the temperature interval over which this occurs is smaller than that previously observed. The explanation for this is twofold. First, these curves were determined right at the critical concentration, whereas the previous measurements were made at concentrations of 30 and 40 mole % C_7F_{16}. Secondly, the purification of the C_7F_{16} was more thoroughly carried out as a result of greater experience with the technique. No systematic attempt was made to define exactly the extent of the temperature range in which there is constant scattering intensity. A preliminary experiment at $\Delta T = 0.70°$ indicated that the intensity showed the same characteristics as those of curves a to d. The lower angle portion of the intensity curves is characterized by breaks toward higher intensity which occur at successively smaller angle, consistent with the observation of Frisch and Brady (22) that these should occur at values of $s^2 \approx \beta^2$. Further, this temperature dependence supplements the preliminary observations of a downward trend in the reciprocal intensity plot of Fig. 5, and accounts as it were for

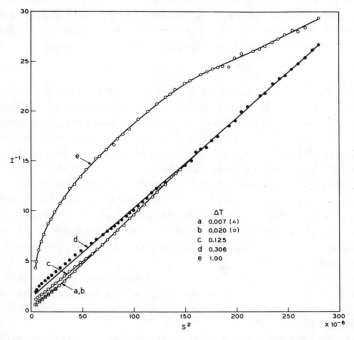

Fig. 7. O.-Z. plots of the data of Fig. 6 corrected for collimation error using the method of Schmidt and Hight.

the missing correlation of the earlier experiments. Consequently it is now possible to attempt a more complete physical interpretation of the data.

In Fig. 7, O.-Z. plots of the reciprocal intensity as a function of s^2 are displayed. We note that the curve for $\Delta T = 1.0°$ exhibits a completely different behavior from that of the rest of the curves. There appears to be a short linear region, followed by a gradual decrease at lower angle; the linear region is the typical O.-Z. type of result, but the rest of the curve cannot be accounted for in this way. The probable explanation is that at this temperature the critical conditions implicit in both the O.-Z. and lattice gas theories, particularly the nature of the long-range correlation function $G(r)$, do not apply, and that this temperature corresponds to the transition region of Fig. 2, where it was conjectured that there was a fundamental change in the correlation scheme of the system. We will content ourselves for the present with this qualitative statement and focus our analysis on the properties of the other curves. It is immediately apparent that as the critical temperature is approached, there are significant deviations from the classical (O.-Z.) behavior. Returning to Eq. 22,

$$I(s)^{-1} \sim (\beta^2 + s^2)^{1-\epsilon/2}$$

we note again that both the O.-Z. theory and the Debye extension of it predict that $\epsilon = 0$. If this prediction is not correct and ϵ is significantly different from 0, the O.-Z. plots should be convex downward for small s^2. The curve for $\Delta T = 0.306°$ is indeed linear over the whole region of the plot, except in the region of smallest s where there is (as in all the curves) a small but significant upward deviation. We attribute this increase either to multiple scattering or to errors resulting from using an approximate Gaussian slope for the beam profile in the slit correction operation. The other curves however do definitely show a marked downward curvature, which increases as ΔT (and consequently β) $\rightarrow 0$. That this is due to an increase in the value of ϵ is demonstrated in Fig. 8, where plots of $I^{-1/(1-\epsilon/2)}$, with ϵ chosen at each temperature as indicated, are shown. Starting from zero for curve d, it increases to 0.04 for curve c, and reaches a constant value of 0.10 for the two lowest temperatures, whose plots are virtually indistinguishable.

These results are interesting for several reasons. Most important, the lattice gas approach appears to be a satisfactory and fruitful method for critical state studies, since it is consistent with the experimental data. Some specific predictions of the theory are found to be at least semi-quantitatively verified. For example, as was conjectured by Stillinger and Frisch (11), Green's value for the r-dependence of the falling off of the correlation function is too high, not only in the two-dimensional case, which was the basis of the Stillinger-Frisch calculation, but also in three dimensions (12) and we conclude from this that his neglect of the wider-sense irreducible graphs is not justified. Further as predicted by Frisch and Brady [cf. Eq. 24 and Ref. (22)], and Fisher, the deviations from O.-Z. behavior occur very close to the critical temperature, whereas in a temperature interval reasonably far away from the critical the classical expression is valid. Finally, the value of the critical exponent is remarkably close to that calculated by Fisher (8) for the simple cubic lattice gas and is in accord with his conjecture that the three dimensional results should be closer to the classical prediction than those derived from the two-dimensional models.

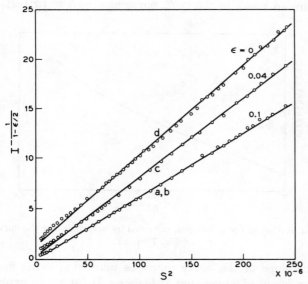

Fig. 8. O.-Z. plots of the data of Fig. 6, corrected for collimation error, with values of ϵ chosen to give straight line plots.

To complete our analysis of the results we now turn our attention to the zero-angle scattering intensity, $I(0)$. According to the lattice gas formulation the ordinate intercepts of the O.-Z. plots should reflect the deviation from classical behavior, resulting in a flattening of the curve as T approaches T_c, with zero slope for $\Delta T = 0$. This flattening results because the exponent in Eq. 22 differs from unity. An expansion of the right-hand side of the equation yields cross terms containing products of β and s. Thus the contribution of β^2 to the intensity is no longer limited to zero angle, but through these terms extends into a finite region of $s^2 > 0$. This plot is shown in Fig. 9, where the flattening out is evident. However, the curve does *not* pass through the origin. The reason for this, we feel, is that the scattering curves do not even now extend into small enough angle to allow a realistic extrapolation to be made.

In this respect we wish to comment briefly on some preliminary data obtained by Lundberg, Richelson, and Collins (24) on the same system from a study of the small-angle scattering of light of 5461 Å, in which they were able to extend the measurements into the range of $s^2 \sim 10^{-9}$. A noteworthy observation they made was that there is a pronounced break in the O.-Z. curve at $s^2 \sim 5 \times 10^{-9}$. Below this

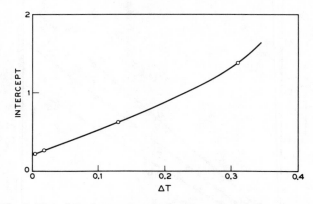

Fig. 9. Plot of the intercepts, obtained by extrapolating the curves of Fig. 7 vs. ΔT.

angle the slope of the curve increases markedly, being greater by roughly a factor of 5 over that at higher s. It cannot be explained by multiple scattering, whose effects have been minimized. Their results in effect supplement the X-ray results and are an extension of the downward curvature into smaller s. Taken in conjunction with Fig. 9, they explain why the intercept remains finite, since the rapid increase in slope they observe is probably a manifestation of the fact that Eq. 22 predicts that at T_c the inverse plot should approach the origin with infinite slope.

Since our system is a mixture of large organic molecules, chosen because of its experimental tractability, it is a legitimate question to ask how valid the results obtained from it would be for the regular hard sphere model on which the Ising lattice calculations are based.

This is difficult to answer as yet because of the lack of rigor in the theoretical calculations, but it seems safe to assume that the long-range $G(r)$ interaction should be relatively insensitive to the details of the local potential around each molecule, whose effects should be more directly manifested in the values of l, which through Eqs. 9 and 14 would affect the slope of the O.-Z. curves. $G(r)$ would be predominantly effective in modifying the shape of the scattering curve in the region of small s where its Fourier transform would be particularly sensitive to its long-range character. We would expect therefore that further study on other systems would confirm that our measured value of ϵ is a general one.

ACKNOWLEDGEMENTS

We gratefully acknowledge the assistance of P. Debye and M. S. Green, both for clarifying discussion and more fundamentally for the former's development of the modern theory of critical phenomena and the latter's basic analysis of the more questionable aspects of the O.-Z. theory. One of us (GWB) is indebted to H. L. Frisch and E. Helfand for their guidance in the theoretical aspects of the problem and particularly to F. H. Stillinger, whose lecture notes were the basis for the cluster and lattice gas discussion of the problem.

BIBLIOGRAPHY

1. Montroll, E. W. and Mayer, J. E., J. Chem. Phys., **9,** 626 (1941); Von Leeuwen, J. M. J., Groeneveld, J. and DeBoer, J., Physica, **25,** 792 (1959).
2. Hill, T. L., *Statistical Mechanics*, Chapter 7, McGraw-Hill Book Company, New York, 1956.
3. Brumberger, H., *Proceedings of the Conference on Phenomena in the Neighborhood of Critical Points*, National Bureau of Standards, Washington, D.C.; N. B. S. Miscellaneous Publication **273,** (1966).
4. Pearson, F. J. and Rushbrooke, G. S., Proc. Roy. Soc. (Edinburgh), **A64,** 305 (1957).
5. See Ref. 2, page 236 and Refs. therein.
6. Zimm, B. H., J. Phys. Chem., **54,** 1306 (1950).
7. Ornstein, L. S. and Zernike, F., Proc. Acad. Sci. (Amsterdam), **17,** 793 (1914).
8. Fisher, M. E., J. Math. Phys., **5,** 944 (1964).
9. Debye, P., J. Chem. Phys., **31,** 680 (1959).
10. Einstein, A., Ann. Physik, **33,** 1275 (1910).
11. Stillinger, F. H. and Frisch, H. L., Physica, **27,** 751 (1961).
12. Green, M. S., J. Chem. Phys., **33,** 1403 (1949).
13. Kaufman, B. and Onsager, L., Phys. Rev., **76,** 1244 (1949).
14. Domb, C. and Sykes, M. F., Proc. Roy. Soc., **A240,** 214 (1957).
15. Brady, G. W., J. Chem. Phys., **32,** 45 (1960).

16. Petz, J. I., Norelco Reporter, **X,4,** 131 (1963).
17. Schmidt, P. W. and Hight, R., Jr., Acta Cryst., **13,** 480 (1960).
18. Hildebrand, J., Fisher, R. and Benesi, M., J. Am. Chem. Soc., **72,** 4348 (1950).
19. Brady, G. W. and Petz, J. I., J. Chem. Phys., **34,** 332 (1961).
20. Brady, G. W. and Frisch, H. L., ibid., **35,** 2234 (1961).
21. Brady, G. W., ibid., **40,** 2747 (1964).
22. Frisch, H. L. and Brady, G. W., ibid., **37,** 1514 (1962).
23. Mueller, C. R. and Lewis, J. E., ibid., **26,** 286 (1957). For criticism of this work see Williamson, A. G., Scott, R. L. and Dunlap, R. D., ibid., **30,** 325 (1959).
24. Private communication.

Small-Angle X-Ray Study of Metallized Catalysts

G. A. SOMORJAI, R. E. POWELL, P. W. MONTGOMERY, AND G. JURA

Department of Chemistry, University of California, Berkeley, California

INTRODUCTION

Since the very beginning of the use of contact catalysts, there has been widespread use of catalytic metal particles dispersed on highly porous supports. Typical supported catalysts include platinum, palladium, nickel, or cobalt supported on carbon, alumina, silica, or alumina-silica. There is a very large literature on the adsorptive and catalytic properties of these systems; but studies of the sizes and shapes of the metal particles themselves are much fewer, primarily because of the lack of effective experimental techniques for such studies.

The techniques which appear to be applicable for such investigation of metal particles in supported catalysts are electron microscopy (1) and the broadening of X-ray diffraction lines (2). The use of magnetic granulometry (3) in studying particle size is limited to small particles of ferromagnetic metals. The technique which seems to be the most versatile, the one utilized in this investigation, is small-angle X-ray scattering.

There is a major practical difficulty in the small-angle investigation of microporous catalysts, namely, the intense scattering arising from the holes in the catalyst support itself. Consider, for example, the metallized catalyst used in the present investigation, platinum supported on η-alumina. The electron density of platinum is $78 \times 21.45/195.09 = 8.58$ faradays per cc; of the crystallites of η-alumina approximately $30 \times 3.7/60 = 1.85$ faradays per cc; and of the holes very nearly zero. Since the absolute intensity of the X-ray scattering is proportional to the square of the difference in electron density, a given volume of holes will scatter about 1/13 the intensity of the same volume of platinum. The scattering from holes is easily large enough to permit the small-angle investigation of the particle sizes in the microporous solids themselves, and there are a number of examples in the literature of studies on finely divided solids, including catalysts (4). Even much smaller differences in electron density are sufficient for small-angle scattering (5): (0.05 faradays/cc).

There is an even more serious difficulty. If the location of the holes were random with respect to the platinum particles, the scattering from the holes could be treated as part of the background correction and subtracted out. But the platinum particles are sitting within the holes, so that their scattering tends to cancel one another. In fact, our preliminary experiments showed that the scattering from platinum-alumina was apparently *less* than that from the alumina itself. It is therefore essential to destroy the holes before studying the scattering from the platinum.

A possible method would be to fill the holes with a liquid of the same electron density as the alumina, namely 1.85 faradays per cc. It has been shown (6) that the hole scattering from a silica-alumina cracking catalyst of electron density 1.245 was diminished about two-fold by saturating it with o-xylene, and diminished more than a hundred-fold by saturating it with n-butyl iodide. In a number of preliminary experiments, we attempted to saturate our alumina with liquids (or conveniently fusible solids) of electron density near 1.85, viz., CHI_3, $SbBr_3$, and $ZnCl_2$-$ZnBr_2$ solid solution. It proved difficult to get these substances into the pores, which apparently they do not wet; and even more difficult to remove the surplus, which when not removed led to a serious (and nonreproducible) attenuation of the X-ray beam. Nevertheless the method does appear capable of successful application, and was abandoned only because the following alternative proved much more successful.

We have found that by pressing the supported catalyst samples in a hydrostatic press (7) under a pressure of 100 kbar or more, substantially all holes in the alumina are reduced to a size giving no small-angle scattering; the gases originally present escape through the material of the press. In the "pressure sintered" sample the scattering by the metal particles is readily measured.

With the development of pressure sintering as a method for the elimination of hole scattering, it becomes possible to study the effect of such variables as the method of original impregnation, the thermal history of the metallized catalyst, or the chemical attack upon the metal particles. In the present investigation the second of these problems was attacked, namely, the effect of thermal history on particle size. The catalyst chosen for investigation was platinum on alumina. It has been possible to follow the growth of the platinum particles and to derive information about the kinetics and mechanism of the process.

EXPERIMENTAL APPARATUS AND PROCEDURE

Metallized Catalysts

The catalysts subjected to study were 5% and 0.5% platinum by weight, dispersed on high-area microporous alumina. The alumina, "Filtrol 90," is characterized by its X-ray diffraction pattern as η-alumina; its crystal structure is not affected by heating to the temperatures used in this investigation, and X-ray diffraction patterns taken on our samples after pressure-sintering showed no change in its crystal structure even after compression to 350,000 atmospheres.

The alumina was impregnated with aqueous chloroplatinic acid, dried, and reduced to metallic platinum by heating in hydrogen.* The catalyst pellets were ground and screened through a 200 mesh/inch screen. Then the samples were heat treated (at 400–700°C) for the desired time (1–96 hours) in an oxidizing atmosphere (air) or a reducing atmosphere (generally illuminating gas, but identical results were obtained with hydrogen).

Every time a platinum-alumina sample was heated, a sample of the pure alumina was heated side by side in the furnace, so that a correction might be made in the X-ray measurement for any changes taking place in the "blank" alumina upon heating. In fact this precaution proved to be unnecessary, since all the alumina samples turned out to give identical X-ray scattering. After heat treatment, the platinum-alumina samples were compressed at 100,000 atmospheres for 15 minutes. The detailed description of the high-pressure press is published elsewhere (7).

Figure 1 shows the X-ray scattering intensity of unheated pure alumina, treated at various pressures. As it shows, the scattering from holes in the alumina has been substantially eliminated at 100,000 atmospheres, so this pressure was chosen for the treatment of all samples. Heating in the range of temperatures used has a negligible effect on the compressibility characteristics of the alumina.

* We wish to thank H. F. Harnsberger and R. J. Houston of the California Research Corp., who kindly prepared the catalysts used in this investigation.

There was no change in the shape of the platinum particles due to compression in the pressure range 5–200 kbars, as was determined from the small-angle X-ray scattering data.

At one stage of the investigation an attempt was made to pressure sinter the platinum-alumina first, then heat treat it. The results were quite disconcerting and the attempt was abandoned: the scattering in the small-angle region increased enormously, producing a background scattering so high that the scattering from the platinum particles could not be determined with any confidence. The cause of

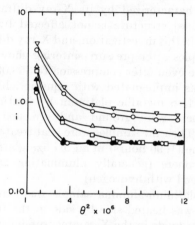

Fig. 1. Scattering intensity of unheated η-alumina at various pressures. i: intensity in units of 256 counts/30 sec; θ: angle in radians. ● air background alone, no sample, ▲ alumina pressed at 100,000 atm., ○ alumina pressed at 300,000 atm., □ alumina pressed at 65,000 atm., △ alumina pressed at 40,000 atm., ○ alumina pressed at 20,000 atm., and ▽ alumina unpressed.

this phenomenon is the high strain to which the alumina is subjected upon compression, which greatly facilitates crystallization. It is probable that the grains rearrange to an order such as they had before compression, thereby introducing a number of holes.

Small-Angle Apparatus

The small-angle X-ray measurements were made using CuKα radiation, a four-slit collimating system, and a sample-to-detector distance of 50 cm. The sample was mounted on a one-mil "Mylar" sheet behind the third slit. The X-ray intensity was measured using scintillation counting with pulse height selection (8). The apparatus has been described in detail elsewhere (9).

The smallest angle which did not strike the primary beam was 0.0733 degrees, or about 4.4 minutes of arc. The largest angle at which these experiments gave scattering intensity distinguishable from the background was about 1.1 degrees. Roughly these angles correspond to particles ranging from 40 Å to 600 Å.

The pulses leaving the analyzer (10) were counted with a conventional scale-of-256 and a mechanical register. The typical counting rate was some 2000 pulses per minute, although it was much higher near the primary beam and much lower at angles near one degree.

The attenuation of the X-ray beam by the pure alumina or 0.5% platinum-alumina samples was negligible. For the 5% platinum there was a small absorption, for which a small correction was applied to intensities measured for it: each intensity was multiplied by 1.106.

The apparatus was calibrated using Bushy Stunt Virus. This virus has previously been studied (11) and its radius is 154 Å. Excellent agreement was found between our result and those of other investigators (9).

WORKING EQUATIONS

From the small-angle X-ray scattering, the datum that can be calculated most directly is the average radius of gyration of the scattering particles. For a detailed discussion of the theory, the reader is referred to Guinier's book (4).

The scattering intensity $i(h)$ for N noninteracting particles is given as

$$i(h) = i_e(h) N \overline{F^2(h)} \qquad 1.$$

where $i_e(h)$ is the intensity of scattering by one electron, $F(h)$ is the structure factor, and $h = (4\pi/\lambda) \sin \theta$, where λ is the X-ray wavelength and 2θ is the scattering angle. At small angles $F(h)$ is approximated by $\overline{F^2(h)} = n^2 \exp(-h^2 r^2/3)$ where r is the radius of gyration and n is the total number of electrons in the particle. For small angles $[(4\pi/\lambda) \sin \theta \simeq (4\pi/\lambda)\theta]$, r is given by

$$r = \frac{1}{2\pi} \sqrt{\frac{3}{\log_{10} e}} \lambda \sqrt{p} \qquad 2.$$

where p is the negative slope of the log $i(h)$ vs. $(\tan 2\theta)^2$ curve. In case of the CuKα radiation $r(\text{Å}) = 0.645 \sqrt{p}$.

The usual plot for obtaining the average radius of gyration is therefore the Guinier plot, $\log i$ vs. h^2, where h is proportional to the angle: $h = (2\pi/\lambda) \tan 2\theta \simeq 4\pi\lambda^{-1} \theta$. The slope gives the desired radius. However, the initial slope must be used, because at larger angles the Guinier exponential approximation begins to fail.

In interpreting the scattering results of this investigation, the Guinier plot was used to determine the average radii of the platinum particles.

PARTICLE SHAPE AND SIZE DISTRIBUTION

For spherical or nearly spherical particles the Guinier approximation holds best. However, there is no "built-in" shape factor in the integral. Furthermore, most of the systems to be investigated by means of small-angle X-ray scattering are heterodisperse, and the distribution of particles is of great interest. Shull and Roess (12) developed a method for calculating small-angle scattering intensities of heterodisperse systems, inserting a distribution function into the scattering integral. It was assumed that all particles are geometrically similar; that is, the number of electrons n is proportional to r, where r is the radius of gyration.

Then

$$i(h) = K \int_0^\infty \overline{F^2(h)} N(r) r^6 \, dr \qquad 3.$$

where $N(r)$ is the number distribution function and $N(r) \, dr$ represents the total number of particles in the size range $r, r + dr$. K is constant. The exact structure factor for spheroids,

$$\overline{F^2(h)} = 9 \left(\frac{\sin rh - rh \cos rh}{r^3 h^3} \right)^2 n^2 \qquad 4.$$

was used instead of the Guinier approximation. The calculated particle sizes agreed within 10% with those obtained by using the Guinier approximation.

In case $N(r)$ is a Maxwellian distribution, $N(r) = r^n e^{-(r/r_0)^2}$ where r_0 and n are constants.

If one writes for a spheroid: $V(r) = (4\pi/3)vr^3$, where v is the axial ratio, and substitutes into the scattering intensity integral, the integral can readily be evaluated at three different limits. For spherical particles $v = 1$, for disc-shaped particles $v \to 0$, and for rod-like particles $v \to \infty$ and $r \to 0$ simultaneously so the product rv approaches

L, the length of a rod with negligibly small radius. Expanding the integral by hypergeometric functions [see Roess and Shull (12)] one gets the scattering from a distribution of particles of different shapes.

By comparison of the experimental data with the theoretical scattering curves, one should be able to get a distribution of the scattering particles and find some information about the general shape of the particles. There are various ways of making graphical comparisons of the data with the theories, one convenient type of plot being that of log $i\theta^2$ vs. log θ^2, which gives curves with distinct maxima, on which the effects of shape and distribution function are fairly readily discernible. In Fig. 2 are plotted the scattering curves for

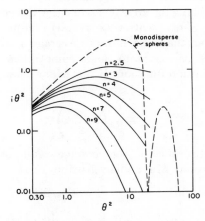

Fig. 2. Scattering curves for monodisperse spheres and Maxwellian distribution of heterodisperse spheres. --- Monodisperse spheres.

monodisperse spheres and for various Maxwellian distributions. Scattering curves for discs and for rods are also available (9, 12).

The experimental scattering data for platinum on alumina were examined in the light of these theoretical curves, and such conclusions as could be drawn will be taken up with the subsequent discussion of the experimental results.

WHAT AVERAGE DOES SMALL-ANGLE SCATTERING MEASURE?

If the scattering particles are not monodisperse, the slope of the Guinier plot will yield an average radius which depends upon the distribution function. If the number distribution function is $N(r)$, the scattered intensity is

$$i \sim \int_0^\infty r^6 N(r) e^{-r^2 h^2/3} \, dr \qquad 5.$$

so the apparent radius, evaluated as the square root of $-3d \ln i/d(h)^2$, is easily seen to be

$$r_{x\text{Av}} = \left(\int_0^\infty r^8 N(r) \, dr \bigg/ \int_0^\infty r^6 N(r) \, dr \right)^{\frac{1}{2}} \qquad 6.$$

That is, it is the root mean radius weighted by the sixth power of the radius (or the volume squared) (12). We shall denote this average radius, the X-ray scattering average radius, by the symbol $r_{x\text{Av}}$. Since this method of averaging greatly favors the larger particles in the distribution, the X-ray scattering average radius will always be considerably larger than the number average radius. To see how much larger, it is necessary to make some assumption about the form of the distribution function. If it is Maxwellian, so that $N(r) = r^n e^{-(r/r_0)^2}$, then

$$r_{x\text{Av}} = r_0 \left(\frac{n+7}{2} \right)^{\frac{1}{2}} \qquad 7.$$

Compare this with some of the average radii, of various kinds, computed for the same Maxwellian distribution:

Most probable radius $\qquad r_0(n/2)^{\frac{1}{2}}$

Number average radius, $r_{n\text{Av}}$ $\qquad r_0 \dfrac{(n/2)!}{\left(\dfrac{n-1}{2}\right)!}$

Root mean square radius, or surface average radius (radius of sphere of average surface), $r_{s\text{Av}}$ $\qquad r_0 \left(\dfrac{n+1}{2} \right)^{\frac{1}{2}}$

Root mean cube radius, or volume average radius (radius of sphere of average volume), $r_{v\text{Av}}$. $\qquad r_0 \left[\left(\dfrac{n+2}{2} \right)! \bigg/ \left(\dfrac{n-1}{2} \right)! \right]^{\frac{1}{3}}$

For a fairly broad distribution, say $n = 3$, $r_{x\text{Av}}$ is seen to be 1.68 times $r_{n\text{Av}}$. For a broader distribution, the discrepancy would be greater, and for a sharper distribution it would be less. In the interpretation of the data reported later, it will be well to bear in mind that the radii reported are, in the first place, radii of gyration and must be multiplied by $(\frac{5}{3})^{\frac{1}{2}} = 1.29$ to obtain the radii of the equivalent spheres; and, in the second place, must be divided by a factor which we cannot give exactly because of our ignorance of the distribution function, but which is of the order of 1.5 or 2, to obtain number average radii.

RESULTS AND DISCUSSION

Detection of 0.5% Platinum on Alumina

Although the body of our platinum-alumina studies were made with 5% platinum catalysts, it was of some interest to see whether

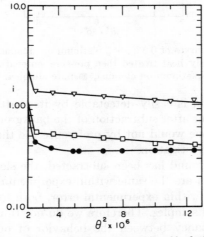

Fig. 3. Scattering curves of unheated, 0.5%, 5% platinum on alumina and pure alumina background after pressure sintering at 100,000 atm. ▽ 5% platinum on alumina, ☐ 0.5% platinum on alumina, and ● pure alumina.

0.5% platinum could be detected satisfactorily, because the technical platforming catalysts are in this lower range of composition. There was some question whether the 0.5% platinum, which on a volume basis is less than 0.1% platinum, would give scattering sufficiently greater than the background.

The answer to the question can be seen from Figs. 3 and 4 which compare the background and the scattering for 0.5% and 5% platinum, for the catalysts unheated and pressure-sintered, then for the catalysts heated at 700°C for 24 hours in air and pressure-sintered.

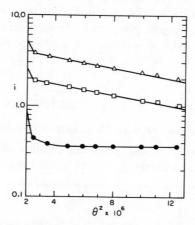

Fig. 4. Scattering curves of 0.5%, 5% platinum on alumina and pure alumina (background previously heat treated then pressure sintered). ▽ 5% platinum on alumina, □ 0.5% platinum on alumina, ● pure alumina.

The 0.5% platinum is easily detectable by its scattering, although the percentage error after subtraction of the background is comparatively large, and one would not by preference use the 0.5% samples for an extended investigation.

After the background has been subtracted, the slopes for the 5% and 0.5% platinum are the same within experimental error, so their radii are the same within experimental error, both for the unheated and for the heated samples. Therefore we do not think that there is any major discrepancy between the behavior of our 5% samples and the more conventional 0.5% catalysts.

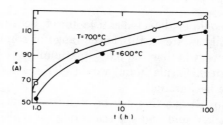

Fig. 5. Change of the platinum particle size as a function of heating time in reducing atmosphere.

Growth of Platinum Particles

Most of the temperature treatments were carried out on 5% platinum alumina catalyst samples, which were then compressed at 100,000 atm.

Figure 5 shows the particle size changes in a reducing atmosphere at two temperatures, 600°C and 700°C and Fig. 6 shows the particle size changes for oxidizing atmospheric heating.

The effect of heat treatment on the particle size is clearly indicated in these data. There is a large increase in the average particle size at these temperatures in the measured time interval. All samples

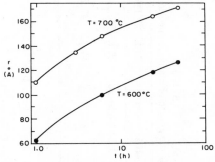

Fig. 6. Change in the platinum particle size as a function of heating time in oxidizing atmosphere.

exhibit a very fast growth in the first few hours, which levels off with time. In reducing atmospheric heating, this leveling off or almost complete stop of the growing process is quite conspicuous. In either reducing or oxidizing atmosphere, a faster change is taking place at the higher temperature. Moreover, it is apparent from the figures that heating in oxidizing atmosphere brings about a very much faster growth process than does heat treatment in reducing atmosphere.

The absolute intensity of scattering (which is proportional to the number of particles whose scattering is detectable) is at first low, increases with heating time, then eventually falls somewhat with long heating times (9). The plainest interpretation is that at first many particles are too small to be within the range of detection by the X-ray method, as heating progresses they grow larger and become detectable, and at very long times a number of them have become too

TABLE I

Apparent activation energies for the growth of platinum particles of different average radii.

R (Å)	94	95	100	104	106	110	115	118	120	122	124	126
E_{act} in reducing atmosphere (kcal)	14.5	14.52	16.30	20.80	24.20	26.6						
E_{act} in oxid. atmosphere (kcal)						44.3	47.2	49.1	51.6	53.8	54.4	54.7

large to be detectable, their scattering being hidden under the primary beam. In other words there is no limiting particle size, but a steady transport of material from small particles to larger ones.

Besides the diminishing rate of growth there is another interesting phenomenon which occurs. The apparent activation energy can be calculated by comparing the elapsed times for a particle to attain a given radius, at two or more temperatures, viz.,

$$E_{\text{activation}} = -R\, d \ln t_r/d(1/T) \qquad 8.$$

where t_r is the time to attain a given radius. The results are shown in Table I.

It can be seen that the apparent activation energy increases notably as the particles become larger. The increase in activation energy is quite adequate to account for the marked decrease in growth rate; we shall return to the question a little later, to show that a reasonable agreement between the experimental data and this theory can be obtained, and also to show the physical plausibility of the radius dependence of growth rate.

The apparent activation energy, computed on the foregoing basis, is roughly 20 kcal for a reducing atmosphere and roughly 52 kcal for an oxidizing atmosphere. It is evident that two quite different mechanisms are involved, and tempting to ascribe the low-energy process to some sort of diffusion mechanism and the high-energy process to the transport through the gas phase by platinum oxide. From the work of Brewer and Elliot (13) the heat of the reaction $Pt + O_2 = PtO_2(g)$ is $\Delta H = +55$ kcal, so the transport by gaseous PtO_2 becomes plausible. However, it is necessary to point out that the apparent activation energy may be relatively unreliable as a measure of the true activation energy, because, as we shall show, the radius-dependence is itself highly temperature dependent.

The growth or precipitation of particles in a solid matrix, far from equilibrium, proceeds at a constant, mostly volumetric rate (14, 15), $r^2\, dr/dt = k$. Here, k is the rate constant. This equation applies to relatively large particles (1 micron or more). Small particles in the 10 Å – 10^3 Å range due to their increased surface energy have solubilities larger than that of the large particles as expressed by the Thompson equation (16)

$$\ln C_r/C_s = 2V\gamma/rRT \qquad 9.$$

Here, C_r is the solubility of particles of radius r, and C_s is the limiting solubility for large particles, V is the molar volume, γ is the surface

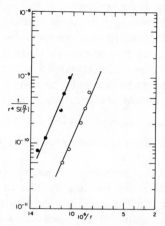

Fig. 7. A plot of the change of the radius of gyration, r, as a function of time for heat treatment at ● 600°C reducing atmosphere and ○ 700°C reducing atmosphere.

tension of the particle in the solid matrix, R is the gas constant and T is the absolute temperature. Therefore the rate of growth of larger particles as long as small particles are present is accelerated by a factor of $\exp(2V\gamma/rRT)$. Under these conditions the growth law becomes

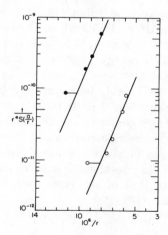

Fig. 8. A plot of the change of the radius of gyration, r, as a function of time for heat treatment at ● 600°C oxidizing atmosphere and ○ 700°C oxidizing atmosphere.

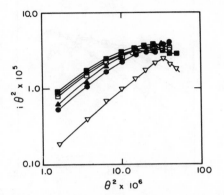

Fig. 9. 5% Platinum on alumina, 600°C in reducing atmosphere. ▽ heated for 1 hour or unheated, ● heated for 3 hours, ▲ heated for 6 hours, □ heated for 24 hours, ○ heated for 48 hours, ■ heated for 96 hours.

$$r^2\, dr/dt = ke^{a/r} \qquad 10.$$

where $a = 2V\gamma/RT$. It is the presence of the surface energy term which accounts for both the decrease of rate with particle size and the increase of activation energy with particle size.

The variables of Eq. 10 are easily separable and the resulting differential equation has been integrated (9) to give

$$kt/a^3 = (r/a)^4 e^{-a/r} S(a/r) \qquad 11.$$

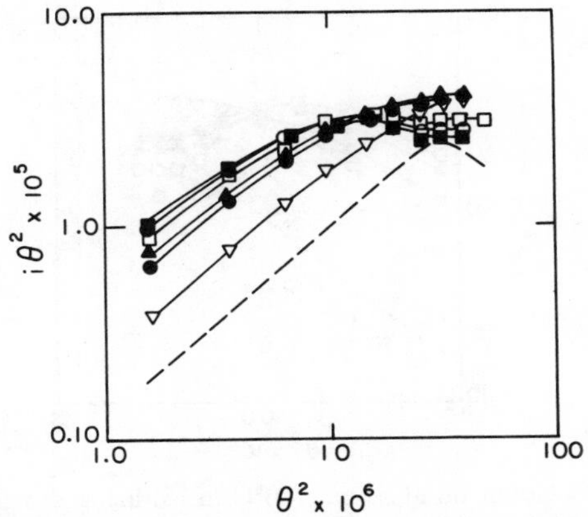

Fig. 10. 5% Platinum on alumina, 700°C in reducing atmosphere. - - - unheated sample, ▽ heated for 1 hour, ● heated for 2 hours, ▲ heated for 6 hours, □ heated for 24 hours, ○ heated for 48 hours, ■ heated for 96 hours.

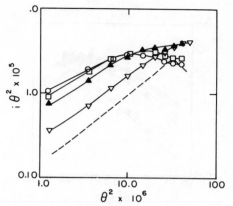

Fig. 11. 5% Platinum on alumina, 600°C in oxidizing atmosphere. ▽ heated for 1 hour, ▲ heated for 6 hours, □ heated for 24 hours, ○ heated for 48 hours, --- unheated sample.

where the function, $S(a/r)$, can be approximated by the simple equation (9)

$$S(a/r) \simeq \frac{1 + a/r}{a/r + 5 + 3(r/a)} \qquad 12.$$

It is worth commenting that the theoretical curve is almost linear when r is plotted against $\log t$ as long as the growth is controlled by the surface energy term.

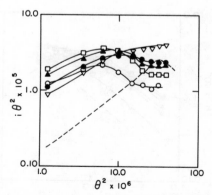

Fig. 12. 5% Platinum on alumina, 700°C in oxidizing atmosphere. ▽ heated for 1 hour, ● heated for 3 hours, ▲ heated for 6 hours, □ heated for 24 hours, ○ heated for 48 hours, --- unheated sample.

The fitting of experimental data to the theoretical curves requires knowledge of the parameters a and k. This can be accomplished by trial in a variety of ways; perhaps the simplest is to plot $\log (t/r^4 S)$ against $1/r$, from which the slope gives a and the intercept gives ka. This kind of plot must be iterated, since S is a function of a/r and requires a prior knowledge of a; however, S is a slowly varying function and an approximate preliminary value of a permits it to be evaluated rather well (9).

The lines drawn through the data points on Figs. 7 and 8 were calculated from the theoretical equation with the foregoing parameters (9). Except for the very shortest times, the agreement is within our experimental error.

The explanation for the great deviation at very short times is that it probably does not represent a real growth of (arithmetic mean) particle size at all, but represents a change in the distribution function of particles, toward a broader distribution. As we shall give evidence to show later, the initial metal particles are fairly near to monodisperse, but rapidly change to a rather broad distribution. Since, as we have pointed out, the X-ray method looks at particles considerably larger than the average particles, the change in distribution will simulate the growth of particle size.

The rate constant k for a reducing atmosphere is given approximately by $5 \times 10^8 \exp(-38,000/RT)$ and for an oxidizing atmosphere by $2.51 \times 10^{17} \exp(-69,000/RT)$ Å$^3/hr$. These activation energies would be the true activation energies for the transport in contact with bulk material, or very large particles. It is clear from the numbers that the two processes, reducing and oxidizing atmosphere growths, are very different. However, there is a rather large possible error in the numerical values and it would be unsafe to draw mechanistic conclusions from their values.

The Size Distribution and Shape of Particles

In order to examine the distribution and shape characteristics of the platinum particles, the data have been replotted as $\log i\theta^2$ vs. $\log \theta^2$, to give the same kind of graphs as were discussed above. The results are displayed in Figs. 9, 10, 11 and 12.

The dotted curves on the figures correspond to the platinum in the unheated samples. As a comparison with the theoretical curve will easily show, these correspond rather well to the curves for monodisperse spheres; so it is a good approximation that the freshly deposited platinum is made up of spherical particles with a narrow distribution range.

Upon heat treatment of the samples, the scattering curves become progressively flatter, indicating a broad distribution of sizes. As time goes on, the intensity at large angles diminishes and at small angles grows, which of course indicates that smaller particles are disappearing while large particles are growing. The curves are similar in shape to those for a Maxwellian distribution of spheres with $n = 3$ or less in $N(r)$.

The very flat distribution is particularly noticeable in the data for an oxidizing atmosphere. The theoretical curves for a flat distribution of spheres are rather similar to those for a distribution of discs, and it is not possible to make a firm decision as to the shape. It is not implausible that in the oxidizing atmosphere the particles redeposit on the alumina in a somewhat plate-like form, since platinum oxide may well be able to wet alumina.

BIBLIOGRAPHY

1. Turkevich, J., J. Chem. Phys., **13,** 235 (1945).
2. For a brief summary of the line-width technique, see Jellinek, M. H., and Fankuchen, I., *Advances in Catalysis,* **I,** 262 (1948).
3. Schuit, G. C. A. and Van Reijen, L. L., *Advances in Catalysis,* **X,** 243 (1958).
4. Guinier, A., et al., *Small-Angle Scattering of X-Rays,* J. Wiley and Sons, New York, 1955; pp. 187–194.
5. Porai-Koshits, E. A. and Andreyev, L. S., J. Soc. Glass Tech., **43,** 235T (1959).
6. Gunn, E. L., paper presented at the April, 1958 meeting of the American Chemical Society, Division of Petroleum Chemistry.
7. Montgomery, P. W., Stromberg, H. and Jura, G., UCRL-9796 (1961).
8. Parrish, W. and Kohler, T. R., Rev. Sci. Instr., **27,** 795 (1956).
9. Somorjai, G. A., Ph.D. Thesis, University of California (1960).
10. Van Rennes, A. B., Nucleonics, **10,** 22 (1952).
11. Beeman, W. W., J. Chem Phys., **19,** 793 (1951).
12. Roess, L. C. and Shull, C. G., J. Appl. Phys., **18,** 295 (1947).
13. Brewer, L. and Elliott, G. R. B., Ph.D. Thesis, University of California (1952).
14. Morin, F. J. and Reiss, H., J. Phys. Chem. Solids, **3,** 196 (1957).
15. Somorjai, G. A., J. Chem. Phys., **35,** 655 (1961).
16. Hardy, H. K. and Heal, T. J., Progr. Metal Phys., **5,** 143 (1954).

A Small-Angle X-Ray Scattering Study of Radiation Damage to Silica Gels*

BENNETT C. LARSON AND HAROLD D. BALE

Physics Department, University of North Dakota, Grand Forks, N. Dak.

INTRODUCTION

Although extensive study of radiation damage to solids has been made in recent years, little of this work has considered irradiation effects on size parameters of colloidal powders or dried gels. Adamson, Ling and Datta's study (1) of irradiation effects on the surface of solids was concerned with low surface area non-porous materials for which they found negligible changes in surface area upon irradiation. Induced changes in the surface area of porous graphite have been reported by Spalaris, Bupp and Gilbert (2). However, these authors concluded that much of their observed effect, measured by gas adsorption, might be due to progressive blocking of pores as bulk expansion took place. Previous irradiation studies of silica gel have been primarily related to adsorption properties and catalytic activity of the gel (3). Weisz and Swegler (4) report negligible changes in the surface area after an irradiation of $7.5 \times 10^{18}\, n_f/cm^2$.

The applicability of small-angle scattering to the study of colloidal systems is well known. Among other things, the technique affords a method of determining both surface area and a characteristic length for the sample. This X-ray method of surface area measurement does not require that pores or voids be accessible to a gas and consequently has an important advantage for radiation damage study over the usual gas adsorption determination.

MATERIALS STUDIED

Four different silicon dioxide materials were studied. The silica designated as sample 1 is a commercial product called "Cab-O-Sil," obtained from Cabot Corporation. The material consists of spherical

* Work supported by Atomic Energy Commission.

particles having an average diameter of 150 Å. It is non-porous and highly dispersed with a bulk density of only 2.2 lb/ft^3. The colloidal material is formed at 1100°C and is stable for long periods of time even at temperatures of 200 to 300°C.

The three silica gels studied, samples 2, 3 and 4, were obtained from Davison Chemical Division of W. R. Grace Co. Their respective surface areas in m^2/gm are 240, 495 and 1320. Gels consist of particles of silica sol that are linked together to form a fairly rigid structure. The nature of the resulting pores depends on the dimension and coordination of the primary particles of the skeleton. A small-pored silica gel with a large surface area consists of close-packed small particles, while a large-pored silica gel with a small surface area consists of large particles loosely arranged. The dried silica gel is stable up to 300°C; however, the presence of water vapor during heat treatment causes a rapid deterioration of the gel structure.

SAMPLE IRRADIATION

The samples were pile irradiated in the 20 megawatt Oak Ridge Research Reactor with a fast neutron dose in excess of 10^{19} n/cm^2. A special method of sample encapsulation was used for the gel samples to ensure the absence of water vapor during irradiation and to minimize the mass present and hence the gamma heating of the sample. For this encapsulation the silica gel was poured directly into the aluminum irradiation can and then heated at 200°C and simultaneously evacuated for one hour. The can was then vacuum sealed and welded. The irradiation periods ranged from one to eleven weeks. Temperature in the cans probably did not exceed 150°C during the irradiation.

EQUIPMENT AND PROCEDURE

The diffractometer used was a four-slit arrangement with 50 cm between neighboring slits (5). The beam path, except in the region of the sample, was evacuated to reduce air absorption and scattering. Monochromatization of the copper radiation was achieved with a Kβ filter and a pulse-height analyzer. Slit widths of 0.005 inch were used to observe the inner portion of the scattering curve while 0.035 inch widths were used to obtain the data at the larger angles. Superposition of these two curves gave a composite scattering curve that extended from 0.0012 radians to the inverse fourth power region at the larger angles. Slit height corrections were made by a method described by Schmidt and Hight (6).

Adsorbed water was removed from the gel before examination by preheating the sample at 200°C for one hour. For the X-ray study the gel was sealed in a flat sample holder of optimum thickness.

THEORY

The extensive linking together of particles in a dried gel would tend to invalidate any interpretation of the data in terms of independent particle scattering. Rather it seems more meaningful to consider the gel structure as an arbitrary arrangement of matter of uniform skeletal density. Appropriate scattering theory for this model has been carefully worked out by others (5); consequently only a brief summary relevant to this study will be presented here. The intensity, $i(h)$, can be expressed in terms of the correlation function, $G(r)$, as

$$i(h) = i_e(h) V \rho^2 c (1 - c) \int_0^\infty G(r) \frac{\sin hr}{hr} 4\pi r^2 \, dr \qquad 1.$$

where $i_e(h)$ is the single electron scattering, V is the illuminated volume, ρ is the electron density, and c is the fraction of the volume occupied. $G(r)$ can be obtained by making a Fourier inversion of Eq. 1; however, this knowledge in itself is not often particularly meaningful. An exception is the exponential correlation function

$$G(r) = e^{-r/a} \qquad 2.$$

Debye, Anderson and Brumberger (7) have shown that if $G(r)$ can be expressed in this form then the corresponding scattering structure is a perfectly random arrangement of matter.

At the larger scattering angles of the small-angle region, where the product hr is large compared to unity, the intensity can be approximated as

$$i(h) \approx \frac{2\pi \rho^2 i_e(h) S}{h^4} \qquad 3.$$

In the region where the scattering varies as h^{-4}, $i(h)$ will be proportional to S, the total surface area of the illuminated volume. In the case of an arbitrary structure, S is the most meaningful parameter that can be determined from the scattering data. In the development of Eq. 3, density fluctuations due to atomic structure were not considered. In the small-angle region, where h is small compared with the reciprocal of the atomic spacings, these fluctuations in density should have little effect, but will become perceptible at larger angles. Thus the product $h^4 i(h)$ will begin to increase when h becomes too large. Also, as mentioned above, the asymptotic law is not valid for too small values of h. Each material will then have a certain range in h over which the scattering is proportional to the surface area provided that the smallest dimensions of the scattering structure are several times larger than the distance of atomic density fluctuations.

Another parameter derived from $G(r)$ is the distance of heterogeneity, l_c:

$$l_c = 2 \int_0^\infty G(r) \, dr = \frac{\int_0^\infty h i(h) \, dh}{2\pi V \rho^2 c(1-c) i_e(h)} \qquad 4.$$

Despite the fact that l_c can be precisely defined (5), it actually gives only an indication of the size of the structure. Determination of S and l_c from Eqs. 3 and 4 requires a value for $i_e(h)$. This can be determined from the normalization relation

$$\int_0^\infty h^2 i(h) \, dh = 2\pi^2 i_e(h) \rho^2 V c(1-c) \qquad 5.$$

Then the surface area per unit mass can be written

$$S/m = \frac{\pi(1-c) \lim [h^4 i(h)]}{d \int_0^\infty h^2 i(h) \, dh} \qquad 6.$$

where d is the skeletal density. Use of the normalization relation, Eq. 5, for surface area determination is only valid if it is experimentally possible to observe the scattering at sufficiently small angles and if the small-angle scattering theory used is applicable to the entire structure. These conditions should be satisfied as long as the structure is continuous over a range of no more than a thousand angstroms. If however, a portion of the sample contains structures several thousand angstroms in size, Eq. 5 will not give the correct value of $i_e(h)$.

Since the theory discussed above pertains to an arbitrary arrangement of matter of uniform skeletal density, it should then also apply to a dilute system of particles. For the case of widely dispersed identical particles, l_c can be more specifically defined than previously as the average length of all the lines that pass through every point in the particle (5).

RESULTS AND DISCUSSION

The irradiation effects on the non-porous material, sample 1, are shown in Fig. 1. After irradiation there is a slight increase in the scattered intensity at the smallest angles and a decrease of intensity in the inverse fourth power region. The latter can be attributed to a decrease in surface area for the sample.

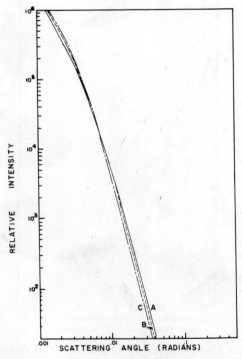

Fig. 1. Sample 1 scattering curves. A, unirradiated; B, irradiation of $2 \times 10^{19} n_f/\text{cm}^2$; C, irradiation of $2.2 \times 10^{20} n_f/\text{cm}^2$.

Sample 2, which is the coarsest gel studied, showed very little change in the scattering curve after one week of irradiation, $2 \times 10^{19} \, n_f/\text{cm}^2$. Figure 2 shows the effects of varied amounts of irradiation on sample 3. Again, as with sample 1, there is a progressive decrease of intensity in the inverse fourth power region with increased irradiation. Several irradiated materials show scattering curves, such as curve B in Fig. 2, that have a small region in which the intensity varies with an inverse power greater than 4. This region, however, is always followed at larger angles by the inverse fourth power region. The surface areas for the irradiated samples 1 and 3 are plotted as a function of total fast neutron irradiation in Fig. 3. It is interesting to note that for sample 3, the surface area decreases exponentially with total irradiation.

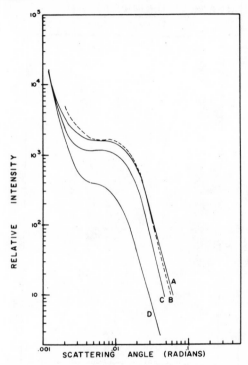

Fig. 2. Sample 3 scattering curves. A, unirradiated; B, irradiation of $2 \times 10^{19} n_f/\text{cm}^2$; C, $8 \times 10^{19} n_f/\text{cm}^2$; D, $2.2 \times 10^{20} n_f/\text{cm}^2$.

The extremely fine structure of sample 4 is very susceptible to irradiation effects, Fig. 4. The scattering curves for sample 4 both show a pronounced rise at the smaller angles. This rise could possibly be attributed to either extensive clustering of primary particles in the gel or more likely to the presence of a fraction of relatively very large particles.

The results of the study are shown in Table I. Two different methods were used to calculate the surface areas from the X-ray data. For the first method a value for $i_e(h)$ was determined for each calcula-

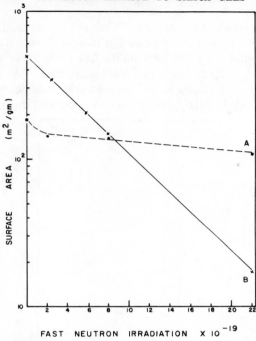

Fig. 3. Surface area as a function of fast neutron irradiation.
A, sample 1; B, sample 3.

TABLE I

| Sample no. | Material | Irradiation n_f/cm^2 | Adsorption* | Surface area, m²/gm | | l_c Å |
				X-ray method 1	X-ray method 2	
1	Cab-O-Sil	—	200	185		200
		2×10^{19}		142	143	215
		8×10^{19}		141	138	218
		22×10^{19}		110	110	242
2	Davison 62	—	340	232	240	96
		2×10^{19}		210	206	98
3	Davison 81	—	600	494	495	35
		2×10^{19}		422	349	38
		8×10^{19}		381	150	44
		22×10^{19}		325	17.2	106
4	Davison 12	—	800	1262	1320	13.3
		2×10^{19}		743	162	25.4

* Manufacturer's values

tion from the normalization relation. In the second method the value of $i_e(h)$ determined for the unirradiated sample 1 was used for all subsequent calculations. The two methods gave similar values for the unirradiated samples but widely differing values for irradiated samples 3 and 4. The values to be accepted are those determined by the second method. Evidently the structure produced during irradia-

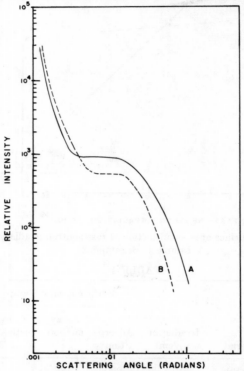

Fig. 4. Sample 4 scattering curves. A, unirradiated; B, irradiation of $2 \times 10^{19} n_f/\text{cm}^2$.

tion of the gels is sufficiently large so that it contributes very little to the observed scattering curve and consequently the integral

$$\int_0^\infty h^2 i(h) \, dh$$

has too small a value for normalization. The value of l_c determined for irradiated samples is also questionable since this calculation involves integrals containing $i(h)$.

The mechanism of damage is a problem of fundamental interest in every study of radiation damage. The changes in surface area that have been observed in this study are basically quite different from the modifications observed in continuous solids. It is difficult to imagine how irradiation-induced point defects or the diffusion of them could contribute to the observed decrease in surface area of a non-porous system of particles such as sample 1. Rather some type of thermal effect, such as a thermal spike (8), seems to have occurred. Estimates indicate that a deposition of 10^5 ev of energy in a single particle of 200 Å diameter would lead to a particle temperature in excess of 1000°C. The duration of such a spike should be considerably longer than a spike formed in a continuous solid. These hot regions could lead to sintering between particles and a smoothening of surfaces. Thus it seems plausible that the observed reduction in surface area in sample 1 can be attributed to thermal spike formation in the particles. Heating sample 1 for one hour at 900°C causes a change in the scattering curve very similar to that observed after pile irradiation.

The gel structure is made up of colloidal particles which are linked or sintered together to form a rigid porous structure. Sample 2 shows very little change upon irradiation. Apparently the heat treatment the material underwent in preparation left it in a condition where further sintering or surface smoothening by thermal spikes was not possible. On the other hand, samples 3 and 4 are more susceptible to damage by thermal spikes, probably because of their finer structure. The irradiation effects in sample 4 are very rapid. After an irradiation of $2 \times 10^{19}\ n_f/\text{cm}^2$ the surface area dropped by a factor of 8 to a value lower than any of the other gels studied. However, l_c increased by less than a factor of 2. One can here only speculate that the irradiation effect is not a gradual coarsening of the gel structure but is rather a complete or almost complete destruction of the gel pore structure in localized regions. These non-porous regions must be sufficiently large to contribute little to the observed small-angle scattering curve. Hence l_c and the surface area would still be primarily determined by the yet undamaged regions. The increase in scattered intensity at the smallest angles after irradiation, Fig. 4, is in agreement with this view.

It was mentioned previously that for sample 3, the surface area is an exponentially decreasing function of the total sample irradiation. This relation is of the form

$$\frac{dS}{dt} = -kS \qquad 7.$$

where dS/dt is the rate of change of surface area and k is a constant. For k to remain constant as the surface area is reduced from 495 to 17 m^2/gm seems to require that as far as the surface area is concerned, a once damaged region cannot experience further damage. Again as in the case of sample 4, it appears that, upon irradiation, regions of the gel structure collapse and form non-porous particles of low surface area. A somewhat similar behavior is described by Ries in Ref. (9), in discussing the heating effect on silica gel. He observed no change in the pore size distribution after heating but only a reduction in the number of the pores.

ACKNOWLEDGEMENT

The authors wish to express their appreciation to the Isotope Sales Office, Oak Ridge National Laboratory, for carrying out the sample irradiations.

BIBLIOGRAPHY

1. Adamson, A. W., Ling, I. and Datta, S. K., Adv. Chem. Ser., **33,** 62 (1961).
2. Spalaris, C. N., Bupp, L. P. and Gilbert, E. C., J. Phys. Chem., **61,** 350 (1957).
3. Kohn, H. and Taylor, E., ibid., **63,** 966 (1959).
4. Weisz, P. and Swegler, E., J. Chem. Phys., **23,** 1567 (1955).
5. Guinier, A. *et al.*, *Small-Angle Scattering of X-Rays*, J. Wiley and Sons, New York, 1955; pp. 16, 75, 81, 95.
6. Schmidt, P. W. and Hight, R., Acta Cryst., **13,** 480 (1960).
7. Debye, P., Anderson, H. R., Jr. and Brumberger, H., J. Appl. Phys., **28,** 679 (1957).
8. Billington, D. S. and Crawford, J. H., Jr., *Radiation Damage in Solids*, Princeton University Press, Princeton, New Jersey, 1961; p. 42.
9. Iler, R. K., *The Colloid Chemistry of Silica and Silicates*, Cornell University Press, Ithaca, New York, 1955; p. 209.

Small-Angle Scattering by Solutions of Complex Ions

ARTHUR HYMAN* AND PHILIP A. VAUGHAN

School of Chemistry, Rutgers, the State University, New Brunswick, New Jersey

INTRODUCTION

The thermodynamic properties of any system depend directly on the statistical distribution of the particles of which it is composed. In solutions of electrolytes the radial distribution of solute particles is of particular interest. It is the purpose of this work to: (1) formulate a procedure for obtaining such information from X-ray scattering patterns, and (2) apply this formulation to experimental data. We have thus determined the principal features of the radial distribution function of the anions in aqueous solutions of H_2PtCl_6.

The problem of interpreting X-ray scattering from solutions is complicated by the presence of solvent, and a general theoretical treatment of this problem which is both adequate and useful has not yet been put forth. Generally, the contribution of solvent to the scattered intensity has been neglected. Nevertheless, the internal structures of some fairly complicated solute species have been determined. The earliest work in this field is due to Brosset (1) and to Vaughan, Sturdivant and Pauling (2). By considering solutes containing heavy atoms, these authors were able to neglect the effects of solvent. They interpreted data at high scattering angles, where the scattering patterns are due principally to the internal structure of molecules. Many such structures have since been determined.

We have in this work investigated the small-angle region of scattering from concentrated solutions of complex ions. At these low scattering angles, the spatial arrangement of solute particles with respect to each other profoundly influences the scattering patterns. In addition, we have observed important effects due to differences in

* Present Address: University of Maryland, Baltimore County, 5401 Wilkens Avenue, Baltimore, Maryland.

electron density between solute and solvent. We have attempted to explain these small-angle scattering phenomena in terms of a model representative of a solution containing only one scattering solute species. An equation has been derived from which the radial distribution of solute particles can be obtained through the application of Fourier's Integral Theorem. This result was then analyzed by means of an empirical distribution function which showed some of the features of the radial distribution obtained by Fourier inversion to be spurious.

THEORY OF SMALL-ANGLE SCATTERING FROM SOLUTIONS

The intensity of X-rays, coherently scattered by a collection of electrons, is given by

$$i(\mathbf{h}) = \int_{\mathbf{r}}\int_{\mathbf{r}'} \rho(\mathbf{r})\rho(\mathbf{r}')e^{i\mathbf{h}\cdot(\mathbf{r}-\mathbf{r}')}\,d\mathbf{r}\,d\mathbf{r}' \qquad 1.$$

in which the intensity, i, is in electron units, $\rho(\mathbf{r})$ is the electron density at a point defined by the vector $\mathbf{r}$, $\mathbf{h}$ is the scattering vector, and the integrations are over the irradiated volume.

We shall specialize 1 to the particular case in which the distribution of scattering matter is that found in a solution. The expression so derived will be applicable at small scattering angles (i.e. small values of $|\mathbf{h}|$). Let us begin by assuming that at any given instant the system may be subdivided into clearly defined regions containing either solute or solvent, so that we may write

$$\rho(\mathbf{r}) = \rho_s(\mathbf{r}) + \rho_m(\mathbf{r}) \qquad 2.$$

in which $\rho_s(\mathbf{r}) = 0$ if $\mathbf{r}$ defines a point in the solute, and $\rho_m(\mathbf{r}) = 0$ if $\mathbf{r}$ defines a point in the solvent. Substitution of 2 into 1 yields

$$\left.\begin{aligned} i(\mathbf{h}) =\ & \int_m\int_m \rho_m(\mathbf{r})\rho_m(\mathbf{r}')e^{i\mathbf{h}\cdot(\mathbf{r}-\mathbf{r}')}\,d\mathbf{r}\,d\mathbf{r}' \\ & + \int_m\int_s \rho_m(\mathbf{r})\rho_s(\mathbf{r}')e^{i\mathbf{h}\cdot(\mathbf{r}-\mathbf{r}')}\,d\mathbf{r}\,d\mathbf{r}' \\ & + \int_s\int_m \rho_s(\mathbf{r})\rho_m(\mathbf{r}')e^{i\mathbf{h}\cdot(\mathbf{r}-\mathbf{r}')}\,d\mathbf{r}\,d\mathbf{r}' \\ & + \int_s\int_s \rho_s(\mathbf{r})\rho_s(\mathbf{r}')e^{i\mathbf{h}\cdot(\mathbf{r}-\mathbf{r}')}\,d\mathbf{r}\,d\mathbf{r}' \end{aligned}\right\} \qquad 3.$$

where $\int_m d\mathbf{r}$ and $\int_s d\mathbf{r}$ denote, respectively, integration over the space occupied by solute and solvent.

We will be primarily interested in the small-angle region in which the scattered intensity is not very sensitive to short-range fluctuations in electron density. It is therefore possible and indeed convenient to assume the solvent electron density to be uniformly distributed. Accordingly, let

$$\rho_s(\mathbf{r}) = c \qquad 4.$$

when $\mathbf{r}$ defines a point in the solvent. Consider, now, the integral $\int_V e^{i\mathbf{h}\cdot\mathbf{r}}\,d\mathbf{r}$, in which the integration is over the entire irradiated volume. This sample-shape scattering is known to be negligible at observable scattering angles, so that we may write with little error

$$\int_V e^{i\mathbf{h}\cdot\mathbf{r}}\,d\mathbf{r} = \int_s e^{i\mathbf{h}\cdot\mathbf{r}}\,d\mathbf{r} + \int_m e^{i\mathbf{h}\cdot\mathbf{r}}\,d\mathbf{r} = 0$$

or

$$\int_s e^{i\mathbf{h}\cdot\mathbf{r}}\,d\mathbf{r} = -\int_m e^{i\mathbf{h}\cdot\mathbf{r}}\,d\mathbf{r} \qquad 5.$$

Combination of 3, 4 and 5 yields for the scattered intensity

$$i(\mathbf{h}) = \int_m \int_m [\rho_m(\mathbf{r})\rho_m(\mathbf{r}') - c\rho_m(\mathbf{r}) - c\rho_m(\mathbf{r}') + c^2] e^{i\mathbf{h}\cdot(\mathbf{r}-\mathbf{r}')}\,d\mathbf{r}\,d\mathbf{r}' \qquad 6.$$

Equation 6 gives the intensity scattered by a static arrangement of molecules. The observed intensity, $\langle i \rangle$, will be given by this equation when each term on the right side is averaged over all spatial distributions and orientations of the molecules. We will consider each term in turn for a system consisting of but one type of solute particle.

Term I: $\qquad \int_m \int_m \langle \rho_m(\mathbf{r})\rho_m(\mathbf{r}') e^{i\mathbf{h}\cdot(\mathbf{r}-\mathbf{r}')} \rangle\,d\mathbf{r}\,d\mathbf{r}'.$

This is just the scattering due to a gas of the molecules and was shown by Menke (3) to be given by $N\langle F^2 \rangle + N^2 \langle F \rangle^2 \langle P \rangle$ under the assumption that the orientation of a given molecule is independent of the orientation of neighboring molecules. In this expression, N is the number of solute particles irradiated;

$$\langle F \rangle = \sum_j f_j \frac{\sin hr_j}{hr_j} \qquad 7.$$

$$\langle F^2 \rangle = \sum_j \sum_k f_j f_k \frac{\sin hr_{jk}}{hr_{jk}} \qquad 8.$$

$h = |\mathbf{h}| = 4\pi(\sin\theta)/\lambda$, θ is one-half the scattering angle, f_j is the atomic scattering factor of the jth atom in a molecule, r_j is the distance from the center* of a molecule to the jth atom in that molecule, r_{jk} is the distance between centers of the jth and kth atoms in the same molecule, $\langle P \rangle$ is the average value of $e^{i\mathbf{h}\cdot\mathbf{u}}$, where $\mathbf{u}$ is a vector connecting molecular centers,* and the summations are over all the atoms in a molecule.

Term II: $\qquad -c \int_m \int_m \langle \rho_m(\mathbf{r}) e^{i\mathbf{h}\cdot(\mathbf{r}-\mathbf{r}')} \rangle\, d\mathbf{r}\, d\mathbf{r}'.$

To evaluate this term we make the substitutions $\mathbf{r}' = \mathbf{R}_v + \mathbf{r}_v$, and when $\mathbf{r}$ ends in the vicinity of atom j in the uth molecule $\mathbf{r} = \mathbf{R}_u + \mathbf{r}_{uj} + \mathbf{x}_{uj}$, where $\mathbf{R}_u$ and $\mathbf{R}_v$ are respectively vectors from an arbitrary origin to the centers of the uth and vth molecules, $\mathbf{r}_{uj}$ is a vector from the center of molecule u to the center of atom j in that molecule, $\mathbf{x}_{uj}$ is a vector whose origin lies at the center of atom j in the uth molecule, and $\mathbf{r}_v$ originates at the center of molecule v. Since the atomic scattering factor of atom j is given by

$$f_j = \int_{\text{atom }j} \rho(\mathbf{x}) e^{i\mathbf{h}\cdot\mathbf{x}}\, d\mathbf{x}$$

Term II becomes

$$-c \sum_u \sum_v \sum_j \langle f_j e^{i\mathbf{h}\cdot\mathbf{r}_{uj}} e^{i\mathbf{h}\cdot(\mathbf{R}_u - \mathbf{R}_v)} \int_{\text{molecule }v} e^{-i\mathbf{h}\cdot\mathbf{r}_v}\, d\mathbf{r}_v \rangle$$

in which the integration is over the volume occupied by molecule v. Calling the terms for which $u = v$ internal terms, we note that there are N such terms, one for each molecule; furthermore they are all equal since we average over all orientations. Making the substitutions

$$F_u = \sum_j f_j e^{i\mathbf{h}\cdot\mathbf{r}_{uj}} \qquad \text{and} \qquad \Phi_u = \int_{\text{molecule }u} e^{i\mathbf{h}\cdot\mathbf{r}_u}\, d\mathbf{r}_u$$

these internal terms may be written $-Nc\langle F_u \Phi_u^* \rangle$, in which the subscripted quantities are computed for the same molecule. There are $N(N-1)$ terms for which $u \neq v$. These external terms are again all equal since we average over all orientations and all possible inter-

---

* The center of a molecule may be chosen arbitrarily subject to the restriction that it be defined similarly when computing $\langle F \rangle$ and $\langle P \rangle$. The assumption of a spherical cavity, to be made later, will imply a choice of origin which will be obvious for symmetrical solute species.

molecular vectors. If we assume that the distribution of intermolecular vectors is independent of molecular orientation, we may write for the contribution of the external terms $-N^2c\langle F_u\Phi_v\rangle\langle P\rangle$, in which we have put $N - 1 \cong N$ since N is large. Assuming further that the orientations of molecules are independent of each other, we have $\langle F_u\Phi_v^*\rangle_{u \neq v} = \langle F_u\rangle\langle \Phi_v^*\rangle = \langle F\rangle\langle \Phi^*\rangle$, and we may write for Term II: $-Nc\langle F_u\Phi_u^*\rangle - N^2c\langle F\rangle\langle \Phi^*\rangle\langle P\rangle$.

Term III: $\quad -c\int_m\int_m \rho_m(\mathbf{r}')e^{i\mathbf{h}\cdot(\mathbf{r}-\mathbf{r}')}\,d\mathbf{r}\,d\mathbf{r}'$

and

Term IV: $\quad c^2\int_m\int_m e^{i\mathbf{h}\cdot(\mathbf{r}-\mathbf{r}')}\,d\mathbf{r}\,d\mathbf{r}'.$

These may be shown by similar arguments to be given, respectively, by $-Nc\langle F_u^*\Phi_u\rangle - N^2c\langle F^*\rangle\langle \Phi\rangle\langle P\rangle$ and $Nc^2\langle|\Phi|^2\rangle + N^2c^2\langle\Phi\rangle\langle\Phi^*\rangle\langle P\rangle$.

If it is further assumed that the volume occupied by a molecule is a sphere of radius r_0, then

$$\langle|\Phi|^2\rangle = \langle\Phi\rangle\langle\Phi^*\rangle = \Phi^2 \qquad 9a.$$

and

$$\langle F_u\Phi_u^*\rangle = \langle F_u^*\Phi_u\rangle = \langle F\rangle\Phi \qquad 9b.$$

with

$$\Phi = (4\pi/h^3)(\sin hr_0 - hr_0\cos hr_0) \qquad 10.$$

Debye (4) has shown $\langle P\rangle$ to be given by

$$\langle P\rangle = \frac{4\pi}{N}\int_0^\infty [D(r) - D_0]r^2\frac{\sin hr}{hr}\,dr \qquad 11.$$

where $D(r)$ is an assumed spherically symmetric function representing the density of solute molecules, the centers of which lie a distance r from the center of a central molecule, and D_0 is the average density of solute molecules in the system. It is further assumed that $D(r)$ rapidly approaches D_0 with increasing r.

In view of the preceding development, we write for the intensity scattered per solute molecule

$$i_N(h) = i(h)/N = [\langle F^2\rangle - 2c\Phi\langle F\rangle + (c\Phi)^2]$$
$$+ 4\pi(\langle F\rangle - c\Phi)^2\int_0^\infty [D(r) - D_0]r^2\frac{\sin hr}{hr}\,dr \qquad 12.$$

in which $\langle F \rangle$, $\langle F^2 \rangle$ and Φ are given respectively by 7, 8 and 10. As usual one may solve for $D(r)$ by applying Fourier's Integral Theorem.

The assumptions under which 12 was derived are listed below:

(1) Solute molecules occupy clearly defined spherical holes.

(2) The electron content of the solvent is uniformly distributed in those regions not occupied by solute.

(3) Sample-shape scattering is not observed.

(4) The orientation of a solute particle is independent of the orientation of any other solute particle as well as the distribution of solute particle centers.

(5) The distribution of solute particle centers, $D(r)$, is spherically symmetric about any given center, with $D(r)$ rapidly approaching D_0 with increasing r.

It is of interest to note that the first (bracketed) term on the right side of 12 was derived solely from a consideration of internal effects, that is interference effects due to electron pairs entirely contained within the hole occupied by one solute molecule. In contrast, the second term on the right side of 12 arises solely from a consideration of electron pairs, each electron of which lies in a different hole. This formal separation of interference effects will be used to advantage in applying 12 to the interpretation of experimental data.

ANALYSIS OF OBSERVED SMALL-ANGLE PATTERNS

Experimental Systems and Methods

Experimental systems particularly suited to analysis by 12 should have certain characteristics. Specifically, such a system should contain only one scattering solute species, preferably with atoms of high electron content relative to the solvent. In addition the solute particles should be sufficiently large so as to prevent important information about $D(r)$ from occurring at such high scattering angles that the assumption of uniform solvent electron density breaks down. Finally, the solute particles should be of such a shape so as to fit reasonably well into spherical holes. These requirements are met fairly well by aqueous solutions of chloroplatinic acid, which were chosen to be studied here. The principal scattering solute specie in these systems is the octahedral doubly-charged chloroplatinate ion ($PtCl_6^=$), with a chlorine-platinum distance of 2.33 Å*(5).

---

* A bond distance of 2.39 Å is reported (2) in solution, but 2.33 Å was used in calculating $\langle F \rangle$ and $\langle F^2 \rangle$, since the crystallographic study (5) is presumed more accurate than that of the solution.

Scattering patterns for 1.0035, 2.0016 and 2.9690 molal solutions of H_2PtCl_6 were measured. A Norelco X-ray generator and diffractometer were used to obtain the data. A copper target was employed along with a curved-crystal monochromator, which selected the $K\alpha$ doublet of Cu (average $\lambda = 1.542$ Å). The scattered radiation was detected by a scintillation counter used in conjunction with a linear amplifier, pulse-height analyzer, and decade counter. Samples of the solution were contained in flat cells with either "Mylar" or "Saran Wrap" windows. The cells varied in thickness from about 0.3 to 0.8 mm, depending on the concentration of the solutions studied. The scattering was measured for values of $(\sin \theta)/\lambda$ in the interval 0.02 to 0.13 Å^{-1} at increments of 0.005 Å^{-1}. The observed data were corrected for polarization, absorption, background and window scatter. The incoherent scatter was calculated from Compton's formula $(Z - f^2/Z)$ along with the Breit-Dirac factor, and was subtracted from the total scatter. The scattering factors used for Cl, H, Pt and O, in all calculations, were obtained respectively from papers by Berghuis et al. (6), McWeeny (7), Thomas and Umeda (8), and James and Brindley (9).

The Internal Contributions to the Scattering

We note that the coherent small-angle scattering as given by 12 is a function of two parameters as well as the angular variable h. In particular, the intensity depends on the solvent electron density c, and the radius of the spherical cavity, r_0. The latter quantity appears via the function $\Phi(h;r_0)$. These two parameters are not independent, being related through the observable bulk density of the solution. It is convenient to eliminate c as a parameter by writing $c = c(r_0)$, the functional relationship being developed by assuming the solvent to be excluded from the holes of volume $(\frac{4}{3})\pi r_0^3$. For aqueous H_2PtCl_6 solutions, c takes the form

$$c(r_0) = \frac{\mathcal{a} \times 10^{-20}}{M_1[(1000 + M_2 m)\, d - (\frac{4}{3})\pi r_0^3 m \mathcal{a} \times 10^{-24}]} \qquad 13.$$

in which c is given in number of solvent electrons per Å^3 of solvent, M_1 and M_2 are respectively the formula weights of H_2O and H_2PtCl_6, m is the molality of the solution with respect to H_2PtCl_6, d is the density of the solution in g/cm^3 and $\mathcal{a}$ is Avogadro's number. In all calculations the following values of the constants were used: $M_1 = 18.016$, $M_2 = 409.98$, and $\mathcal{a} = 6.023 \times 10^{23}$.

We refer now to Fig. 1, in which is presented, for three concentrations of solution, the normalized and corrected coherent intensity as a function of $(\sin \theta)/\lambda$. For each system studied, the angular range of observation has been divided into two parts. The lower- and higher-angle parts have been labeled, respectively, Region A and Region B. Both sections, however, fall into what may be called the small-angle region of scattering. The dotted curves represent $\langle F^2 \rangle$ as calculated for the chloroplatinate ion via 8. It is of interest to note that in both Regions A and B, $\langle F^2 \rangle$ gives an inadequate description of the experimental scatter. Furthermore, it is apparent that no change in the normalization of the data will secure a fit with $\langle F^2 \rangle$.

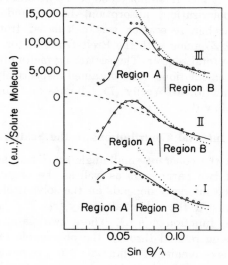

Fig. 1. Comparison of corrected experimental data with theoretical curves. —— $i_p(h;\mathbf{p})$, --- $i_{\text{INT}}(h;r_0)$, $\cdots$ $\langle F^2 \rangle$, ○ corrected experimental intensity. I, II and III refer respectively to 1.0035 m, 2.0016 m and 2.9690 m solutions.

After computing $\langle F^2 \rangle - 2c\Phi\langle F \rangle + (c\Phi)^2$ [henceforth denoted as $i_{\text{INT}}(h;r_0)$] with a trial value of r_0, it became evident that for each system studied there existed a significant portion of the small-angle region for which, with proper normalization, the experimental data could reasonably be made to fit $i_{\text{INT}}(h;r_0)$. This angular range is what we have called Region B. The trial value of r_0 was chosen so as to make the corresponding spherical hole neatly contain the chloroplatinate ion. Specifically, r_0 was set equal to the Pt-Cl bond distance plus one van der Waals radius of chlorine, yielding a trial value of 3.73 Å.

In view of these facts, we attempted a least-squares fit of the data to $i_{INT}(h;r_0)$ in Region B. The parameters that were varied were r_0 and a normalization constant, k. We minimized the quantity $\Sigma_i[ki_e(h_i) - \xi(h_i;r_0)]^2$ with respect to k and r_0, where $i_e(h_i)$ is the experimental intensity, corrected for absorption and polarization, observed at the point $h = h_i$, and $\xi(h_i;r_0)$ is given by $i_{INT}(h_i;r_0)$ plus the calculated incoherent scatter at the point i. This process yielded two equations:

$$\Sigma_i [ki_e(h_i) - \xi(h_i;r_0)][c(r_0)\Phi(h_i;r_0) - \langle F(h_i)\rangle]$$
$$\times \partial/\partial r_0[c(r_0)\Phi(h_i;r_0)] = 0 \quad 14.$$

and

$$k = \frac{\Sigma_i i_e(h_i)\xi(h_i;r_0)}{\Sigma_i i_e^2(h_i)} \quad 15.$$

in which $c(r_0)$ is given by 13. The quantity k was eliminated from 14 and 15; the resulting equation was then solved numerically for r_0. The least-squares radii so obtained are listed in Table I.

TABLE I
Some of the principal results obtained for the chloroplatinic acid solutions.

Solution molality	r_0 (Å)	$D_0 \times 10^4$ (molecules/Å^3)	Position of first maximum (Å)		Solvent electron density,* (electrons/Å^3)	Nearest neighbor distance (Å) if close-packed
			$D(r)$	$D_e(r;\mathbf{p})$		
1.0035	3.52	5.48	13.5	Absent	0.3367	13.7
2.0016	3.62	10.04	9.5	9.7	0.3477	11.2
2.9690	3.55	13.74	8.5	8.4	0.3460	10.1

* For pure water at 25°C, $c = 0.3333$.

Since the solute particles do not, in reality, occupy spherical holes, r_0 must ultimately be regarded as an empirical parameter. Nevertheless, its value should be some measure of particle size. The results of the least-squares analysis of Region B yield over the concentration range of about 1 m to 3 m, an average r_0 of 3.56 Å with a standard deviation of 0.04 Å. This is within 5% of the value calculated from the sum of the Pt-Cl bond distance plus a van der Waals radius of chlorine. A further test of the applicability of the model is that the radii of the spherical holes be independent of solution concentration. The least-squares values of r_0 indicate that this is the case.

The quantity $i_{\text{INT}}(h;r_0)$ is represented in Fig. 1 by a dashed curve. Let us, for the moment, confine our attention to Region B. Since the value of Φ falls off rapidly with increasing angle, we note that in the high-angle end of Region B, $i_{\text{INT}} \approx \langle F^2 \rangle$. Here the experimental points fit both these curves reasonably well. At smaller angles, $\langle F^2 \rangle$ deviates from i_{INT}; the experimental points are seen to follow i_{INT} throughout the remaining part of Region B. It is difficult to account quantitatively for the errors in the corrected experimental intensities. Probably the chief sources of error are those due to counting and correcting for absorption. The former can be estimated by assuming a Poisson distribution of photons arriving at the counter. The latter errors arise principally because the sample is not perfectly flat, whereas the absorption correction is for a flat sheet specimen. These errors cannot easily be estimated. The root mean square deviation of the corrected values of the observed intensities i_c from i_{INT} was computed as $\langle [(i_c - i_{\text{INT}})/i_{\text{INT}}]^2 \rangle^{\frac{1}{2}}$, yielding values of this quantity for the 1.0035 m, 2.0016 m and 2.9690 m solutions respectively equal to 0.021, 0.040 and 0.026. The corresponding deviations expected from counting errors alone are respectively 0.012, 0.013 and 0.013. We cannot decide on the basis of these experiments whether the difference between the two sets of deviations is due to a theoretical inadequacy or rather to unaccountable errors, absorption or otherwise. One may, however, safely conclude that the model predicts the scattering pattern very much better than just the $\langle F^2 \rangle$ term alone (dotted curve in Fig. 1) and that any adequate description of the scattering in this angular range must include the effects of solvent.

The External Contributions to the Scattering

In view of the excellent fit of the data to i_{INT} in Region B, it appears that external contributions to the scattered intensity are negligible in this angular range. Thus one may assume that the information regarding the spatial distribution of solute molecule centers lies essentially in Region A of the scattering pattern. Not unexpectedly, one notes that Region B covers a progressively smaller range of angles as the solution concentration increases. As the solute particles are packed closer together, the intermolecular interference effects appear at higher scattering angles.

Rearrangement of 12 and the application of Fourier's Integral Theorem gives

$$4\pi r^2[D(r) - D_0] = \frac{2r}{\pi} \int_0^\infty \frac{[i_c(h) - i_{\text{INT}}(h;r_0)]}{(\langle F \rangle - c\Phi)^2} h \sin hr \, dr \qquad 16.$$

The integral in 16 can be evaluated from the value of r_0 obtained in the previous section and corrected values of the experimental intensity, $i_c(h)$. The choice of an upper limit of integration for computational purposes is, in this case, clearly indicated. In accordance with the preceding interpretation of the data, the integrand vanishes outside the Region A. A natural choice of the upper limit is, therefore, the

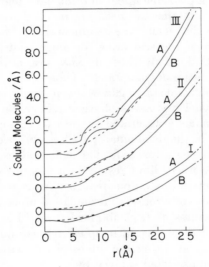

Fig. 2. Curves of $4\pi r^2 D_e(r)$, A, and $4\pi r^2 D(r)$, B. The dashed curves represent the functions $4\pi r^2 D_0$. I, II and III refer respectively to the 1.0035 m, 2.0016 m and 2.9690 m solutions.

largest value of h in Region A. The functions $4\pi r^2 D(r)$ were so computed and are displayed in Fig. 2.

There are certain features that one may reasonably expect the radial distribution of solute particles to incorporate. In particular, $D(r)$ should vanish in the neighborhood of the origin, since deep interpenetration of molecules is forbidden. Furthermore, it is expected that the more concentrated solutions will yield more sharply defined features in this function. Both of these characteristics seem to be present in the calculated radial distribution. The areas under the curves $4\pi r^2 D(r)$ are negligible for r less than about 6 Å, a value somewhat less than the sphere diameter $2r_0$ of 7.12 Å. In all cases a maximum follows the initial region of negligible D, and as r increases, the remaining features occur with decreasing amplitude. The first maxima appear respectively in the 1.0035 m, 2.0016 m and 2.9690 m solutions at 13.5 Å, 9.5 Å and 8.5 Å, indicating a decrease in the most probable anion-anion nearest-neighbor distance with increasing concentration.

An Alternative Analysis of the External Contributions

The fact that a plausible upper limit of the integral in 16 was used does not necessarily free the resulting radial density function from error other than that incurred experimentally. Although i_{INT} gives a good description of the scattering for angles outside of Region A, external contributions to the intensity are nevertheless present throughout the entire angular range. Thus $i_c(h) - i_{\text{INT}}(h;r_0)$ cannot be precisely zero outside of Region A, even if $i_c(h)$ is free of error and the proposed model represents physical reality. Small deviations of i_c from i_{INT} are in fact exaggerated by the h-weighting in the integrand of 16; in addition, the denominator $(\langle F \rangle - c\Phi)^2$ becomes very small for large h and indeed vanishes for particular values of h. The imposition of the finite upper limit results in a loss of resolution and the possible introduction of spurious features in the radial density function. It is therefore of interest to formulate a procedure to determine which of the features of $D(r)$ obtained from 16 are experimentally meaningful. This has been attempted by assuming an empirical density function and fitting the intensity calculated from it to the data. The function selected is given by

$$D_e(r;\mathbf{p}) = D_0(2\pi)^{-\frac{1}{2}} \int_{-\infty}^{[r-2(R_0+\sigma_0)]/\sigma_0} e^{-u^2/2}\, du + \sum_{i=1}^{n} B_i e^{-(r-R_i)^2/2\sigma_i^2} \qquad 17.$$

where the argument $\mathbf{p}$ indicates dependence on the parameters R_0, σ_0, B_1, R_1, σ_1, $\cdots$, B_n, R_n, σ_n. The first term on the right side of 17 makes no appreciable contribution to $D_e(r;\mathbf{p})$ at small values of r. There then follows a sharp increase in the value of this term which proceeds asymptotically to approach D_0 in a manner controlled by the parameters R_0 and σ_0. Superimposed on this term are the remaining Gaussian-shaped maxima (or minima for $B_i < 0$).

Substitution of 17 for $D(r)$ in 12 and evaluation of the integrals, subject to the restriction that $\exp\{-[r - 2(R_0 + \sigma_0)]^2/2\sigma_0^2\} \approx 0$ for $r < 0$, yields for the intensity

$$i_p(h;\mathbf{p}) = i_{\text{INT}}(h;r_0) + 4\pi h^{-1}(\langle F \rangle - c\Phi)^2 T(h;\mathbf{p}) \qquad 18.$$

where

$$T(h;\mathbf{p}) = D_0 e^{-(h\sigma_0)^2/2}\{2(R_0 + \sigma_0)h^{-1}\cos[2h(R_0 + \sigma_0)] \\
- (\sigma_0^2 + h^{-2})\sin[2h(R_0 + \sigma_0)]\} \\
+ (2\pi)^{\frac{1}{2}}\sum_{i=1}^{n} \sigma_i B_i e^{-(h\sigma_i)^2/2}(h\sigma_i^2 \cos hR_i + R_i \sin hR_i) \qquad 19.$$

An attempt was made to fit the right side of 18 to the corrected, normalized intensity data. The number of terms in the summation in 19 was kept to the minimum necessary for each solution concentration. That is, additional terms that produced no significant improvement in the fit were dropped. Employed for this purpose was the least-squares procedure of Newton and Gauss as modified by Strand et al. (10). The initial set of parameters was chosen so as to make $D_e(r;\mathbf{p})$ approximate $D(r)$ as obtained by Fourier inversion. The best set of parameters is listed in Table II. Included in Figs. 1 and 2 respectively are the functions $i_p(h;\mathbf{p})$ and $4\pi r^2 D_e(r;\mathbf{p})$.

TABLE II
The parameters of $D_e(r;\mathbf{p})$.

H_2PtCl_6 solution	$R_0(\text{Å})$	$\sigma_0(\text{Å})$	$B_1 \times 10^4$ (molecules/Å^3)	$R_1(\text{Å})$	$\sigma_1(\text{Å})$	$B_2 \times 10^4$ (molecules/Å^3)	$R_2(\text{Å})$	$\sigma_2(\text{Å})$
1.0035 m	3.33	0.001	—	—	—	—	—	—
2.0016 m	2.81	0.588	1.36	9.71	0.659	—	—	—
2.9690 m	3.43	0.539	4.85	8.41	1.72	−2.39	12.44	0.557

The most notable feature of this analysis is the reduction in the number of parameters necessary as the solution concentration decreases. Thus the 2.9690 m solution required an eight-parameter $D_e(r;\mathbf{p})$ to explain the observed intensities. Both a relative maximum and minimum were required in the density function. The 2.0016 m and 1.0035 m solutions required five and two parameters respectively, $D_e(r;\mathbf{p})$ having a relative maximum in the former case. In all cases $D_e(r;\mathbf{p})$ first becomes non-zero for r between 6.3 Å and 6.6 Å, again somewhat less than the value of $2r_0$ of 7.12 Å. It is of interest to note that for the 1.0035 m solution $D_e(r;\mathbf{p})$ is essentially the distribution of an infinitely dilute gas of the ions with only a hard sphere interaction. This fact illustrates the relative insensitivity of scattered X-rays to the distribution of just the anions, even at what one normally considers fairly high concentrations, and points up the need for high-quality data in investigations of this kind.

A comparison of $D(r)$ with $D_e(r;\mathbf{p})$ shows that the radial distribution functions obtained by Fourier inversion contain more features at large values of r than are necessary to explain the observed data. For this reason, no significance is ascribed to these features. These extra features are most likely artifacts of the procedure used to obtain $D(r)$, and arise from the facts that the data are imperfect and that, of necessity, a finite upper limit was imposed on the integral in 16. Of the remaining features the following observations may be made. Referring to Table I, which summarizes some of the important results of this analysis, it is to be noted that the position of the first maximum in $D(r)$ appears at lower values of r as the solution concentration increases. If the $PtCl_6^=$ ions were distributed as a gas of hard spheres, the position of these maxima would be independent of concentration and equal to the sphere diameter (11). This variation in peak position manifests the longer-range forces (principally electrostatic) at play in these systems. Included in Table I are the nearest-neighbor distances expected from a close-packed arrangement of anions consistent with the observed solution densities. The most probable nearest-neighbor positions are seen to be intermediate to those of a hard sphere gas and the close-packed arrangement.

On the Applicability of the Method

The small-angle region of scattering contains, and has been made to yield, information regarding the radial distribution of solute particles. It has been necessary, however, to introduce certain assumptions which limit the types of systems that may be considered by this technique. Of primary importance is the fact that only one scattering solute species may be present in the solutions studied. In the particular systems studied here, nothing has been learned of the disposition of the cations. Secondly, let us consider the assumption of uniform solvent electron density. It has been pointed out that this approximation is valid only at small scattering angles, since in this region short-range fluctuations in electron density do not strongly influence the scattering patterns. Thus, the preceding analysis is valid only for systems in which the solute particles are of sufficient size so that the distance of closest approach of solute particles is considerably greater than the shortest solvent-particle separation. If this condition is not met, features in the scattering curve related to external interference effects will appear at scattering angles too large for the uniform solvent density approximation to apply and the solute-solute interactions will be inextricably tied up with solvent-solvent and solvent-solute interactions. Thus, for example, aqueous solutions of simple ions may not be considered by this technique without considerable modification of the theory.

BIBLIOGRAPHY

1. Brosset, C., Arkiv Kemi, **1,** 353 (1949).
2. Vaughan, P. A., Sturdivant, J. and Pauling, L., J. Am. Chem. Soc., **72,** 5477 (1950).
3. Menke, H., Phys. Z., **33,** 593 (1932).
4. Debye, P., J. Math. and Phys., **4,** 133 (1925).
5. Ewing, F. J. and Pauling, L., Z. Krist., **68,** 223 (1928).
6. Berghuis, J., Ibertha, J., Haanappel, M. and Potters, M., Acta Cryst. **8,** 478 (1955).
7. McWeeny, R., ibid., **4,** 513 (1951).
8. Thomos, L. H. and Umeda, K., J. Chem. Phys., **26,** 293 (1957).
9. James, R. W. and Brindley, G. W., Z. Krist., **78,** 470 (1931).
10. Strand, T. G., Kohl, D. A. and Bonham, R. A., J. Chem. Phys., **39,** 1307 (1963).
11. Kirkwood, J. G., Maun, E. K. and Alder, B. J., ibid., **18,** 1040 (1950).

X-Ray Scattering Study of Bone Mineral

E. D. EANES AND A. S. POSNER

The Hospital for Special Surgery, Cornell University Medical College,
New York, N.Y. 10021

INTRODUCTION

Bone is a connective tissue whose extracellular substance consists of a hydroxyapatite-like mineral deposited in a hydrated protein matrix. The submicroscopic crystals of the calcium phosphate mineral phase are uniformly dispersed throughout the protein creating a large mineral-matrix interface of considerable physiological importance. In the studies reported herein, small-angle diffraction techniques were employed to measure the area of this interface. It was felt that small-angle X-ray scattering would be an effective means of measuring this area because the method involved minimal treatment in preparing the bone samples for measurement. In previous investigations, using gas-solid adsorption techniques, drastic heating and dehydration procedures had to be employed to prepare the samples for specific surface measurements (1). The value reported in these earlier studies, 100 square meters per gram, represents the specific surface of the crystals of bone after dehydration and chemical removal of the protein (1). Whether these surface area measurements obtained by low-temperature gas adsorption reflect the extent of the *in vivo* mineral-matrix interface is open to question. The human bone specimens used in the present study were defatted by alcohol and ether extraction, ground and sieved; 44-74 micron fractions were used for the low-angle measurements. This method of sample preparation, although it dehydrates bone partially, leaves the *in vivo* crystal protein interfaces essentially intact.

For purposes of interpreting the small-angle diffraction pattern, bone tissue can best be described as a binary system composed of a phase of hydroxyapatite microcrystals dispersed in a homogeneous matrix. The dispersing medium is a mixture of collagen fibers, ground substances, dissolved body salts, and water. Even though the principal constituent of this medium, collagen, is sufficiently structured to produce a characteristic low-angle X-ray pattern, it was found from scattering studies on decalcified, powdered, randomly-oriented bone matrix that the intensity of this scattering is negligible compared to that of whole bone. Further, from similar studies on bones treated with ethylenediamine to remove the organic fraction, it was found that the X-ray scattering profile at low angles from whole bone is predominantly due to the size, shape and spatial distribution of the apatite crystals. Consequently, the area of the interface being measured in this study on whole bone is essentially equal to the sum of the areas of the non-contiguous crystal surfaces.

EXPERIMENTAL METHODS

The small-angle X-ray diffraction patterns were recorded electronically on a commercially available small-angle scattering apparatus of the Kratky design (2). Aperture lengths were chosen so that the corresponding scattering profile would approximate the shape of a curve produced by a beam of infinite height under the same conditions. Monochromatization of the X-ray beam was achieved by using Ni-filtered $CuK\alpha$ radiation and a proportional counter detector coupled with a pulse height analyzer. Fixed time, manual step-scanning of the diffraction curve was employed to collect the intensity data. The intensity at each step was recorded to within a probable error of 1.0% or less.

EVALUATION OF RESULTS

Porod (3) established that an absolute measure of the specific surface, defined as the ratio of the dispersed phase interface to the volume of the dispersed phase, can be obtained from the invariant, Q, of the system and the outer (i.e., higher-angle) portion of the low-angle scattering curve. The invariant, dependent only upon the total volume of the dispersed phase, can be evaluated from the integral

$$\int_0^\infty I(M)M \, dM$$

where $I(M)$ is the scattered intensity and M is related to the scattering angle 2θ by the expression $\tan 2\theta = M/a$ where a is the sample-detector distance. In the outer portion, or tail, of the scattering curve the quantity $I(M)M^3$ becomes constant and proportional to the total surface area of the dispersed phase. According to Porod (3), it follows that, S_{sp}, the specific surface, can be evaluated from the relation:

$$S_{sp} = \frac{8\pi W_2 I(M) M^3}{\lambda a Q} \qquad 1.$$

where λ is the wavelength of the X-radiation employed and W_2 is the volume fraction of the dispersing phase. The value, 0.572, used for the W_2 of bone was calculated from bone density data (4). The

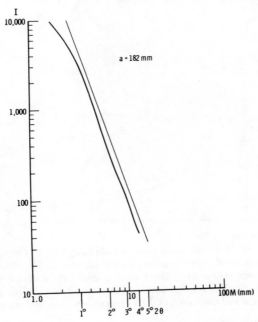

Fig. 1. Log $I(M)$ vs. log M for a sample of low fluoride human rib bone. $I(M)$ is the scattered intensity for a primary beam of infinite height, and M is the distance (in mm) from the origin along the line of register. The solid straight line to the right of the scattering curve represents a line of slope -3.

above relations are valid for scattering curves not corrected for slit-length collimation error. To determine background level and to establish the validity of Porod's equations for our experimental system, the data were plotted according to the following equation:

$$I(M)M^3 = K_1 + K_2M^3 \qquad 2.$$

A straight-line plot was obtained for the data in the tail of our experimental curve, thus validating the use of Porod's equations for slit systems of infinite height and the constant term K_2, representing the background error, was subtracted from the observed scattering curve. Figure 1 shows a $\log I(M)$ vs. $\log M$ plot of background-corrected data from a human bone sample.

In the experimental evaluation of the invariant, Q, the value of the integrand had to be estimated for the extreme inner and outer portions of the diffraction pattern by extrapolation of the experimental curve. The function, K_1/M^3, was assumed to be a valid

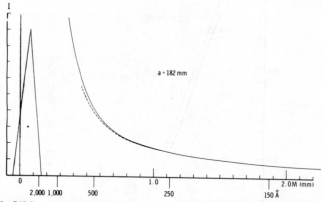

Fig. 2. $I(M)$ vs. M for the same bone sample as in Fig. 1. The solid line represents the experimental curve uncorrected for slit width. The dashed line is the corresponding corrected curve. The cross-sectional profile of the incident beam is shown at the left. The angular positions corresponding to M are given in terms of Bragg spacings and expressed in Å. The quantity, a, is the sample-to-detector distance.

extrapolation of the integrand for the tail of the scattering curve. The inner part of the integrand could not be so easily delineated, however. As shown in Fig. 2, the lower-angle portion of the experimental scattering curve for bone exhibits a pronounced concavity. This sharp rise in the curve at low angles can probably be attributed to the fact that the assemblage of bone crystals is polydisperse in size and shape. Results obtained by electron microscopy support this conclusion (5). When log $I(M)$ of the bone data is plotted against M^2 the concavity persists making an extrapolation to zero angle based on the assumptions inherent in Guinier's law (6) uncertain. In addition, a plot of $MI(M)$ against M could not be extrapolated to low angles since the experimental portion of the integrand failed to converge to zero.

In the present experiment, the following extrapolation method was employed: The background-corrected experimental intensity data were corrected for finite slit width aberration by using the expression given by Kratky, Porod and Skala (7):

$$I(M) = 2\hat{I}(M) - \left[\frac{\hat{I}(M + \epsilon) + \hat{I}(M - \epsilon)}{2}\right] \qquad 3.$$

where ϵ is the positive square root of the second moment about the mean, M_0, of the incident beam profile and $\hat{I}(M)$ and $I(M)$ are, respectively, the scattered intensity before and after correction for beam width. In Fig. 2, the dashed line represents the corrected curve while the solid line represents the original uncorrected scattering data for bone. Figure 2 shows that the experimental slit width actually allowed for a resolution of 1870 Å. It was decided to cut off the corrected curve arbitrarily at 600 Å because of the difficulty of estimating the slit correction for the steeply rising portion of the scattering curve beyond this point.

Following the approximation suggested by Guinier and Fournet (8), to extend the angular range over which the Guinier formula can be applied, the slope (p) of the (log of corrected intensities) vs. M^2 curve was plotted as a function of M^2. The p vs. M^2 plot was then linearly extrapolated to zero from the two innermost values of p obtained from the experimental data. The extrapolated p's were then used to calculate the values of $I(M)M$ needed to evaluate Q to its lower limit. Both numerical and graphical methods were used to compute Q, with results agreeing to within 0.2%.

From the low-angle scattering data the value of S_{sp} for a low-fluoride human rib bone was found to be 185 square meters per gram of bone mineral. A value of 152 square meters per gram, uncorrected for slit width, was previously reported by the writers (9). The equivalent value per gram of whole bone would be 111 square meters. Though considerably higher than the values found from gas-solid adsorption data (1), these values agree with the surface of 160 square meters per gram of mineral calculated from the electron microscopic data on bone crystallite size given by Johansen and Parks (5). The drastic treatment required in preparing bones for gas adsorption measurements could cause crystal growth, effect an increase in the numbers of continuous surfaces through clumping, or cause the formation of closed pores whose inner surfaces are inaccessible to the adsorbing gases. Each of these factors would tend to decrease the surface available for gas adsorption accounting for the lower specific surface reported from adsorption data.

Some degree of uncertainty exists as to the accuracy of the value of S_{sp} obtained by small-angle scattering. Because the scattering curve of bone increases steeply at very small angles, the linear extrapolation procedure may underestimate the contribution of this unknown part of the curve to Q, resulting in subsequent overevaluation of S_{sp}. In addition, the uncertainty of the scattering at low angles prevents gathering information about the possible presence of large crystals or aggregates. Thus the value 185 square meters per gram must be considered as an upper limit to the specific surface of the mineral portion of bone. In this regard, absolute intensity measurements of the low-angle data could be a more accurate approach to the values of specific surface as the difficulties of evaluating the invariant, Q, are avoided.

Previously reported wide-angle X-ray diffraction data (10) have shown that the ingestion and subsequent accumulation of fluoride in osseous tissue is accompanied by significant alterations in the crystal size of bone apatite. Low-angle diffraction was employed to measure the corresponding changes that occurred in the area of the mineral-matrix interface. To measure these changes in surface area with increased fluoride content, it was not necessary to follow the laborious procedure discussed above for S_{sp}. Since the values of Q would be identical, if one compares the X-ray scattering from a number of bone samples containing the same volume (or mass) of mineral, the ratio of the specific surfaces of the samples can be obtained from the ratios of the $I(M)M^3$ values. In the present study all measurements were taken at the same scattering angle in the Porod region of the scattering curve and the surface area of each sample relative to the same reference (a low fluoride sample) was calculated from the relation:

$$S_{sp}/S_{sp}(ref) = I(M)/I(M)(ref) \qquad 4.$$

where the term (ref) identifies the reference sample. The quantity $S_{sp}/S_{sp}(ref)$ is defined as the relative surface, S_{rel}, of the sample. The experimental details of this low-angle study on fluoride containing bone appear elsewhere (11).

The data for twelve human rib samples, varying in percent fluoride, are shown in Fig. 3. The figure shows a plot of the relative surface values, S_{rel}, obtained from the X-ray low-angle scattering data as a function of the percent fluoride in the mineral portion of the bones.

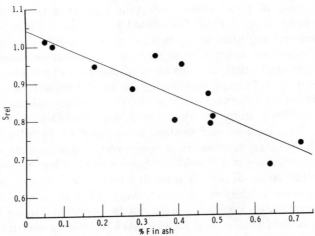

Fig. 3. S_{rel} vs. percent fluoride (ash weight basis) for a series of human rib specimens. The solid line is the calculated least squares (regression) line. The slope (k) of this line is -0.46, i.e. a 0.1% change in fluoride level reduces S_{sp} by 4.6%. The corresponding correlation coefficient (r) is 0.87. A Student's "t" test for linearity by r showed that the straight-line relationship is significant at a probability level <0.001.

The decrease in relative surface with rise in fluoride content is consistent with the increase in resolution observed in the wide-angle X-ray diffraction work previously noted (10). The rib samples were chosen from a collection of specimens obtained at *postmortem* from mature individuals who lived for at least ten years in one of four geographical areas where the municipal water supply contained 0.1, 1.0, 2.0 and 4.0 ppm. fluoride.

The wide-angle X-ray studies (10) confirmed the substitution of fluoride entering the bone into the crystalline hydroxyapatite structure. Hence, there is a chemical stabilization of the bone apatite by (a) the formation of a less soluble fluoride-substituted hydroxyapatite, and (b) the reduction of reactive surfaces per unit weight of bone crystals. The exact mechanism by which fluoride effects an increase in the size of bone apatite crystals is not clear. The final size a crystal achieves under a given set of environmental conditions is the result of the actions of two essentially competitive factors: the rate at which new crystals in the system are being formed (nucleation) and the rate at which already formed crystals are increasing in size (growth). The total volume occupied by the mineral relative to the protein phase in bone remains constant regardless of the fluoride content of bone (11). Thus, fluoridated bone must contain fewer of the enlarged bone apatite crystals in order to maintain the same mineral-to-protein volume ratio as found in non-fluoridated bone. This would imply that a suppression of the nucleation rate accompanies a fluoride-induced increase in the crystal growth rate. However, the higher supersaturation in the precipitating solution (blood serum) resulting from the decrease in the solubility of the final product (i.e., fluor-hydroxyapatite as opposed to hydroxyapatite) should lead to an increase in homogeneous nucleation as well as crystal growth rate in a fluoridated bone system. Since it is believed that the formation of bone is regulated by a heterogeneous and not a homogeneous nucleation process, however (12), the above effect of supersaturation on nucleation may not be operative. In all events, it seems evident that crystal growth takes precedence relative to the nucleation process in fluoride-containing bones.

Note added in proof:

Results obtained in this laboratory since submitting the manuscript have shown that bone mineral is composed of a non-crystalline calcium phosphate in addition to the poorly crystallized hydroxyapatite (13, 14). In adult bovine and rat bone, this non-crystalline phase comprises as much as 40% of the total mineral (15). Fortunately, a close chemical similarity exists between the two phases so that the magnitude of the discontinuity in electronic density across their respective mineral-matrix interfaces is nearly the same. The S_{sp} reported in this paper can be taken, then, as the average of the specific surfaces of the two phases. The disparity in the S_{sp}'s of the separate components *in vivo* is not known as yet, but from S_{sp} measurements on the synthetic analogues, there is reason to believe that the S_{sp} of the non-crystalline phase may be appreciably less than that of the apatite (16).

ACKNOWLEDGEMENT

This work was supported in part by PHS Grant DE-01945 from The National Institute of Dental Research.

BIBLIOGRAPHY

1. Holmes, J. M., Davies, D. H., Meath, W. J. and Beebe, R. A., Biochemistry, **3,** 2019 (1964).
2. Kratky, O., Z. Elektrochem., **58,** 49 (1954).
3. Porod, G., Kolloid-Z., **124,** 83 (1951); **125,** 51 (1952).
4. Gong, J. K., Arnold, J. S. and Cohn, S. H., Anat. Rec., **149,** 319, 325 (1964).
5. Johansen, E. and Parks, H. F., J. Biophys. Biochem. Cytol., **7,** 743 (1960).
6. Guinier A., Ann. Phys., **12,** 161 (1939).
7. Kratky, O., Porod, G. and Skala, Z., Acta Phys. Austriaca, **13,** 76 (1960).
8. Guinier, A. *et al.*, *Small-Angle Scattering of X-Rays*, Chap. 4, J. Wiley and Sons, New York, 1955.
9. Posner, A. S. and Eanes, E. D., Polymer Preprints, **6,** 202 (1965).
10. Posner, A. S., Eanes, E. D., Harper, R. A. and Zipkin, I., Arch. Oral Biol., **8,** 549 (1963).
11. Eanes, E. D., Zipkin, I., Harper, R. A. and Posner, A. S., ibid., **10,** 161 (1965).
12. Fleisch, H., Clin. Orthopaedics, **32,** 170 (1964).
13. Harper, R. A. and Posner, A. S., Proc. Soc. Exp. Biol. Med., **122,** 137 (1966).
14. Eanes, E. D., Harper, R. A., Gillessen, I. and Posner, A. S., *"Fourth European Symposium on Calcified Tissues, Abridged Proceedings,"* Excerpta Medica International Congress Series, **120,** 24 (1966).
15. Termine, J. D. and Posner, A. S., Science, **153,** 1523 (1966).
16. Eanes, E. D. and Posner, A. S., Trans. N.Y. Acad. Sci., **28,** 233 (1965).

Index

Actin, 157
Activity isotherm, 427, 430, 435
Age-hardening, 295, 366
Aging, 289, 292, 296, 297, 299, 303, 304, 307, 308, 312, 366–369
Al-Ag, 291, 292, 294, 296, 298, 299, 301, 302, 304–308, 319, 320, 329, 332
Al-Cu, 283, 293, 294, 364
Al-Ga, 363
Alloy(s), 278, 289, 291, 293, 294, 296, 298, 299, 302–308, 323, 332, 335–337, 345, 363–370, 382, 401–403, 408, 411, 415, 422
Al-Mg-Zn, 308
Alumina, 449–453, 455, 457, 458, 463, 464
Aluminum, 291, 292, 361, 362, 364, 366, 382
Al-Zn, 291, 292, 294, 296, 299, 300–302, 304–308, 319, 320, 331, 364–367, 369, 370, 402, 406, 411, 416–418, 420, 421
Amorphous regions, 151, 153, 154, 186, 192, 196
Annealing, 146–148, 150–154, 188, 193, 293, 295–297, 303, 308, 309, 311, 314, 349, 362, 364, 365, 368, 385
Apoferritin, 111, 113, 114
Arginine, 269
Au-Ni, 335, 337, 339–342, 344, 346, 348, 349
Axial ratio, 6, 8, 9, 209, 210, 277, 283, 287, 306, 384, 454

Background, 63, 81, 83, 95, 107, 122, 142, 198, 199, 205, 247, 281, 291, 297, 310, 313, 315, 332, 345, 358–360, 434, 437, 441, 442, 450, 452, 453, 457, 458, 483, 495–497
Bence-Jones protein, 215

Bond length, 33, 38
Bone, 493–500
Burgers vectors, 277, 392–394, 396

Calibration sample, radioactive, 73
Carbon black, 176
Catalyst(s), metallized, 449–451
Cellulose, 152
  nitrate, 106–111
Chain
  element, statistical, 34, 39–42, 44, 53, 60
  fold, 134, 140, 141, 145, 151, 152, 196, 272
  Gaussian, 106
Characteristic
  function, 18–20
  length, 277, 286, 467
Chloroplatinic acid, 477, 482, 483, 485, 489
Close-packed array, 125, 126
Clustering function, 435, 436, 438
Cohesive energy density, 430, 438
Coil(s), 33–35, 39, 40, 42, 44, 47, 48, 50, 51, 53, 55, 56
  Gaussian, 108, 110, 111
  random, 33–35, 40, 56
Coincidence correction, 91
Collagen, 157, 493, 494
Collimating system (see collimation system)
Collimation
  error, 3, 63, 88, 94–97, 114, 116, 128, 202, 203, 205, 206, 246, 248, 249, 253, 255, 258, 259, 418–420, 435, 442–445, 469, 495–497
  line, 277, 280, 284
  point, 277, 280, 284, 304
  system, 64, 68, 69, 71, 75–82, 84, 93,

121, 124, 128, 197–200, 246, 300, 375, 434, 452. 468
Collimator (see collimation system)
assembly (see collimation system)
Colloid(al), 17
  order, 196
  particles, 475
  regions, 14
  solutions, 278
  structure, 173, 186
  substance, 6
  system, 1, 12, 14, 467, 468
Compressibility, 205, 427, 433, 451
Concentration fluctuations, 411, 412
Conformation, 34, 40–46, 54, 55, 150, 152, 158, 170, 177, 214, 216–218, 273, 275
Convolution, 142, 170, 178, 179, 389, 404, 407, 409
Coordination shell, 406
Copper
  phthalocyanine, 102
  single-crystals, 309, 310, 311–314, 382, 383, 385, 392, 398
Correlation
  distance, 430
  function, 1, 2, 3, 10–12, 15, 177, 179, 180, 336, 337, 348, 402, 405–408, 411–413, 426–428, 432, 433, 435, 438, 441, 444, 469
  function, semi-fine, 406, 407, 412, 413
  length, 415, 421, 422, 444
  range, 401, 413, 427
Critical
  concentration, 296, 416, 435, 442
  opalescence, 401–403, 407, 417, 428
  point, 21, 24, 26–29, 31, 171, 296, 300, 401–403, 406, 407, 411–414, 416, 417, 421, 423, 427, 428, 430, 431
  region, 425–427, 436–439
  temperature, 296, 416, 420, 426, 427, 435, 436, 438, 444, 445
Cross-section factor, 49–51, 56
Cryostat, 355–357
Crystal
  grooved, 122, 123
  plastically deformed, 146, 150, 151, 386

Crystallinity, 150, 153
Cu-Au, 337, 369, 370
Cu-Be, 293
CuNiFe, 337
Cylinders, 17, 30, 31, 160, 215, 218, 391

Degree
  of association, 99, 101, 102, 104–106, 365, 366
  of polymerization, 109, 110
Demixtion, 371, 401, 402, 407, 411, 412, 416, 417
Density fluctuations, 401, 402, 405. 406, 411–413, 436, 469, 470
Diamond, 373–378, 380
Direct beam (see Primary beam)
Dislocation(s), 277, 279, 282, 283, 290, 307, 308, 314, 386, 389, 390
  density tensor, 386, 397
  loops, 277, 290, 291, 311, 313, 315, 389–391, 393–396
  rings, 383–386, 391, 392
Disorder, 134–136, 139–141, 196, 335, 338, 344, 369, 382
  intensity, 336, 340
Distance of heterogeneity, 470
Distribution,
  function, molecular, 405
  pair, 204, 403
  radial, 158, 159, 403, 404, 406, 426, 477, 490
  single-particle, 404
  of intersects, 2, 4, 5, 12, 15
DNA, 221–223, 230, 231, 233, 235, 237, 240, 241
Dye solutions, 78, 79, 99, 106

Elastic modulus, 307
Electron density, 2, 18, 36, 49, 50, 112, 117, 126, 137, 140, 201, 208–211, 215, 245, 246, 250, 252, 253, 261, 267, 269, 271, 274, 277, 281–284, 289, 294, 295, 299, 302, 307, 320, 351, 352, 383, 408, 409, 449, 469, 478, 479, 482, 483, 485, 490, 500
  difference(s), 78, 79, 101, 116, 145, 217, 278, 290, 426, 449, 450, 478

INDEX 505

distributions(s), 115, 169, 223, 244, 248, 256, 258, 270, 281
  effective, 246, 252, 254, 255, 256, 259–262
  fluctuations, 209, 281, 479, 490
Electropolishing, 291
Ellipsoid, 1, 6–9, 14, 113, 179, 209, 210, 215, 268, 269, 283, 285–287, 306, 308
Etching, 291, 340
Ewald sphere (shell), 341, 345, 376–378, 380, 381
Excluded volume, 35, 36

Feather keratin (see Keratin, feather)
Ferritin, 111, 112, 115, 116
Fiber(s), 73, 79, 131–135, 137–142, 150–153, 159, 182, 184, 185, 188, 237, 239
Fibrils, 134, 137–139, 151, 152, 187–190, 192, 193. 196
Filaments, 18, 19, 21, 24, 28–31, 157, 197
Fluorescence radiation, 84
Focus
  line, 64, 65, 67–71, 197, 198, 201
  square, 64–66, 68–71
Fourier
  integral, 19, 21, 31
  integral theorem, 478, 482, 487
  inversion, 113, 159, 245, 247, 411, 414, 426, 469, 478, 489, 490
  series, 338
  synthesis, 270
  transform, 3, 20, 115, 142, 169, 179, 191, 246, 249–254, 256–261, 263, 277, 282, 336, 386, 387, 389, 407, 408, 410, 413, 433, 447
Four-point diagram, 137, 138, 140, 193
Frenkel defect, 378

Geiger counter, 72, 73, 217, 417
Germanium crystals, 122, 124, 128
γ-Globulin, 213–218
Guinier
  approximation, 284, 287, 301–303, 313, 454
  camera, 417
  law, 383, 497

plot, 3, 4, 6, 7, 97, 101, 103, 105, 107, 207, 217, 218, 310, 439, 454, 455
Helix, 37, 38, 47, 48, 52, 53, 150, 214
Hemoglobin, 205, 211
Heterogeneity, 278, 282, 283, 352, 371
Hexagonal packing, 221, 222, 231, 235, 237, 240
Histone, 221, 223, 230, 233, 240, 241
Huang scattering, 338, 339, 370
Hydroxyapatite, 493, 499, 500

Intensity
  absolute, 1, 48, 56, 64, 88, 92, 101, 103, 104, 118, 201, 203–205, 222, 296, 310, 333, 352, 353, 361, 459
  by filter attentuation, 88, 89
  by gaseous scattering, 88, 201, 203
  by rotating sector, 48, 49, 89, 90
  by standard sample, 92, 93
  integrated, 285, 287, 289, 296, 297, 299, 319–321, 332–334, 384, 385
  primary, 74, 81, 103, 202, 205, 223, 409
  reduced, 408, 410, 411
  rods, 293, 294
  smeared, 3, 203
  zero (angle), 99–101, 103, 108, 109, 202, 205, 391, 395, 396, 427, 445
Interparticle interference, 96, 101, 125, 127–129, 174, 205, 206, 247, 277, 284, 287, 288, 295, 301–304, 328, 329
Intersects, 2, 4, 5, 8–10, 12
Interstitial, 373, 379
Invariant, 3, 98, 100, 494, 496, 498
Ising model, 425, 432, 446

Johansson crystal, 72, 188

KCl, 104–106, 382
Keratin
  α, 157
  β, 182–184
  feather, 157, 158, 162, 165
Kiessig camera, 343, 349, 374, 375
Kossel lines, 309

Lamella(e), 141, 145, 150, 151, 184, 353, 355, 361
Latex, 122, 125, 127–129
Lattice
  crystalline, 135, 145, 177–179, 273, 293, 352, 410
  cylindrical, 157, 159, 160, 162, 166
  defects, 152, 352, 362
  dilatation, 279
  hexagonal, 157, 159, 160, 193
  macro-, 170, 174, 177, 179, 182, 186, 187, 189, 191, 192, 194
  layer, 159, 160
  paracrystalline, 135, 136, 142, 159, 178–181, 185, 189, 193
  parallel-arc, 162, 164–166
  reciproçal, 132, 181, 292, 293, 322, 323, 329, 331, 332, 335–337, 370
  strains, 279
Laue
  monotonic intensity, 321, 324, 325, 331–333
    scattering, 327, 329, 352
    terms, 329, 331, 332, 336
    units, 344, 347, 348
  photograph, 340, 342, 374–376, 380, 382
  scattering, 364, 365, 367, 369, 370
LiF, 345, 353, 361, 362
Light-scattering, 33, 35, 36, 51, 53, 102, 118, 401, 402, 421, 446
Local order, 335
Lysine, 269

Masks, 131, 133, 134, 137–139
Mass distribution, radial, 114
Matrix, 58–60, 278, 293, 299, 300, 306, 307, 309, 320, 321, 324, 326, 327–329, 331–333, 337, 364, 461, 462, 493, 494, 498, 500
Metals, deformed, 279, 314
Metmyoglobin, 205–207
Mg-Pb, 308
Mg-Zn, 308
Mica, 200, 434
Micelle
  DNH, 221, 230, 231, 238, 240

fringed, 137, 140, 151
iron (of ferritin), 112, 113
Miscibility gap, 296, 298, 299, 300, 302–304, 319, 323, 333, 334, 340, 342, 344, 345, 348, 349
Moiré pattern, 148
Molecular weight, 36, 101–103, 106–110, 117, 118, 147, 152, 201, 203, 204, 213, 215, 243, 244, 246, 252, 258, 260, 262, 267
Monochromatic beam (see monochromatization)
Monochromatic radiation (see monochromatization)
Monochromatization, 71, 72, 89, 121, 128, 129, 188, 292, 342, 345, 353, 355, 358, 374, 468, 483, 494
  by Bragg reflection, 71, 89
  crystal, 121, 128
  filter difference method, 71, 72
  total reflection, 71, 72
Monochromator (see monochromatization)
Monte Carlo method, 45, 57, 107, 108
Mosaic crystal, 147, 148, 196, 340
Muscle, 157
Myoglobin, 197, 200, 202, 205, 206, 208, 210, 211, 267, 268, 270–275
Myosin, 157

NaCl-AgCl, 319, 320
Naphthol Orange, 102–106
Ni-Al, 309, 332
Nucleoprotein(s), 201, 221, 224–229, 240
Nylon, 131, 132, 134, 140, 151

Order
  intensity, 321, 329
    integrated, 323, 329, 330
  parameter(s), 327, 330, 331, 336, 338, 340, 348, 369
  short-range, 321, 332, 340, 346, 349, 365, 369
Ornstein-Zernike, 401, 413, 419, 420–423, 427, 428, 431–433, 439–441, 443–447
Optical
  analogs, 131, 179, 180

diffraction pattern, 131, 132, 134, 138, 140, 160, 163–165

Packing, dense, 10, 14, 152, 176
Papain, 213, 216, 217
Paracrystal(s), 170, 177, 181, 182
Paracrystalline model, 131, 140, 187, 192
Parallelopiped, 1, 11, 178, 181
Parameter
  differential, 1, 2, 4, 10, 14
  integral, 1–5, 8, 14
Particle(s)
  angular, 11, 13, 15
  anisometric, 7
  irregular, 6
  mass, 99, 100
  shape, 1, 2, 5–9, 15, 17–19, 57, 99–101, 130, 170, 174, 215, 283, 287, 289, 303, 306, 449, 452, 455, 465
    distribution, 283, 454, 455, 465
  size, 1, 2, 8, 99–101, 117, 121, 125–128, 170, 215, 283, 288, 299, 331, 449, 450, 454, 459, 461, 463, 465, 485
    distribution, 7, 8, 283, 288, 289, 302, 303, 454, 465 466
  smooth, 11, 13, 15
  spherical, (see sphere)
  weight, 103, 104
Patterson function, 272, 274
Perfluoroheptane-i-octane, 426, 435–437
Persistence
  length, 46, 47, 54, 55, 58, 106–108, 110, 111, 118, 428, 430, 436, 439
  of curvature, 36, 38
Phase
  transformation, 151
  transition, 406, 425, 431
Phenanthrene, 380–382
Plane of registration, 49, 63, 65–67, 69, 76–78, 94, 98
Platinum, 449–452, 454, 455, 457–460, 463–466
Point defect(s), 308, 351, 353, 358–362, 370, 373, 475
Poisson's ratio, 277, 384, 398
Polydispersity, 1, 5–9, 15, 96, 99, 110, 173–175, 187, 306, 496

Polyethylene, 93, 131, 140, 145, 146, 148, 151–153, 170, 184, 186–192, 194, 195
Polymer(s), 33, 38, 131, 145, 148, 150, 151, 153, 194, 196, 278
  coils, 33, 34, 49
  single crystal mats, 146, 148
Polymethyl methacrylate, 50, 51
  isotactic, 33, 36–38, 49, 51, 53–55
  syndiotactic, 33, 36–38, 49, 51–53
  tactic, 48
Polyoxymethylene, 151, 152
Polypeptide, 157, 158, 267
Polystyrene, 122, 127, 129
Polyvinyltoluene, 122
Precipitate, 294, 308, 309, 326, 364
Precipitation, 323, 366
Pre-precipitation, 293, 364, 366, 368
Pressure sintering, 450, 451, 457, 458
Primary beam, 3, 49, 63, 65–72, 74, 75, 80–83, 88–95, 122, 126, 129, 188, 193, 223, 279, 292, 309, 314, 353, 355, 358, 373, 374, 376, 379, 382, 409, 453, 461, 495
  finite length, 95
  infinite length, 94, 95, 494
  profile, 94, 435, 444
  uniform, 65–68, 70, 94, 95
  energy, 63, 64, 71, 81, 88, 91, 92, 277, 279, 358
Proportional counter, 72, 73, 89, 122, 199, 434
Protamine, 221
Protein(s)
  fibrous, 157, 161, 166
  globular, 197, 201
  solution, 78, 79, 100, 102, 224–229

Quartz, 200, 201
Quench(ing), 153, 279, 291, 293, 294, 296, 303, 307, 309, 311–313, 320, 338, 340, 342, 344–346, 348, 349, 351, 364, 366, 368, 370, 383

Radiation damage, 467, 475
Radius
  of gyration, 3, 5, 6, 33, 35, 37, 53, 60, 98, 99, 101, 106, 107, 110, 111,

113, 115, 118, 174, 200–202, 205–208, 210, 211, 215, 217–219, 223, 230, 238, 245, 246, 256, 257, 262, 268–270, 274, 275, 277, 311, 313, 384, 385, 395, 396, 453, 454, 457, 462
  of preferred curvature, 47, 48, 52
Range of intermolecular forces, 428, 430
Reciprocal space, 280, 322, 323, 331, 332, 338, 341, 344, 345, 348, 376, 379
Reflection
  equatorial, 157, 158, 189, 190, 193, 237
  meridional, 151, 184, 186, 193, 237
  multiple, 72, 121–124, 128, 129
  superstructure, 337
Reversion, 295, 308, 319, 366–369
RNA, 243, 244, 248, 252–255, 257, 258, 260–263
Rocking curve, 123, 129
Rod(s), 18, 35, 215, 217, 218, 223, 230, 231, 454, 455
Rotating anode, 64, 197

Scattering
  anomalous, 373, 374, 380, 381
  Bragg, double, 279, 291, 309, 314, 353, 381, 385
  Compton, 353, 359, 361, 363, 369, 377, 483
  critical, 401, 411, 413, 415, 418–423
  curve
    filaments, 18
    smeared, 10, 14
  diffuse, 63, 72, 310, 329, 332, 335, 337–339, 341, 344–346, 351, 365, 373–379, 381, 382, 402, 408, 411
  disorder, 335–338, 340, 347, 348
  equatorial (see Reflection, equatorial)
  equivalent, 113, 117
  factor, 352, 359, 483
    atomic, 321, 324, 325, 331, 336, 352, 353, 480
  meridional (see Reflection, meridional)

  multiple, 99, 121, 125, 130, 373, 381, 431, 444, 446
  parasitic, 63, 70, 75, 77, 79, 80–82, 84, 198, 199, 202, 205, 355, 381, 382, 419
  power, atomic, 408, 410, 411, 426
  Rayleigh, 171, 172, 174
  TDS (see scattering, temperature diffuse)
  temperature diffuse, 142, 310, 338, 348, 373
  thermal, 359–363, 369, 380, 382
Segregation, 278, 282, 293, 294, 297, 299, 369
Shape
  factor, 278, 287, 302, 305, 306, 311, 313, 384, 454
  function, 142, 169, 181
Shear stress, 308
Silica gel, 467, 468, 476
Silicon crystals, 122–124
Slit
  arrangement (see collimation system)
  assembly (see collimation system)
  collimation (see collimation system)
  correction (see collimation error)
  effective, 122, 124, 127, 129
  height correction (see collimation error)
  infinite, 124, 126, 127, 496
  system (see collimation system)
Solid(s)
  microporous, 449
  solutions, 278, 282, 283, 291–293, 296, 319, 320, 321, 323, 324, 333, 335, 337, 340, 345, 362–364, 366, 369, 370, 450
Solutions of complex ions, 477
Spectral distribution of local fluctuations, 407, 413
Sphere(s), 2, 5, 6, 7, 8, 14, 31, 99, 112, 113, 115, 122, 125–129, 210, 218, 244, 249–252, 255, 258, 260–262, 264, 283, 285–287, 288, 289, 295, 301, 302, 306, 310, 313, 323, 383, 384, 454, 455, 457, 466, 467, 481, 488, 490
Spherulite(s), 145, 150

Spinodal, 340, 345
  decomposition, 340, 348, 349
Strain field, 279, 282, 283, 306, 307, 385
Streaks, 296, 308, 341, 343–345, 348, 373, 375–378, 380, 381
Stress tensor, 386
Structure
  amplitude, 272
  factor, 142, 245, 246, 273, 453, 454
Sucrose, 204
Surface
  area, 268, 269, 273, 285, 467–473, 475, 476, 493, 494, 498
  curvature, 10
  internal, 10, 15, 98, 99
  particle, 2
  relative, 499
  specific, 10, 13, 98, 269, 468, 470, 493–495, 497, 498, 500
  structure, 2, 10, 15
Stencil function, 160

Theta
  solution, 35, 36, 49
  solvent, 55, 56
Theory of elasticity, 290, 385, 391
TiO_2, 95, 96, 99
Twinning, 151
Two-phase
  field, 293, 294
  model, 304, 319, 324, 329, 330

  region, 293
  system, 78

Vacancies, 279, 291, 307–311, 315, 351, 352, 359–361, 368, 373, 379, 383, 385, 395
  agglomeration, 279, 291, 383, 385, 396
  clusters, 279, 309, 310, 313, 361
Valence angle, 33, 34, 38, 53, 54, 59
Virial coefficient, second, 36
Virus, 197, 214, 243–245, 247, 249, 252, 256, 257, 260–263, 264
  broad bean mottle (BBMV), 243, 244, 258, 259, 261, 262
  bromegrass mosaic (BMV), 243, 244, 253, 254, 255, 261–263
  R17, 243, 244, 255–258, 261–263
  squash mosaic (SMV), 243, 244, 258–262, 264
  turnip yellow mosaic (TYMV), 243
  wild cucumber mosaic (WCMV), 243, 244, 247–252, 260–262, 264

Weighting function, 202, 203

Zone(s), 321, 326–333, 336, 366–368, 385
  depleted, 383, 385
  Guinier-Preston, 278, 283, 293, 300, 301, 319, 320, 325, 331, 332, 366
  hardening, 295
  spherical, 294, 295, 301, 302, 304, 306, 308, 319, 320, 385